The Structure and Function of Nucleic Acids

The Structure and Function of Nucleic Acids

Balaji Yadav Maddina

RANDOM PUBLICATIONS
NEW DELHI (INDIA)

The Structure and Function of Nucleic Acids

ISBN 978-93-5111-803-9

Published in 2016 in India by

RANDOM PUBLICATIONS

4376-A/4B, Gali Murari Lal, Ansari Road
New Delhi-110 002
Phone : +9111-43580356, 011-23289044, 011-43142548
e-mail: sales@randompublications.com,
info@randompublications.com, randomexports@gmail.com

Reprinted 2021

Type Setting by : Friends Media, Delhi-110089
Digitally Printed at: Replika Press Pvt. Ltd.

Preface

Nucleic acids are biopolymers, or large biomolecules, essential for all known forms of life. Nucleic acids, which include DNA (deoxyribonucleic acid) and RNA (ribonucleic acid), are made from monomers known as nucleotides. Each nucleotide has three components: a 5-carbon sugar, a phosphate group, and a nitrogenous base. If the sugar is deoxyribose, the polymer is DNA. If the sugar is ribose, the polymer is RNA. When sugar and a nitrogenous base get combined they form a nucleotide. Nucleotides are also known as phosphate nucleotides. Nucleic acids are among the most important biological macromolecules (others being amino acids/proteins, sugars/carbohydrates, and lipids/fats). They are also found in abundance in all living things, where they function in encoding, transmitting and expressing genetic information-in other words, information is conveyed through the nucleic acid sequence, or the order of nucleotides within a DNA or RNA molecule. Strings of nucleotides strung together in a specific sequence are the mechanism for storing and transmitting hereditary, or genetic information via protein synthesis. Nucleic acids were discovered by Friedrich Miescher in 1869. Experimental studies of nucleic acids constitute a major part of modern biological and medical research, and form a foundation for genome and forensic science, as well as the biotechnology and pharmaceutical industries.

– Author

Contents

1

Nucleic Acids

The first isolation of what we now refer to as DNA was accomplished by Johann Friedrich Miescher *circa* 1870. He reported finding a weakly acidic substance of unknown function in the nuclei of human white blood cells, and named this material "nuclein". A few years later, Miescher separated nuclein into protein and nucleic acid components.

In the 1920's nucleic acids were found to be major components of chromosomes, small gene-carrying bodies in the nuclei of complex cells. Elemental analysis of nucleic acids showed the presence of phosphorus, in addition to the usual C, H, N & O. Unlike proteins, nucleic acids contained no sulfur. Complete hydrolysis of chromosomal nucleic acids gave inorganic phosphate, 2-deoxyribose (a previously unknown sugar) and four different heterocyclic bases.

To reflect the unusual sugar component, chromosomal nucleic acids are called deoxyribonucleic acids, abbreviated DNA. Analogous nucleic acids in which the sugar component is ribose are termed ribonucleic acids, abbreviated RNA. The acidic character of the nucleic acids was attributed to the phosphoric acid moiety.

The two monocyclic bases shown here are classified as pyrimidines, and the two bicyclic bases are purines. Each has at least one N-H site at which an organic substituent may be attached. They are all polyfunctional bases, and may exist in tautomeric forms.

Base-catalyzed hydrolysis of DNA gave four nucleoside products, which proved to be N-glycosides of 2'-deoxyribose combined with the heterocyclic amines.The base components are coloured green, and the sugar is black. As noted in the 2'-deoxycytidine structure on the left, the numbering of the sugar carbons makes use of primed numbers to distinguish them from the heterocyclic base sites. The corresponding N-glycosides of the common sugar ribose are the building blocks of RNA, and are named adenosine, cytidine, guanosine and uridine (a thymidine analog missing the methyl group). From this evidence, nucleic acids may be formulated as alternating copolymers of phosphoric acid (P) and nucleosides (N), ~

P – N – P – N'– P – N''– P – N'''– P – N ~

At first the four nucleosides, distinguished by prime marks in this crude formula, were assumed to be present in equal amounts, resulting in a uniform structure, such as that of starch. However, a compound of this kind, presumably common to all organisms, was considered too simple to hold the hereditary information known to reside in the chromosomes.

This view was challenged in 1944, when Oswald Avery and colleagues demonstrated that bacterial DNA was likely the genetic agent that carried information from one organism to another in a process called "transformation". He concluded that *"nucleic acids must be regarded as possessing biological specificity, the chemical basis of which is as yet undetermined."* Despite this finding, many scientists continued to believe that chromosomal proteins, which differ across species, between individuals, and even within a given organism, were the locus of an organism's genetic information.

It should be noted that single celled organisms like bacteria do not have a well-defined nucleus. Instead, their single chromosome is associated with specific proteins in a region called a "nucleoid". Nevertheless, the DNA from bacteria has the same composition and general structure as that from multicellular organisms, including human beings.

Views about the role of DNA in inheritance changed in the late 1940's and early 1950's. By conducting a careful analysis of DNA from many sources, Erwin Chargaff found its composition to be species specific. In addition, he found that the amount of adenine (A) always equaled the amount of thymine (T), and the amount of guanine (G) always equaled the amount of cytosine (C), regardless of the DNA source. As set forth in the following table, the ratio of (A+T) to (C+G) varied from 2.70 to 0.35. The last two organisms are bacteria.

NUCLEOSIDE BASE DISTRIBUTION IN DNA

Organism	Base Composition (mole %)	Base Ratios	Ratio (A+T)/(G+C)				
	A	G	T	C	A/T	G/C	
Human	30.9	19.9	29.4	19.8	1.05	1.00	1.52
Chicken	28.8	20.5	29.2	21.5	1.02	0.95	1.38
Yeast	31.3	18.7	32.9	17.1	0.95	1.09	1.79
Clostridium perfringens	36.9	14.0	36.3	12.8	1.01	1.09	2.70
Sarcina lutea	13.4	37.1	12.4	37.1	1.08	1.00	0.35

In a second critical study, Alfred Hershey and Martha Chase showed that when a bacterium is infected and genetically transformed by a virus, at least 80percent of the viral DNA enters the bacterial cell and at least 80percent of the viral protein remains outside. Together with the Chargaff findings this work established DNA as the repository of the unique genetic characteristics of an organism.

THE CHEMICAL NATURE OF DNA

The polymeric structure of DNA may be described in terms of monomeric units of increasing complexity. In the top shaded box of the following illustration, the three relatively simple components mentioned earlier are shown.

Condensation polymerization of these leads to the DNA formulation outlined above. Finally, a 5'- monophosphate ester, called a nucleotide may be drawn as a single monomer unit, shown in the shaded box to the right. Since a monophosphate ester of this kind is a strong acid (pK_a of 1.0), it will be fully ionized at the usual physiological pH. Isomeric 3'-monophospate nucleotides are also known, and both isomers are found in cells.

They may be obtained by selective hydrolysis of DNA through the action of nuclease enzymes. Anhydride-like di- and tri-phosphate nucleotides have been identified as important energy carriers in biochemical reactions, the most common being ATP (adenosine 5'-triphosphate).

Table. Names of DNA Base Derivatives

Theymine	2′-Dexythymidine	2Deoxyth Nucleoside 5′-Nucleotide
Adenine	2'-Deoxyadenosine	2'-Deoxyadenosine-5'-monophos-phate
Cytosine	2'-Deoxycytidine	2'-Deoxycytidine-5'-monophos-phate
Guanine	2'-Deoxyguanosine	2'-Deoxyguanosine-5'-monophos-phateYmidine-5′-monophos-phate

A complete structural representation of a segment of the DNA polymer formed from 5'-nucleotides. Several important characteristics of this formula should be noted.

- First, the remaining P-OH function is quite acidic and is completely ionized in biological systems.
- Second, the polymer chain is structurally directed.
- Third, although this appears to be a relatively simple polymer, the possible permutations of the four nucleosides in the chain become very large as the chain lengthens.
- Fourth, the DNA polymer is much larger than originally believed. Molecular weights for the DNA from multicellular organisms are commonly 109 or greater.

Information is stored or encoded in the DNA polymer by the pattern in which the four nucleotides are arranged. To access this information the pattern must be "read" in a linear fashion, just as a bar code is read at a supermarket checkout. Because living organisms are extremely complex, a correspondingly large amount of information related to this complexity must be stored in the DNA. Consequently, the DNA itself must be very large, as noted above. Even the single DNA molecule from an *E. coli* bacterium is found to have roughly a million nucleotide units in a polymer strand, and would reach a millimeter in length if stretched out. The nuclei of multicellular organisms incorporate chromosomes, which are composed of DNA combined with nuclear proteins called histones. The fruit fly has 8 chromosomes, humans have 46 and dogs 78 (note that the amount of DNA in a cell's nucleus does not correlate with the number of chromosomes).

The DNA from the smallest human chromosome is over ten times larger than *E. coli* DNA, and it has been estimated that the total DNA in a human cell would extend to 2 meters in length if unraveled. Since the nucleus is only about 5 m in diameter, the chromosomal DNA must be packed tightly to fit in that small volume. In addition to its role as a stable informational library, chromosomal DNA must be structured or organized in such a way that the chemical machinery of the cell will have easy access to that information, in order to make important molecules such as polypeptides.

Furthermore, accurate copies of the DNA code must be created as cells divide, with the replicated DNA molecules passed on to subsequent cell generations, as well as to progeny of the organism. The nature of this DNA organization, or secondary structure, will be discussed in a later section.

RNA, A DIFFERENT NUCLEIC ACID

The high molecular weight nucleic acid, DNA, is found chiefly in the nuclei of complex cells, known as eucaryotic cells, or in the nucleoid regions of procaryotic cells, such as bacteria. It is often associated with proteins that help to pack it in a usable fashion. In contrast, a lower molecular weight, but much more abundant nucleic acid, RNA, is distributed throughout the cell, most commonly in small numerous organelles called ribosomes.

Three kinds of RNA are identified, the largest subgroup being ribosomal RNA, rRNA, the major component of ribosomes, together with proteins. The size of rRNA molecules varies, but is generally less than a thousandth the size of DNA. The other forms of RNA are messenger RNA, mRNA, and transfer RNA, tRNA. Both have a more transient existence and are smaller than rRNA.

All these RNA's have similar constitutions, and differ from DNA in two important respects.The sugar component of RNA is ribose, and the pyrimidine base uracil replaces the thymine base of DNA. The RNA's play a vital role in the transfer of information (transcription) from the DNA library to the protein factories called ribosomes, and in the interpretation of that information (translation) for the synthesis of specific polypeptides.

THE SECONDARY STRUCTURE OF DNA

In the early 1950's the primary structure of DNA was well established, but a firm understanding of its secondary structure was lacking. Indeed, the situation was similar to that occupied by the proteins a decade earlier, before the alpha helix and pleated sheet structures were proposed by Linus Pauling. Many researchers grappled with this problem, and it was generally conceded that the molar equivalences of base pairs (A & T and C & G) discovered by Chargaff would be an important factor. Rosalind Franklin, working at King's College, London, obtained X-ray diffraction evidence that suggested a long helical structure of uniform thickness.

Francis Crick and James Watson, at Cambridge University, considered hydrogen bonded base pairing interactions, and arrived at a double stranded helical model that satisfied most of the known facts, and has been confirmed by subsequent findings.

BASE PAIRING

Careful examination of the purine and pyrimidine base components of the nucleotides reveals that three of them could exist as hydroxy pyrimidine or purine tautomers, having an aromatic heterocyclic ring. Despite the added stabilization of an aromatic ring, these compounds prefer to adopt amide-like structures.

A simple model for this tautomerism is provided by 2-hydroxypyridine. A compound having this structure might be expected to have phenol-like characteristics, such as an acidic hydroxyl group. However, the boiling point of the actual substance is 100° C greater than phenol and its acidity is 100 times less than expected (pKa = 11.7). These differences agree with the 2-pyridone tautomer, the stable form of the zwitterionic internal salt.

Note that this tautomerism reverses the hydrogen bonding behaviour of the nitrogen and oxygen functions,the N-H group of the pyridone becomes a hydrogen bond donor and the carbonyl oxygen an acceptor.

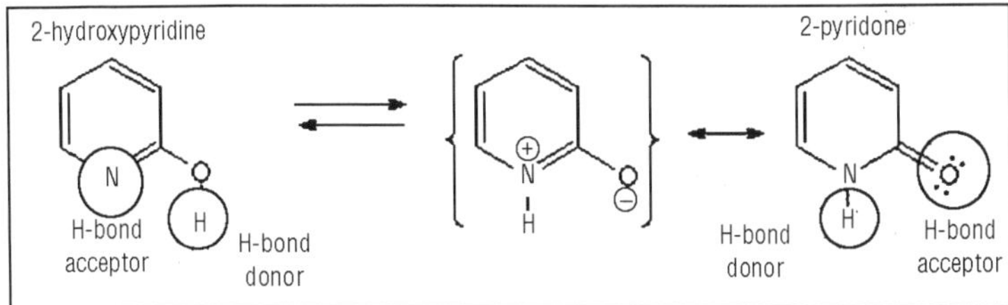

The additional evidence for the pyridone tautomer, consists of infrared and carbon nmr absorptions associated with and characteristic of the amide group.

Once they had identified the favored base tautomers in the nucleosides, Watson and Crick were able to propose a complementary pairing, via hydrogen bonding, of guanosine (G) with cytidine (C) and adenosine (A) with thymidine (T). This pairing, which is shown in the following diagram, explained Chargaff's findings beautifully, and led them to suggest a double helix structure for DNA.

Before viewing this double helix structure itself, it is instructive to examine the base pairing interactions in greater detail. The G#C association involves three hydrogen bonds (coloured pink), and is therefore stronger than the two-hydrogen bond association of A#T. These base pairings might appear to be arbitrary, but other possibilities suffer destabilizing steric or electronic interactions. The C#T pairing on the left suffers from carbonyl dipole repulsion, as well as steric crowding of the oxygens. The G#A pairing on the right is also destabilized by steric crowding (circled hydrogens).

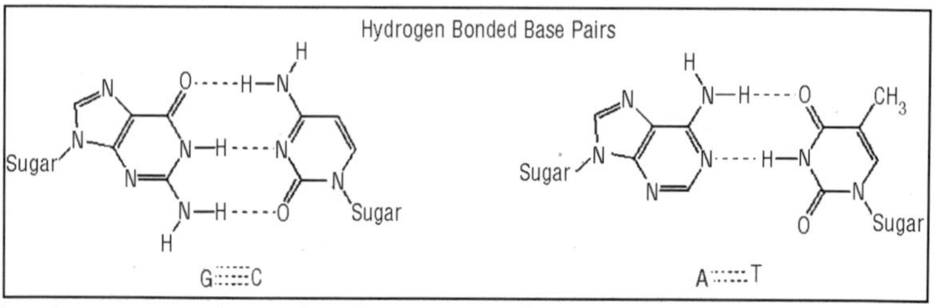

Hydrogen Bonded Base Pairs

G:::::C A:::::T

A simple mnemonic device for remembering which bases are paired comes from the line construction of the capital letters used to identify the bases. A and T are made up of intersecting straight lines. In contrast, C and G are largely composed of curved lines. The RNA base uracil corresponds to thymine, since U follows T in the alphabet.

The Double Helix

After many trials and modifications, Watson and Crick conceived an ingenious double helix model for the secondary structure of DNA. Two strands of DNA were aligned anti-parallel to each other.Complementary primary nucleotide structures for each strand allowed intra-strand hydrogen bonding between each pair of bases.Coiling these coupled strands then leads to a double helix structure.

THE DOUBLE HELIX STRUCTURE FOR DNA

The helix shown here has ten base pairs per turn, and rises 3.4 Å in each turn. This right-handed helix is the favored conformation in aqueous systems, and has been termed the B-helix. Two alternating grooves are evident, a wide and deep major groove, and a shallow and narrow minor groove.

Other molecules, including polypeptides, may insert into these grooves, and in so doing perturb the chemistry of DNA. Other helical structures of DNA have also been observed, and are designated by letters.

DNA REPLICATION

A double helix structure for DNA, Watson and Crick stated, *"It has not escaped our notice that the specific pairing we have postulated immediately suggests a possible copying mechanism for the genetic material.".* The essence of this suggestion is that, if separated, each strand of the molecule might act as a template on which a new complementary strand might be assembled, leading finally to two identical DNA molecules. Indeed, replication does take place in this fashion when cells divide, but the events leading up to the actual synthesis of complementary DNA strands are sufficiently complex that they will not be described in any detail.

As depicted in the following drawing, the DNA of a cell is tightly packed into chromosomes. First, the DNA is wrapped around small proteins called histones.

These bead-like structures are then further organized and folded into chromatin aggregates that make up the chromosomes. An overall packing efficiency of 7,000 or more is thus achieved. Clearly a sequence of unfolding events must take place before the information encoded in the DNA can be used or replicated.

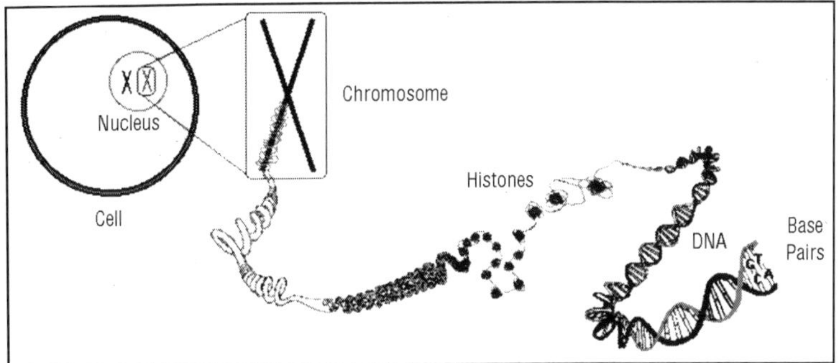

Once the double stranded DNA is exposed, a group of enzymes act to accomplish its replication. These are described briefly here:

- *Topoisomerase*: This enzyme initiates unwinding of the double helix by cutting one of the strands.
- *Helicase*: This enzyme assists the unwinding. Note that many hydrogen bonds must be broken if the strands are to be separated..
- *SSB*: A single-strand binding-protein stabilizes the separated strands, and prevents them from recombining, so that the polymerization chemistry can function on the individual strands.
- *DNA Polymerase*: This family of enzymes link together nucleotide triphosphate monomers as they hydrogen bond to complementary bases. These enzymes also check for errors (roughly ten per billion), and make corrections.
- *Ligase*: Small unattached DNA segments on a strand are united by this enzyme.

Polymerization of nucleotides takes place by the phosphorylation reaction described by the following equation.

Di- and triphosphate esters have anhydride-like structures and are consequently reactive phosphorylating reagents, just as carboxylic anhydrides are acylating reagents.

Since the pyrophosphate anion is a better leaving group than phosphate, triphosphates are more powerful phospho-rylating agents than are diphosphates.

The DNA polymerization process that builds the complementary strands in replication, could in principle take place in two ways. Referring to the general equation above, R1 could represent the next nucleotide unit to be attached to the growing DNA strand, with R2 being this strand. Alternatively, these assignments could be reversed.

In practice, the former proves to be the best arrangement. Since triphosphates are very reactive, the lifetime of such derivatives in an aqueous environment is relatively short. However, such derivatives of the individual nucleosides are repeatedly synthesized by the cell for a variety of purposes, providing a steady supply of these reagents. In contrast, the growing DNA segment must maintain its functionality over the entire replication process, and can not afford to be changed by a spontaneous hydrolysis event. As a result, these chemical properties are best accomodated by a polymerization process that proceeds at the 3'-end of the growing strand by 5'-phosphorylation involving a nucleotide triphosphate.

The polymerization mechanism described here is constant. It always extends the developing DNA segment toward the 3'-end. when a nucleotide triphosphate attaches to the free 3'-hydroxyl group of the strand, a new 3'-hydroxyl is generated. There is sometimes confusion on this point, because the original DNA strand that serves as a template is read from the 3'-end toward the 5'-end, and authors may not be completely clear as to which terminology is used.

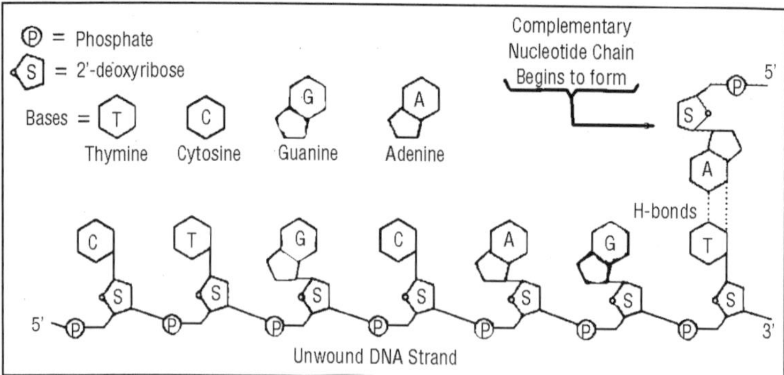

Because of the directional demand of the polymerization, one of the DNA strands is easily replicated in a continuous fashion, whereas the other strand can only be replicated in short segmental pieces. Separation of a portion of the double helix takes place at a site called the replication fork. As replication of the separate strands occurs, the replication fork moves away, unwinding additional lengths of DNA. Since the fork in the diagram is moving toward the 5'-end of the red-coloured strand, replication of this strand may take place in a continuous fashion (building the new green strand in a 5' to 3' direction). This continuously formed new strand is called the leading strand.

In contrast, the replication fork moves toward the 3'-end of the original green strand, preventing continuous polymerization of a complementary new red strand. Short segments of complementary DNA, called Okazaki fragments, are produced, and these are linked together later by the enzyme ligase. This new DNA strand is called the lagging strand. When you consider that a human cell has roughly 109 base pairs in its DNA, and may divide into identical daughter cells in 14 to 24 hours, the efficiency of DNA replication must be extraordinary.

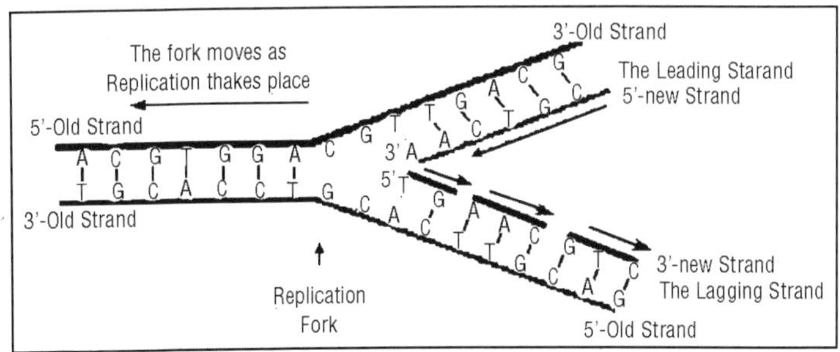

The procedure described above will replicate about 50 nucleotides per second, so there must be many thousand such replication sites in action during cell division. A given length of double stranded DNA may undergo strand unwinding at numerous sites in response to promoter actions.

The unraveled "bubble" of single stranded DNA has two replication forks, so assembly of new complementary strands may proceed in two directions. The polymerizations associated with several such bubbles fuse together to achieve full replication of the entire DNA double helix

REPAIR OF DNA DAMAGE AND REPLICATION ERRORS

One of the benefits of the double stranded DNA structure is that it lends itself to repair, when structural damage or replication errors occur. Several kinds of chemical change may cause damage to DNA:

- Spontaneous hydrolysis of a nucleoside removes the heterocyclic base component.
- Spontaneous hydrolysis of cytosine changes it to a uracil.
- Various toxic metabolites may oxidize or methylate heterocyclic base components.
- Ultraviolet light may dimerize adjacent cytosine or thymine bases.

All these transformations disrupt base pairing at the site of the change, and this produces a structural deformation in the double helix. Inspection-repair enzymes detect such deformations, and use the undamaged nucleotide at that site as a template for replacing the damaged unit. These repairs reduce errors in DNA structure from about one in ten million to one per trillion.

RNA and Protein Synthesis

The genetic information stored in DNA molecules is used as a blueprint for making proteins. Why proteins? Because these macromolecules have diverse primary, secondary and tertiary structures that equip them to carry out the numerous functions necessary to maintain a living organism. As noted in the protein, these functions include:

- Structural integrity (hair, horn, eye lenses etc.).
- Molecular recognition and signaling (antibodies and hormones).
- Catalysis of reactions (enzymes)..
- Molecular transport (hemoglobin transports oxygen).
- Movement (pumps and motors).

The critical importance of proteins in life processes is demonstrated by numerous genetic diseases, in which small modifications in primary structure produce debilitating and often disasterous consequences. Such genetic diseases incluse Tay-Sachs, phenylketonuria (PKU), sickel cell anemia, achondroplasia, and Parkinson disease. The unavoidable conclusion is that proteins are of central importance in living cells, and that proteins must therefore be continuously prepared with high structural fidelity by appropriate cellular chemistry.

Early geneticists identifed genes as hereditary units that determined the appearance and/ or function of an organism (i.e. its phenotype). We now define genes as sequences of DNA that occupy specific locations on a chromosome. The original proposal that each gene controlled the formation of a single enzyme has since been modified as: one gene = one polypeptide.

The intriguing question of how the information encoded in DNA is converted to the actual construction of a specific polypeptide has been the subject of numerous studies, which have created the modern field of Molecular Biology.

THE CENTRAL DOGMA AND TRANSCRIPTION

Francis Crick proposed that information flows from DNA to RNA in a process called transcription, and is then used to synthesize polypeptides by a process called translation. Transcription takes place in a manner similar to DNA replication.

A characteristic sequence of nucleotides marks the beginning of a gene on the DNA strand, and this region binds to a promoter protein that initiates RNA synthesis.

The double stranded structure unwinds at the promoter site., and one of the strands serves as a template for RNA formation. The RNA molecule thus formed is single stranded, and serves to carry information from DNA to the protein synthesis machinery called ribosomes. These RNA molecules are therefore called messenger-RNA (mRNA).

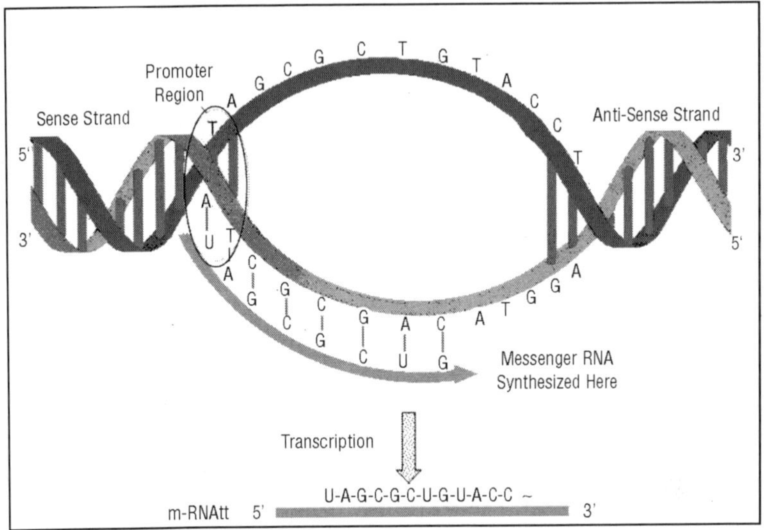

An important distinction must be made here. One of the DNA strands in the double helix holds the genetic information used for protein synthesis. This is called the sense strand, or information strand. The complementary strand that binds to the sense strand is called the anti-sense strand, and it serves as a template for generating a mRNA molecule that delivers a copy of the sense strand information to a ribosome.

The promoter protein binds to a specific nucleotide sequence that identifies the sense strand, relative to the anti-sense strand. RNA synthesis is then initiated in the 3' direction, as nucleotide triphosphates bind to complementary bases on the template strand, and are joined by phosphate diester linkages. A characteristic "stop sequence" of nucleotides terminates the RNA synthesis. The messenger molecule is released into the cytoplasm to find a ribosome, and the DNA then rewinds to its double helix structure.

In eucaryotic cells the initially transcribed m-RNA molecule is usually modified and shortened by an "editing" process that removes irrelevent material. The DNA of such organisms is often thousands of times larger and more complex than that composing the single chromosome of a procaryotic bacterial cell. This difference is due in part to repetitive nucleotide sequences (ca. 25 per cent iu the human genome). Furthermore, over 95 per cent of human DNA is found in intervening sequences that separate genes and parts of genes.

The informational DNA segments that make up genes are called exons, and the noncoding segments are called introns. Before the mRNA molecule leaves the nucleus, the nonsense bases that make up the introns are cut out, and the informationally useful exons are joined together in a step known as RNA splicing. In this fashion shorter mRNA molecules carrying the blueprint for a specific protein are sent on their way to the ribosome factories.

The Central Dogma of molecular biology, which at first was formulated as a simple linear progression of information from DNA to RNA to Protein, is summarized in the following illustration. The replication process on the left consists of passing information from a parent DNA molecule to daughter molecules. The middle transcription process copies this information to a mRNA molecule. Finally, this information is used by the chemical machinery of the ribosome to make polypeptides.

As more has been learned about these relationships, the central dogma has been refined to the representation displayed on the right. The dark blue arrows show the general, well demonstrated, information transfers noted above. The relatively simple RNA-protein structures of a virus may induce RNA replication or a reverse transcription from RNA to DNA (magenta arrow), but these are exceptional, albeit significant transfers.

The AIDS virus, for example, functions by the action of reverse transcriptase enzymes accompanying its RNA. Direct translation of DNA information into protein synthesis (orange arrow) has not yet been observed in a living organism. Finally, proteins appear to be an informational dead end, and do not provide a structural blueprint for either RNA or DNA.

TRANSLATION

Translation is a more complex process than transcription. This would, of course, be expected. After all, the coded messages produced by the German Enigma machine could be copied easily, but required a considerable decoding effort before they could be read with understanding.

In a similar sense, DNA replication is simply a complementary base pairing exercise, but the translation of the four letter (bases) alphabet code of RNA to the twenty letter (amino acids) alphabet of protein literature is far from trivial. Clearly, there could not be a direct one-to-one correlation of bases to amino acids, so the nucleotide letters must form short words or codons that define specific amino acids. Many questions pertaining to this genetic code were posed in the late 1950's:

- How many RNA nucleotide bases designate a specific amino acid?
  If separate groups of nucleotides, called codons, serve this purpose, at least three are needed. There are $4^3 = 64$ different nucleotide triplets, compared with $4^2 = 16$ possible pairs.
- Are the codons linked separately or do they overlap?
  Sequentially joined triplet codons will result in a nucleotide chain three times longer than the protein it describes. If overlapping codons are used then fewer total nucleotides would be required.

- If triplet segments of mRNA designate specific amino acids in the protein, how are the codons identified?
  For the sequence ~CUAGGU~ are the codons CUA & GGU or ~C, UAG & GU~ or ~CU, AGG & U~?
- Are all the codon words the same size?

In Morse code the most widely used letters are shorter than less common letters. Perhaps nature employs a similar scheme.

Physicists and mathematicians, as well as chemists and microbiologists all contributed to unravelling the genetic code. Although earlier proposals assumed efficient relationships that correlated the nucleotide codons uniquely with the twenty fundamental amino acids, it is now apparent that there is considerable redundancy in the code as it now operates. Furthermore, the code consists exclusively of non-overlapping triplet codons.

Clever experiments provided some of the earliest breaks in deciphering the genetic code. Marshall Nirenberg found that RNA from many different organisms could initiate specific protein synthesis when combined with broken E.coli cells (the enzymes remain active). A synthetic polyuridine RNA induced synthesis of poly-phenylalanine, so the UUU codon designated phenylalanine. Likewise an alternating ~CACA~ RNA led to synthesis of a ~His-Thr-His-Thr~ polypeptide. The following table presents the present day interpretation of the genetic code. Note that this is the RNA alphabet, and an equivalent DNA codon table would have all the U nucleotides replaced by T. Methionine and tryptophan are uniquely represented by a single codon. At the other extreme, leucine is represented by eight codons. The average redundancy for the twenty amino acids is about three. Also, there are three stop codons that terminate polypeptide synthesis.

The translation process is fundamentally straightforward. The mRNA strand bearing the transcribed code for synthesis of a protein interacts with relatively small RNA molecules (about 70-nucleotides) to which individual amino acids have been attached by an ester bond at the 3'-end. These transfer RNA's (tRNA) have distinctive three-dimensional structures cosisting of loops of single-stranded RNA connected by double stranded segments.

This cloverleaf secondary structure is further wrapped into an "L-shaped" assembly, having the amino acid at the end of one arm, and a characteristic anti-codon region at the other end.

The anti-codon consists of a nucleotide triplet that is the complement of the amino acid's codon(s). Models of two such tRNA molecules are shown to the right. When read from the top to the bottom, the anti-codons depicted here should complement a codon in the previous table.

A cell's protein synthesis takes place in organelles called ribosomes. Ribosomes are complex structures made up of two distinct and separable subunits (one about twice the size of the other). Each subunit is composed of one or two RNA molecules associated with 20 to 40 small proteins. The ribosome accepts a mRNA molecule, binding initially to a characteristic nucleotide sequence at the 5'-end.

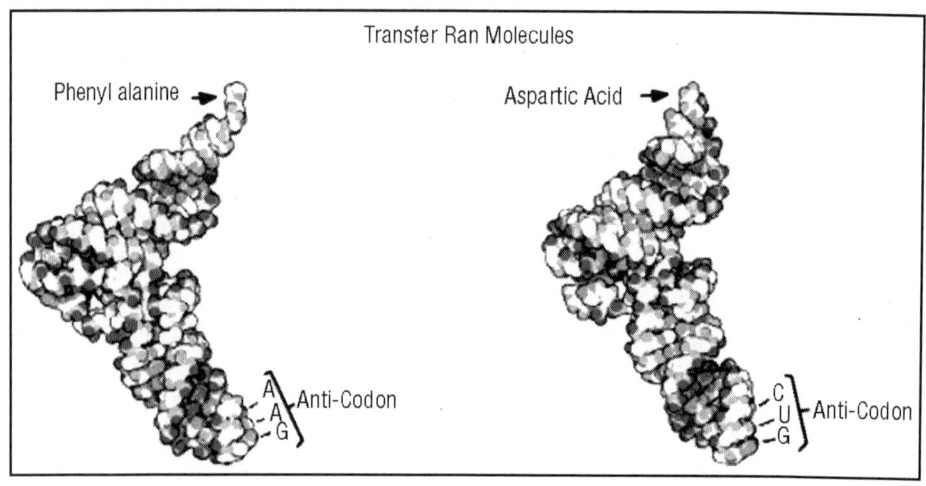

Fig. TRNA Molecules.

This unique binding assures that polypeptide synthesis starts at the e codon. A tRNA molecule with the appropriate ant-codon then attaches at the starting point and this is followed by a series of adjacent tRNA attachments, peptide bond formation and shifts of the ribosome along the mRNA chain to expose new codons to the ribosomal chemistry.

ANALYSIS OF STRUCTURAL SIMILARITIES AND DIFFERENCES BETWEEN DNA AND RNA

We know that living organisms have the ability to reproduce and to pass many of their characteristics on to their offspring. From this we may infer that all organisms have genetic substances and an associated chemistry that enable inheritance to occur. It is instructive to consider the essential requirements such genetic materials must fullfill.

- *Information:* Biologically useful information, especially instructions for protein sysnthesis, must be incorporated in the material.
- *Stability:* The inherited information must be stable (unchanged) over the lifetime of the organism if accurate copies are to be conveyed to the offspring. Infrequent changes may take place.
- *Reproduction:* A method of faithfully replicating the information encoded in the material, and transmitting this copy to the offspring must exist.
- *Mutability:* Despite the inherent stability noted above, the material must be capable of incorporating stable structural change, and passing this change on to succeeding generations.

Since this genetic substance has been identified as the nucleic acids DNA and RNA, it is instructive to examine the manner in which these polymers satisfy the above requirements.

INFORMATION STORAGE

The complexity of life suggests that even simple organisms will require very large inheritance libraries. Although the four nucleotides that make up of DNA might appear to be too simple for this task, the enormous size of the polymer and the permutations of the monomers within the chain meet the challenge easily. DNA has four letters in its alphabet (A, C, G & T), so the number of words that can be formed increase exponentially with the number of letters per word. Thus, there are 42 or 16 two letter words, and 43 or 64 three letter words. Assuring the stability of information encoded by the DNA alphabet presents a serious challenge. If the letters of this alphabet are to be strung together in a specific way on the polymer chain, chemical reactions for attaching (and removing) them must be available. Simple carboxylic ester or amide links might appear suitable for this purpose step-growth polymerization, but these are used in lipids and polypeptides, so a separate enzymatic machinery would be needed to keep the information processing operations apart from other molecular transformations. The overall stability of such covalent links presents a more serious problem. Under physiological conditions (aqueous, pH near 7.4 & 27 to 37° C) esters are slowly hydrolyzed. Amides are more stable, but even a hydrolytic cleavage of one bond per hour would be devastating to a polymer having tens of thousands to millions such links.

Furthermore, short difuctional linking groups, such as carbonates, oxylates and malonates show enhanced reactivity, and their parent acids are unstable or toxic. Phosphate is an ubiquitous inorganic nutrient. Mono, di and triesters of the corresponding acid (phosphoric acid) are all known. Because of their acidity (pKa H" 2), the mono and diesters are negatively charged at physiological pH, rendering them less susceptible to nucleophilic attack. The influence of negative charge on the rate of nucleophilic hydrolysis of some representative esters is shown in the table on the right. Clearly, a polymer in which monomer units are joined by negatively charged diphosphate ester links should be substantially more stable than one composed of carboxylate ester bonds. The negative charge found on all biological phosphate derivatives serves other purposes as well.

Table. Hydrolysis at 35°C and pH 7

Ester	Rate of Hydrolysis	Relative Rate
Ethyl Acetate $CH_3CO_2C_2H_5$	1.0×10^{-2}	5×10^{6}
Trimethyl Phosphate $(CH_3O)_3PO$	3.4×10^{-4}	2×10^{5}
Dimethyl Phosphate $(CH_3O)_2PO_2(-)$	2.0×10^{-9}	1.0

- The diphosphate ester links that join the nucleotides units of DNA are formed by phosphorylation reactions involving nucleotide triphosphate reagents. These reagents are the phosphoric acid analogs of carboxylic acid anhydrides, a functional group that would not survive the aqueous environment of a cell. The high density of negative charge on the triphosphate function not only solubilizes the organic moiety to which it is attached, but also reduces the rate at which it is hydrolyzed.
- Living cells must conserve and employ their chemical reagents within a volume defined and enclosed by a membrane barrier. These lipid bilayer membranes have hydrophobic interiors, which resist the passage of ions. Indeed, special trans-membrane structures called ion channels exist so that controlled ion transport across a membrane may take place. Small neutral organic molecules, such as adenosine, cytidine and guanosine, may pass through lipid membranes, albeit at a reduced rate, but their mono, di and triphosphate derivatives are more tightly sequestered in the cell.

SUGAR MOIETY IN DNA

Common perhydroxylated sugars, such as glucose and ribose, are formed in nature as products of the reductive condensation of carbon dioxide we call photosynthesis. The formation of deoxysugars requires additional biological reduction steps, so it is reasonable to speculate why DNA makes use of the less common 2'-deoxyribose, when ribose itself serves well for RNA. At least two problems associated with the extra hydroxyl group in ribose may be noted. First, the additional bulk and hydrogen bonding character of the 2'-OH interfere with a uniform double helix structure, preventing the efficient packing of such a molecule in the chromosome. Second, RNA undergoes spontaneous hydrolytic cleavage about one hundred times faster than DNA.

This is believed due to intramolecular attack of the 2'-hydroxyl function on the neighbouring phosphate diester, yielding a 2',3'-cyclic phosphate. If stability over the lifetime of an organism is an essential characteristic of a gene, then nature's selection of 2'-deoxyribose for DNA makes sense. Structural stability is not a serious challenge for RNA. The transcripted information carried by mRNA must be secure for only a few hours, as it is transported to a ribosome.

Once in the ribosome it is surrounded by structural and enzymatic segments that immediately incorporate its codons for protein synthesis. The tRNA molecules that carry amino acids to the ribosome are similarly short lived, and are in fact continuously recycled by the cellular chemistry.

THE THYMINE VS. URACIL ISSUE

Structural formulas for the three pyrimidine bases, cytosine, thymine and uracil are shown on the right. The carbon atoms that are part of these compounds may be categorized as follows. All of these compounds are apparently put together from a three-carbon malonate-like precursor and a single high oxidation state carbon species.

Such biosynthetic intermediates are well established. Thymine is unique in having an additional carbon, the green methyl group. Biosynthesis of this compound must involve additional steps, thus adding constructional complexity to the DNA molecules in which it replaces uracil.

The reason for the substitution of thymine for uracil in DNA may be associated with the repair mechanisms by which the cell corrects damage to its DNA. One source of error in the code is the slow hydrolysis of heterocyclic enamines, such as cytosine and guanine, to their corresponding lactams.

This changes the structure of the base, and disrupts base pairing in a manner that can be identified and then repaired. However, the hydrolysis product from cytosine is uracil, and this mismatched species must somehow be distiguished from the uracil-like base that belongs in the DNA. The extra methyl group serves this role nicely.

THE VITAMINS

Vitamin, any of the organic (carbon-containing) compounds that the body requires in small amounts to maintain health and function properly. Children additionally need vitamins to grow. The body gets most of its vitamins from the foods we eat. A healthy diet with plenty of fruits and vegetables should provide nearly all of the vitamins a person needs. Scientists have classified 13 compounds as vitamins. They have given most of these vitamins letter or letter plus number names, such as A, B_{12}, and D. Most vitamins are produced by plants. Some, such as vitamin D, are produced only by animals. A few vitamins are made by the body itself. For example, bacteria in the digestive tract help produce vitamin K, and the skin uses sunlight to produce vitamin D.

Vitamins are also manufactured for sale as supplements for people who need additional vitamins to meet their body's requirements.

For example, doctors often prescribe vitamin supplements for pregnant and nursing women to provide the additional nutrients needed by a rapidly growing fetus or infant. Older adults may not meet their vitamin requirements through food because the body's ability to absorb vitamins is impaired with age. In the United States, since 1940, the Food and Nutrition Board of the National Research Council has published recommended dietary allowances (RDA) for vitamins, minerals, and other nutrients. Expressed in milligrams (mg) or international units (IU) for adults and children of normal health, these recommendations provide useful guidelines for daily intake. Such guidelines are useful not only for professionals in nutrition but also for the growing number of families and individuals who eat irregular meals and rely on prepared foods, most of which are now required to carry nutritional labeling.

VITAMINS ARE VITAL

The word *vitamine,* later shortened to *vitamin,* was coined by Polish-American chemist Casimir Funk in the early 20th century. Funk was searching for the then-unknown substance in foods that prevents such diseases as beriberi, rickets, and scurvy. After experiments on pigeons, Funk guessed that an amine—a compound containing nitrogen and hydrogen—in foods was responsible.

He called this substance a *vital* (necessary for life) *amine,* or vitamine. Scientists later discovered that not all vitamins were amines, however.

Vitamins help the body carry out essential biochemical processes. Vitamins generally combine with proteins to create enzymes that promote chemical reactions. These enzymes play an important role in metabolism—the reactions that break down the fats, carbohydrates, and proteins in food so that the body can use them for energy and cell repair.

The enzymes also promote reactions involved in the formation of bone, hormones, blood cells, nervous-system chemicals, and genetic material. Without vitamins, many of these reactions would slow down or cease. The different vitamins are not chemically related, and most differ in their actions in the body.

Deficiency of particular vitamins can lead to various diseases. Too little vitamin C, for example, can cause rickets, a disease in which the bones fail to develop properly. Bowlegs or "knock knees" are signs of rickets. Rickets and the other vitamin-deficiency diseases that interested Casimir Funk have largely disappeared in the developed world as the result of fortified foods and improved nutrition. But these diseases still occur in the developing world, especially where malnutrition is common. Interest in vitamins grew during the 1990s, especially in those vitamins that act as antioxidants—vitamins A, C, and E. Antioxidants neutralize molecules known free radicals that cause cell damage. Free radicals have been linked to a number of disorders, including Alzheimer's disease, arteriosclerosis, cancer, diabetes, and Parkinson disease. Researchers theorized that increased doses of antioxidants might also prevent and even cure such diseases.

These claims have so far not panned out, however, as scientific studies have failed to demonstrate the preventive abilities of antioxidants. Research on antioxidants and other vitamins continues.

WATER-SOLUBLE VITAMINS

Water-soluble vitamins include eight well-recognized members of the vitamin B complex: thiamine, riboflavin, pyridoxine, pantothenic acid, niacin, biotin, folic acid and vitamin B_{12}; the water-soluble essential nutritional factors: choline, inositol, ascorbic acid; and vitamins with less-defined activity for fish: p-aminobenzoic acid, lipoic acid and citrin.

Table. Water-Soluble Vitamin Test Diet H-440

Complete test diet (g)	Vitamin mix Mineral mix (mg) (mg)				
Vitamin-free casein	38	Thiamine HCl	5	USP XII No. 2	plus
Gelatin	12	Riboflavin	20	$AlCl_3$	15
Corn oil	6	Pyridoxine HCl	5	$ZnSO_4$	300
Cod liver oil	3	Choline chloride	500	CuCl	10
White dextrin	28	Nicotinic acid	75	$MnSO_4$	80
Cellulose mixture	9	Calcium	50	KI	15
			pantothenate		
Cellulose	8	Inositol	200	$CaCl_2$	100
Vitamins	$\frac{1}{9}$	Biotin	0.5	per 100 g of salt mixture	
		Folic acid	1.5		
Mineral mix	4	L-Ascorbic acid	100		
Water	200	Vitamin B_{12}	0.01		
Total diet as fed	300	Menadione (K)	4		
		Tocopherol acetate (E)	40		

The first eight, though required only in small amounts in the diet, play major roles in growth, physiology and metabolism.

Choline, inositol and ascorbic acid are required in appreciable quantities in the diet and sometimes are not referred to as vitamins but as major dietary nutrients. A typical vitamin test diet for fish control over the water-soluble vitamins is listed in Table.

- Diet preparation: Dissolve gelatin in cold water. Heat with stirring on water bath to 80°C. Remove from heat. Add with stirring - dextrin, casein, minerals, oils, and vitamins as temperature decreases. Mix well to 40°C. Pour into containers; move to refrigerator to harden. Remove from trays and store in sealed containers in refrigerator until used. Consistency of diet adjusted by amount of water in final mix and length and strength of beating.
- Delete two parts a -cellulose and add two parts CMC for preliminary feeding
- Dissolve a -tocopherol in oil mix
- Add vitamin B_{12} in water during final mixing

THIAMINE

Experimental avian polyneuritis, a condition resembling beriberi, was produced by Eijkman in Java in 1886 and the anti-beriberi factor was crystallized and named "vitamine". The term "thiamine" was introduced when the chemical nature of the factor was established. Thiamine was isolated from rice polishings in 1926 and synthesized in 1936.

Chemistry

Thiamine hydrochloride is a water-soluble, colourless, monoclinic, crystalline compound. It is comparatively stable to dry heat but is rapidly broken down in neutral or alkaline solutions and is split by sulphites into constituent pyrimidine and thiazole moieties. It has a characteristic yeast-like odour. The pyrimide ring is relatively stable, but the thiazole ring is easily opened by hydrolysis.

Several derivatives are stable to heat and appear to be more completely soluble in weak alkaline solutions than thiamine itself and still show biological activity in animals. These derivatives include thiamine propyl disulphide, benzoylthiamine disulphide, dibenzoylthiamine, and benzoyl-thiamine monophosphate. Both thiamine hydrochloride and thiamine mononitrate have been successfully used as the active vitamin in test diets for fish nutrition studies.

Positive Functions

Thiamine functions in all cells as the coenzyme cocarboxylase, thiamine pyrophosphate, which participates in the oxidative decarboxylation of pyruvic acid to acetate for entry into the tricarboxylic acid (TCA) cycle.

THIAMINE PYROPHOSPHATE

Thiamine pyrophosphate is also a coenzyme of the transketolase system by which direct oxidation of glucose occurs in the cytoplasm of cells via the pentose phosphate pathway. Erythrocyte (RBC) levels of metabolites of this system have been used to indicate thiamine status in experimental animals, including salmon and trout.

Thiamine is essential for good appetite, normal digestion, growth and fertility. It is needed for normal functioning of the nervous tissue and the requirement is determined by the caloric density of the diet.

Deficiency Syndrome

Deficiency signs in salmonids include impaired carbohydrate metabolism, nervous disorders, poor appetite, poor growth, and increased sensitivity to shock. A trunk-winding symptom in eels has been reported, together with haemorrhage at the base of the fins. Skin congestion and subcutaneous haemorrhage occurs in carp fed thiamine-deficient diets. Typical symptoms observed in salmonids, carp and catfish are listed in Table. Thiamine deficiency has also been reported in marine flatfish started on clam neck diets stored long enough for thiaminase present to hydrolyze the thiamine in the ration. Typical nervous paralysis occurred with rapid mortality from physical shock.

Table. Vitamin Deficiency Syndromes

Vitamin	Symptoms in salmon, trout, carp, catfish
Thiamine	Poor appetite, muscle atrophy, convulsions, instability and loss of equilibrium, oedema, poor growth
Riboflavin	Corneal vascularization, cloudy lens, haemorrhagic eyes - photophobia, dim vision, incoordination, abnormal pigmentation of iris, striated constrictions of abdominal wall, dark colouration, poor appetite, anaemia, poor growth
Pyridoxine	Nervous disorders, epileptiform fits, hyperirritability, ataxia, anaemia, loss of appetite, oedema of peritoneal cavity, colourless serous fluid, rapid postmortem rigor mortis, rapid and gasping breathing, flexing of opercles
Pantothenic acid	Clubbed gills, prostration, loss of appetite, necrosis and scarring, cellular atrophy, gill exudate, sluggishness, poor growth
Inositol	Poor growth, distended stomach, increased gastric emptying time, skin lesions
Biotin	Loss of appetite, lesions in colon, colouration, muscle atrophy, spastic convulsions, fragmentation of erythrocytes, skin lesions, poor growth
Folic acid	Poor growth, lethargy, fragility of caudal fin, dark colouration, macrocytic anaemia
Choline	Poor growth, poor food conversion, haemorrhagic kidney and intestine
Nicotinic acid	Loss of appetite, lesions in colon, jerky or difficult motion, weakness, oedema of stomach and colon, muscle spasms while resting, poor growth
Vitamin B_{12}	Poor appetite, low haemoglobin, fragmention of erythrocytes, macrocytic anaemia
Ascorbic acid	Scoliosis, lordosis, impaired collagen formation, altered cartilage, eye lesions, haemorrhagic skin, liver, kidney, intestine, and muscle
p-Aminobenzoic acid	No abnormal indication in growth appetite, mortality

Requirements

In determining dietary thiamine requirements of fish, some considerations must be placed upon the composition of dietary ingredients in the ration. Fat content of the diet may affect not only caloric intake but also the thiamine requirement because cocarboxylase participates in the oxidation of that through a-ketoglutarate. Therefore, fish on a high fat diet and low thiamine intake might take longer to develop deficiencies and will give an erroneous requirement.

Sources and Protection

Common sources for thiamine are dried peas, beans, cereal bran, and dried yeast. Fresh glandular tissue is also a good source for thiamine and other members of the vitamin B water-soluble complex, but is seldom used in modern commercial fish diets. Thiamine can be easily lost by holding diet ingredients too long in storage or by preparing the diet under slightly alkaline conditions or in the presence of sulphide. Wet or frozen diets pose a problem because moisture content increases the chance of enzymatic (thiaminase) hydrolysis and subsequent destruction of thiamine. Obviously, wet or moist diet, preparations containing any fresh fish or shellfish tissue must be used immediately.

Antimetabolites and Inactivation

Acetylcholine is an antagonist to thiamine and pyrithiamine. Oxythiamine and normal butylthiamine are specific antimetabolites.

Several thiaminases occur which destroy thiamine. These rupture the thiazole ring at the sulphur bond making the residue inactive. Freshwater fish tissues have high thiaminase activity as do tissues from clams, shrimp and mussels. Thiaminases have also been found in beans and mustard seed and in several micro-organisms.

Thiaminase activity is low in most saltwater fish tissues, however, and the enzyme is inactivated by heating or prolonged pasteurization. Thiamine present in fresh Torula yeast is relatively unavailable to fish, but the yeast becomes an excellent thiamine source after rupturing the cells by steam treatment or by dehydration.

Clinical Assessment

Clinical assessment of thiamine status have been made by measuring erythrocyte transketolase activity. Typical saturated levels for thiamine activity in sea salmon range from 15 to 20 g of thiamine/g of wet liver tissue. Fingerling chinook or coho salmon reared in 10 or 15° C water on test diets containing the listed amount or more of thiamine hydrochloride: assayed at 8-10 g of thiamine/ g of wet liver tissue. These liver storage levels and normal erythrocyte transketolase activity in the absence of any deficiency sign will indicate an adequate thiamine intake for that fish population.

RIBOFLAVIN

Following the recognition that the original water-soluble B consisted of more than one factor, a new growth-promoting factor consisting of yellow-green pigments was isolated in 1879. This new factor was designated G by some and

83 by others. The second water-soluble vitamin discovered was given its specific chemical name riboflavin. Lactoflavin, hepatoflavin, and ovoflavin were also shown to be identical with the pure riboflavin. Riboflavin occurs in the free form only in the eye, whey and urine.

Table. Vitamin Requirements for Growth

Vitamin (mg/kg dry diet)	Rainbow trout	Brook trout	Brown trout	Atlantic salmon	Chinook salmon	Coho salmon
Thiamine	10–12	10–12	10–12	10–15	10–15	10–15
Riboflavin	20–30	20–30	20–30	5–10	20–25	20–25
Pyridoxine	10–15	10–15	10–15	10–15	15–20	15–20
Pantothenate	40–50	40–50	40–50	–	40–50	40–50
Niacin	120–150	120–150	120–150	–	150–200	150–200
Folacin	6–10	6–10	6–10	5–10	6–10	6–10
Cyanocobalamin	–	–	–	–	0.015–0.02	0.015–0.02
myo-Inositol	200–300	–	–	–	300–400	300–400
Choline	–	–	–	–	600–800	600–800
Biotin	1–1.5	1–1.5	1.5–2	–	1–1.5	1–1.5
Ascorbate	100–150	–	–	–	100–150	50–80
Vitamin A	2 000–2 500 I.U.	–	–	–	–	–
Vitamin E [2/]	–	–	–	–	40-50	–
Vitamin K	*	*	*	-	*	*

- Fish fed at reference temperature with diets at about protein requirement
- Requirement directly affected by amount and type of unsaturated fat fed
- Denotes a requirement, the level of which has not been established

Table. Vitamin Requirements for Growth

Vitamin (mg/kg dry diet)	Carp	Channel catfish	Eel	Sea bream	Turbot	Yellowtail
Thiamine	2–3	1–3	2–5	–	2–4	–
Riboflavin	7–10	–	–	–	–	–
Pyridoxine	5–10	–	–	2–5	–	–
Pantothenate	30–40	25–30	–	–	–	–
Niacin	30–50	–	–	–	–	–
Folacin	–	–	–	–	–	–
Cyanocobalamin	–	–	–	–	–	–
myo-Inositol	200–300	–	–	300–500	–	–
Choline	500–600	–	–	–	–	–
Biotin	1–15	–	–	–	–	–
Ascorbate	30–50	30–50	–	–	–	–
Vitamin A	1 000–2 000 I.U.	–	–	–	–	–
Vitamin E [2/]	80–100	–	–	–	–	–
Vitamin K	–	–	–	–	–	-

- Fish fed at reference temperature with diets at about protein requirement

- Requirement directly affected by amount and type of unsaturated fat fed
- Denotes a requirement, the level of which has not been established

PANTOTHENIC ACID

A chick dermatitis was cured by Elvehjem and Köehn in 1935 using a factor containing b -alanine. The term pantothenic acid was given earlier by Williams to a yeast growth factor which was later recognized as the anti-dermatitis factor.

Pantothenic acid was synthesized by Stiller in 1940. Phillips observed clubbed gills in trout fed pantothenic acid-deficient diets in 1945. Rucker detected the same condition in salmon fed low pantothenic acid diets.

CHEMISTRY

Pantothenic Acid

Pantothenic acid may be considered as dihydroxydime-thylbutyric acid bonded to β -alanine. The free acid is a yellow, viscous oil and therefore the compound generally used in fish diet preparation is the calcium salt. This salt is a white crystalline powder readily soluble in water, mild acid, and is almost insoluble in organic solvents.

It is stable to oxidizing and reducing agents and to autoclaving, but is labile to dry heat, hot alkali, or hot acid. Pantothenol has almost as much activity as pantothenic acid for growth of chicks. Pantothenic acid acetate, benzoate, and diphosphate esters are biologically active for animals but not for lactic acid bacteria.

The optical isomer L-pantothenic acid, appears physiologically inert. Some organisms may utilize a portion of the molecule. Bacteria appeared to require only the dihydroxydimethylbutyric acid, and some yeasts utilize only g-alanine. Animals, however, need the entire pantothenic acid molecule or its reduced alcohol form.

Positive Functions

Pantothenic acid is part of acetyl coenzyme A which occurs in many enzymatic processes involving 2-carbon compounds. It has been shown to be required by all animal species studied and by many micro-organisms.

The acetyl coenzyme A system is involved in the acetylation of aromatic amines and choline; condensation reactions for synthesis of acetate, fatty acids, and citrate; the oxidation of pyruvate and acetaldehydc; and is essential for the development of the central nervous system. The 2-carbon fragment called 'active acetate', or acetyl coenzyme A, is an essential intermediate in metabolism. It is involved in acylation of acetate, succinate, benzoate, propionate and butyrate. Pantothenic acid is involved in adrenal function and for the production of cholesterol. Coenzyme A is also involved in many other steps of intermediate metabolism of carbohydrates, fats and proteins.

Deficiency syndromes

Under standard test conditions with deficient diet fed in 10-15°C water systems, salmon and trout exhaust pantothenic acid stores within 8-12 weeks. Fish stop feeding and close examination of gill filament show proliferation of epithelial surface plus swelling and clubbing together of the filaments and lamellae.

The opercules become distended and the surface of the gills is often covered with an exudate. Fish become prostrate or sluggish. Necrosis, scarring and cellular atrophy of the tender gill elements occur and anaemia develops after prolonged deficiency. Dietary gill disease has been adequately described and correlated with pantothenic acid deficiency.

The same type of symptom has been observed in salmon, trout, eel, carp and catfish. After replacement of pantothenic acid in the diet, recovery is rapid for those fish still feeding and gross deficiency symptoms disappear after about four weeks on the recovery diet although evidence of necrosis and scarring remains.

Sources and Protection

Good sources for pantothenic acid are cereal bran, yeast, liver, kidney, heart, spleen and lung. Fish flesh is a relatively rich source, although the content is only about 20 per cent of pantothenic acid found in animal glandular tissue. Royal jelly probably contains the greatest amount as it contains over 500 μ g of pantothenol/g dry weight.

The calcium or sodium salt of pantothenic acid is relatively stable and can be incorporated into either moist or dry fish diets. Some loss is incurred during autoclaving and excessive heat should therefore be minimized during diet preparations.

Since the free acid is labile to heat and also to acid and alkali, some loss can be expected during moist diet preparation or during storage. Certain cereal brans may have pantothenic acid bound in a form unavailable to fish because of the low digestibility coefficient and should not be relied upon as the sole pantothenic acid source in the diet.

Antimetabolites and Inactivation

Since pantothenic acid affects the respiration of many types of cells, compounds like 6-mercaptopurine, 2,6-diaminopurine, and 8-azaguanine, which inhibit growth of tumors are antagonistic toward pantothenic acid. Pantoyltaurine is an antimetabolite of pantothenic acid and has been used to accelerate deficiency syndromes in experimental animals.

Methyl-ω -pantothenic acid was reported to interfere with the formation of acetyl coenzyme A and accelerates deficiency syndromes in animals. This compound inhibits sulfanilamide acetylation in pigeon liver homogenates but did not prevent citric acid formation.

Pantothenic acid can be used to overcome these inhibitory effects, including the reversal of mitosis blockage by 7-mecaptopurine in animals. High levels of calcium pantothenate in the diet also has a transient effect on the migratory urge of salmon.

Clinical Assessment

Assay of pantothenic acid content in the diet may be misleading unless care is exercised in proper hydrolysis of the raw materials being assayed since pantothenic acid is only slowly liberated by normal hydrolytic procedures. Complete hydrolysis with enzyme preparations will liberate all of pantothenic acid from biologically active material of glandular tissues, fish flesh, yeast, and bran.

Salmon feeding actively in the oceans showed liver pantothenic acid content of 18-20 μ/g of fresh tissue. Young chinook and coho salmon fingerlings reared in fresh water at 12-15°C showed maximum liver storage at about 14-16 m g of pantothenic acid/g of fresh liver tissue. Load tests and acetylation reactions of sulfanilamide have been used in man and other experimental animals but have not yet been extended for assessment of pantothenic acid status in fish.

NIACIN

Nicotinic acid was synthesized in 1873 but was left on the shelf as an organic compound unrelated to the severe pellagra afflictions occurring throughout the world at that time. Sixty years later the compound was shown to be present in coenzymes I and II and two years thereafter Elvehjem cured 'black tongue' in dogs with the vitamin. Niacin was postulated to be part of factor H for fish in 1937, but deficiency symptoms were not adequately described until reported in trout in 1947.

CHEMISTRY

Nicotinic Acid

Niacin is the preferred nomenclature. Nicotinic acid amide or niacinamide is the common form in which the vitamin is physiologically active.

Niacin is a white, crystalline solid, soluble in water and alcohol. It is stable in the dry state and may be autoclaved for short periods without destruction. It is also stable to heat in mineral acids and alkali. Niacin is both a carboxylic acid and an amine and forms quaternary ammonium compounds because of its basic nature.

Acidic characteristics include salt formation with alkali. Niacin can be esterified easily, then converted to amides. Niacinamide is a crystalline powder soluble in water and ethanol and the dry material is stable up to about 60°C.

In aqueous solutions it is stable for a short period when autoclaved. It is the form in which the vitamin is normally found in niacinamide adenine dinucleotide (NAD), and in niacinamide adenine dinucleotide phosphate (NADP).

POSITIVE FUNCTIONS

The major function of niacin in NAD and NADP is hydrogen transport in intermediary metabolism. Most of these enzyme systems function by alternating between the oxidized and reduced state of the coenzymes NAD-NADH and NADP-NADPH. Oxidation-reduction reactions may be anaerobic as when pyruvate acts as the hydrogen acceptor and lactate is formed, or the reactions may be coupled to electron transport systems with oxygen as the ultimate hydrogen acceptor; such aerobic reactions occur in respiration.

Both NAD and NADP are involved in the synthesis of high energy phosphate bonds which furnish energy for certain steps in glycolysis, in pyruvate metabolism, amino acid and protein metabolism, and in photosynthesis.

An interrelationship between thiamine and niacin exists since both vitamins are coenzymes in intermediary metabolism whereby food is oxidized to furnish energy for physiological functions, to maintain homeostasis, and for body temperature maintenance in homeotherms.

DEFICIENCY SYNDROME

Stores of niacin are more slowly exhausted under experimental conditions than are some of the other vitamins resulting in less defined and more slowly developing symptoms.

Niacin deficiencies in fish were experimentally induced in the late forties and early fifties by using basal diets which had a low niacin content. Loss of appetite and poor food conversion were the first signs noted. Then the fish turned dark and went off feed, followed by the appearance of lesions in the colon, erratic motion, oedema of the stomach and colon and muscle spasms while fish were apparently resting.

A predisposition to sunburn in fish confined in the open in shallow ponds or raceways was described. Common carp showed a congestion of the skin with subcutaneous hemorrhages. Common symptoms of niacin deficiency in most fish studied were muscular weakness and spasms, coupled with poor growth and poor food conversion.

Requirements

In homeotherms on a balanced test ration the niacin requirement is generally estimated to be about ten times that of the thiamine requirement. These rations generally contain considerable carbohydrate material to furnish energy to maintain body temperature. In fish the requirement appears to be twenty to thirty times that of the thiamine needs determined for the same test conditions and test rations.

This difference may be due to low carbohydrate content of young fish diets and the higher protein content of these rations. Conversion of tryptophan to niacin occurs in the mammalian liver and possibly also in the liver of fish. This conversion may account for the slow development of the niacin-deficiency syndrome in fish. However, after 10-14 weeks on diets devoid of niacin, deficiency symptoms did occur in several species of fish. The symptoms were reduced by replacement of niacin in the ration even when high protein diets containing an excess of tryptophan were fed. Too much niacin inhibits growth.

Sources and Protection

Niacin is found in most animal and plant tissues. Rich sources are yeast, liver, kidney, heart, legumes, and green vegetables. Wheat contains more niacin than corn and the vitamin is also found in milk and egg products. The vitamin is very stable since it is generally found in coenzyme form in raw materials. Niacin added to the diet as a supplement remains relatively unaltered during diet manufacture, processing, and storage.

Antimetabolites and Inactivation

Pyridine-3-sulphonic acid and 3-acetylpyridine are compounds structurally related to niacin and are antimetabolites for this vitamin in animals and in micro-organisms. Additional niacin can overcome the anti-metabolic effect. Deficiency symptoms in rats may be induced by 6-amino niacinamide. The symptoms are reversed by addition of ten times more niacinamide than the antimetabolite. Thioacet-amide has been reported to be a niacin antagonist in fish.

Clinical Assessment

Salmon feeding actively in the oceans showed liver niacin content of 70-80 μ g of niacin/g of fresh tissue. About half of this amount is present in fingerling salmon raised in fresh water at 12-15°C and fed test rations containing 40-50 per cent protein and niacin supplements 500-750 mg/kg dry diet. Urinary metabolities of niacin have been measured in other animals on standard niacin load in test rations containing a standard tryptophan load. The technique is well developed to measure the N^1-methyl derivative in mammalian urine. These data have not been reported for fish but metabolism chambers are available for collecting branchial and urinary wastes from large fish intubated with different diet material.

BIOTIN

Certain foods, notably liver and kidney, were found to be protective against a form of dermatitis resulting from the consumption of egg white by rats, in the early thirties. It was soon learned that a specific component of egg albumin, avidin, rendered dietary biotin unavailable, hence producing the symptoms.

Biotin was variously called coenzyme R and 'vitamin H'. It was isolated by Du Vigneaud in 1941 and synthesized by workers of Merck and Company in 1943. Biotin was once thought to be part of factor H for fish. Blue slime patch disease due to biotin deficiency was reported in trout.

CHEMISTRY

Biotin

Biotin is a monocarboxylic acid slightly soluble in water and alcohol and insoluble in organic solvents. Salts of the acid are soluble in water. Aqueous solutions or the dry material are stable at 100 C and to light. The vitamin is destroyed by acids and alkalis and by oxidizing agents such as peroxides or permanganate. Biocytin is a bound form of biotin isolated from yeast, plant, and animal tissues. Other bound forms of the vitamin can generally be liberated by peptic digestion. Oxybiotin has partial vitamin activity but oxybiotin sulphonic acid and other analogues are antimetabolites inhibiting the growth of bacteria.

Avidin, a protein found in raw egg white, binds biotin and makes it unavailable to fish and other animals. Heating to denature the protein makes the bound biotin available again to the fish. Biocytin or e -biotinyl lysine (the epsilon amino group of lysine and the carboxyl of biotin being combined in a peptide bond) is hydrolyzed by the enzyme biotinase making the protein-bound biotin available.

POSITIVE FUNCTIONS

Biotin is required in several specific carboxylation and decarboxylation reactions, including the carboxylation of pyruvic acid to form oxaloacetic acid. It is part of the coenzyme of several carboxylating enzymes fixing CO_2 such as propionyl coenzyme A involved in the conversion of propionic acid to succinic acid in methylmalonyl coenzyme A.

Biotin is also involved in the conversion of acetyl CO_2 to malonyl coenzyme A in the formation of long chain fatty acids. It has possible involvement in citrulline synthesis and may have effects on purine and pyrimidine synthesis. It is involved in the conversion of unsaturated fatty acids to the stable cis form in the synthesis of biologically active fatty acids.

DEFICIENCY SYNDROME

Some signs of biotin deficiency in salmonids are skin disorders, muscle atrophy, lesions in the colon, loss of appetite, and spastic convulsions. Haematology discloses fragmentation of erythrocytes. Poor growth is a common symptom and has been reported for salmonids, common carp, goldfish (Carassius auratus) and eel. Blue slime patch disease in brook trout deficient in biotin appears typical for this species. Fish reared in 10 -15 °C water exhaust biotin stores in 8-12 weeks and the first signs are anorexia, poor food conversion, and general listlessness before the more acute deficiency symptoms become detectable.

Sources and Protection

Rich sources of biotin are liver, kidney, yeast, milk products, and egg yolks. Nut meats contain good supplies of biotin. The diet should be protected from strong oxidizing agents or conditions which promote oxidation of ingredients. Raw egg white should not be incorporated into moist fish diets. Cooking will inactivate the avidin.

Antimetabolites and Inactivation

Raw egg white has already been discussed which irreversibly binds biotin and makes it unavailable to young fish. Many biotin homologues with different side chain lengths inhibit the growth of bacteria. Oxybiotin, in which sulphur is replaced by oxygen, has about the same biological activity as the natural biotin. On the other hand, oxybiotin sulphonic acid inhibits biotin activity.

Clinical Assessment

Measurement or urinary excretion of biotin in animals does not prove a good clinical method since biotin is synthesized by several organisms in the gut. In fish, this technique needs to be explored and may well be valuable as a clinical tool since the gastrointestinal tract of many freshwater fish contains only small quantities of bacteria.

Biotin is one of the most expensive vitamins to add to fish rations. Salmon feeding actively in the oceans had liver biotin concentrations of 10-12 mg of wet liver tissue. The concentrations in the liver of young salmon fingerlings in fresh water fed test diets containing an excess of biotin were between 6-8 mg of the vitamin/g of tissue. Fish with these levels of biotin in the liver should probably be in sound biotin nutritional status.

FOLIC ACID (FOLACIN)

A factor effective in curing metaloblastic anaemia in monkeys was found in yeast and liver concentrates and was designated vitamin M in 1935. In 1939 an anti-anaemic factor was found in liver and called vitamin B_C. These were later shown to be the same substance with folic acid as the active ingredient. Folic acid was synthesized in 1946 and was soon used in fish diets for preventing purified diet anaemia.

CHEMISTRY

Folic Acid

Folic acid crystallizes into yellow spear-shaped leaflets which are soluble in water and dilute alcohol. It can be precipitated with heavy metal salts. It is stable to heat in neutral or alkaline solution, but unstable in acid solution. It deteriorates when exposed to sunlight, or during prolonged storage. Several analogues have biological activity including pteroic acid, rhizopterin, folinic acid, xanthopterin and several formyltetrahydropteroyl-glutamic acid derivatives.

These have closely allied ring structures and many have been isolated as derivatives in various animals or microbiological preparations. One simple form, xanthopterin, present in the pigments of insects, is shown below and is of special interest because of early work with this compound as the anti-anaemic factor H for fish.

Positive Functions

Folic acid is required for normal blood cell formation and is involved as a coenzyme in one-carbon transfer mechanisms.

In the presence of ascorbic acid, folic acid is transformed into the active (5-formyl-5, 6,7,8) tetrahydrofolic acid. Folic acid is involved in many one-carbon metabolism systems such as serine and glycine interconversion, methionine-homocysteine synthesis, histidine synthesis, and pyrimidine synthesis.

Several coenzyme forms of the active vitamin have been isolated. Folic acid is involved in the conversion of megaloblastic bone marrow to normoblastic type. It has a role in blood glucose regulation and improves cell membrane function and hatchability of eggs.

Deficiency Syndrome

Macrocytic normochromic anaemia occurs in several experimental animals, including fish fed diets devoid of the folic acid. Increasing numbers of senile cells are observed as the deficiency progresses until only a few old and degenerating cells are found in the blood of deficient fish.

Anterior kidney imprints disclose only adult cells and no preforms present. Other signs observed have been poor growth, anorexia, general anaemia, lethargy, fragile fins, dark skin pigmentation, and infarction of spleen.

SOURCES AND PROTECTION

Yeast, green vegetables, liver, kidney, glandular tissue, fish tissue, and fish viscera are good sources of folic acid. Insects contain xanthopterin which has folic acid activity. At one time the yellow pigment of xanthopterin was identified as the fish anti-anaemic factor H, but subsequent experiments showed only partial activity and that folic acid itself was a much more potent antimacrocytic anaemia factor. Insects may contribute significantly to the folic acid requirements of wild fish, but in scientific fish husbandry artificial diets are more reliable sources. Activity is lost during extended storage and when material is exposed to sunlight.

Therefore, dry feeds should be carefully protected during manufacture and moist diet rations should be carefully preserved.

Both types of fish diets should be fed soon after manufacture to assure minimal loss of folic acid activity.

ANTIMETABOLITES AND INACTIVATION

One antagonist of folacin is 4-aminopteroylglutamic acid or aminopterin. This material, when incorporated in the diet of guinea pigs and rats, induces anaemia and leucopenia and has been used to treat leukemia in man. Amethopterin (4-amino-N^{10}-methylpteroylglutamic acid) can also be used to induce deficiency by inhibiting purine, pyrimidine, and nucleic acid protection; viz., inhibition of nucleic acid synthesis results in macroytic anaemia.

CLINICAL ASSESSMENT

Haematology is used as a simple clinical tool to assess haemopoiesis in fish. Anterior kidney imprints easily disclose normal distribution of immature cells and preforms undergoing reticulosis. Salmon actively feeding in the ocean and young salmon fingerlings raised in fresh water and fed diets rich in folacin show liver storage of 3-4 μ g of folic acid/g of wet tissue. Microbiological assay is preferred for assessment of total folic acid in dietary raw materials because the total biological activity it measures includes all the various coenzyme forms and folic acid analogues.

Assessment of the dietary intake of folic acid is important for intensive cold water fish husbandry. In pond culture, aquatic and terrestrial insects, algae, etc., may also be available and folic acid in the supplementary feed may not be as critical. Since folic acid is labile in storage, excess amounts are generally added to manufactured feed in anticipation of storage losses. However, prudent fish husbandry dictates rapid use of manufactured rations with minimum storage.

Routine periodic haematology of fish assures proper nutritional status for maximum production and sound health. The author has noted in several series of experiments that when fish diseases occur through inadvertent contamination of the water supply, those groups of fish partially or completely deficient in folic acid were among the first to show acute disease symptoms. Therefore, folic acid must also play an important role in resistance to disease.

VITAMIN B_{12}

The antipernicious anaemia factor found to be contained in liver was isolated and crystallized in the mid forties. This substance, named vitamin B_{12} by its discoverers, was later to be recognized as essential for growth of chicken fed diets entirely of plant origin and was designated animal protein factor (APF). When anaemic salmon were injected with crystalline B_{12} in combination with folic acid and xanthopterin positive haemopoiesis occurred within a few days, and the salmon showed rapid recovery from the anaemia.

CHEMISTRY

Vitamin B_{12} or cynacobalamin has the following approximate chemical structure.

Cyanocobalamin

The molecule has a planar group and a nucleotide group lying nearly at right angles to one another. This cobalt-containing vitamin has a net charge of one at the central cobalt atom to which is attached a replaceable cyano group. Vitamin B_{12} is stable to mild heat in neutral solution, but is rapidly destroyed by heating in dilute acid or alkali.

Crude concentrates are more unstable and rapidly lose activity. The compound is similar to the porphyrins in its spatial configuration with a central cobalt atom linked to four reduced pyrrole rings in the haeme series. Replacing the cyanide ion with a variety of anions produce derivatives which have comparable biological activities; viz., hydroxocobalamin, nitritocobalamin, chlorocobalamin and sulphatocobalamin.

Positive Functions

Cyanocobalamin is involved with folic acid in haemopoiesis. It is required by many micro-organisms and is a growth factor for many animals. The animal protein factor present in fish and animal by-products was not recognized until crystalline vitamin B_{12} was injected into anaemic chinook salmon fingerlings in 1949 and positive haemopoiesis was observed. A coenzyme incorporating vitamin B_{12} is involved in the reversible isomerization of methyl-malonyl coenzyme A to succinyl coenzyme A and in the isomerization of methylaspartate to glutomate. Cyanocobalamin is involved in the coenzyme for the methylation of homocystine to form methionine.

It is also involved in several other one-carbon reactions and in the synthesis of labile methyl compounds. One vitamin B_{12} containing coenzyme acts in methylation of the purine ring during thymine synthesis. Vitamin B_{12} is also involved in cholesterol metabolism, in purine and pyrimidine biosynthesis, and in the metabolism of glycols.

Deficiency Syndrome

Deficiency signs in young pigs, chicks, and rats show abnormal blood elements, poor growth, and pernicious anaemia. An intrinsic factor is necessary for good absorption of the vitamin from the gut. This factor is a low molecular weight mucoprotein which normally occurs in gastric juice, and especially in hog gut mucosa. Pernicious anaemia in chinook and coho salmon is characterized by fragmented, erythrocytes with many aberrant forms present. Haemoglobin levels are inconsistent and erythrocyte counts have a range extending from frank anaemia to a near normal blood pattern. Cyanocobalamin stores in fish tissues are slowly exhausted and only after 12-16 weeks on test do the symptoms appear in deficient salmon populations. Poor appetite, poor growth, poor food conversion, and some dark pigmentation can be observed before frank anaemia is detected.

Sources and Protection

Rich sources of vitamin B_{12} are found in fish meal, fish viscera, liver, kidney, glandular tissues, and slaughter house wastes. Since vitamin B_{12} is labile on storage, and in mild acid solution is easily destroyed by heating, care must be exercised in diet preparation containing flesh or meat scraps.

Antimetabolites and inactivation.

The vitamin B_{12} coenzymes are very unstable under light, which rapidly decomposes the coenzymes. Photo-sensitivity is increased in dilute acid solutions.

Clinical Assessment

Generalized anaemia with fragmentation of erythrocytes and extremely variable haemoglobin levels and erythrocyte counts indicate possible B_{12} deficiency. Prompt response in individual fish is obtained by injecting B_{12} alone or in combination with folic acid in the ratio of 1 part vitamin B_{12} to 100 parts folic acid. Careful interpretation of haemotological data will enable one to distinguish one form of anaemia from the other.

ASCORBIC ACID

Experimental work to cure scurvy with fruit juice was described by Lind in 1753, but nearly 200 years elapsed before the exact chemical compound responsible for reducing the symptoms was defined. The isolation of ascorbic acid was due to Szent-Györgyi of Hungary and C.G. King of the U.S.A. Vitamin C synthesis was accomplished in 1933 after the chemical structure of ascorbic acid was established by British and Swiss workers.

McCay and Tunison reported scoliosis in brook trout fed formalin-preserved meat in 1934 and McLaren observe haemorrhages in trout fed rations low in ascorbic acid. It was not until the sixties that a critical need for L-ascorbic acid by trout and salmon was demonstrated.

CHEMISTRY

L-ascorbic acid is a white, odourless, crystalline compound, soluble in water but insoluble in fat solvents. It is readily oxidized to dehydroascorbic acid, the less biologically potent form. Ascorbic acid is very stable in acid solution because of the preservation of the lactone ring, but in alkaline solution hydrolysis occurs rapidly and vitamin activity is lost.

It is very heat labile and prone to atmospheric oxidation, especially in the presence of copper, iron, or several other metallic catalysts. The reduced form is the most biologically active but several derivatives or salts are obtainable which have varying degrees of ascorbate activity.

Positive Functions

L-ascorbic acid acts as a biological reducing agent in hydrogen transport. It is involved in many enzyme systems for hydroxylation; i.e., hydroxylation of tryptophan, tyrosine, or proline. It is involved in the detoxification of aromatic drugs and also acts in the production of adrenal steroids.

Ascorbic acid is necessary for the formation of hydroxy proline which is a constituent of collagen, a component of intercellular material in bones and soft tissues. Ascorbic acid plays a synergistic role with vitamin E as intracellular antioxidants and free radical traps. The conversion of folic acid to folinic acid requires vitamin C.

Ascorbic acid is involved in the formation of chondroitin sulphate and intercellular ground substance. Labelled ascorbic acid fed to fish previously deficient in the vitamin was shown to be rapidly mobilized and fixed in areas of rapid collagen synthesis and became concentrated in the thick collagen of the skin and in cartilagenous bones, as well as in the glands of the anterior kidney. Ascorbic acid is also involved in erythrocyte maturation.

Deficiency Syndrome

Scurvy with impaired collagen formation, perifollicular haemorrhages, loose teeth and poor bone formation, anaemia, and oedema have been reported in other animals. Deficiency signs in fish are generally related to impaired collagen formation. Fish soon show hyperplasia of jaw and snout. The same symptoms have been observed in trout, salmon, yellowtail, carp, guppies and char.

Histologically, hypertrophy of the adrenal tissues and haemorrhage at the bases of fins have been observed in coho salmon. Deficiency signs cease to develop and new growth becomes normal upon replacement of ascorbic acid in the ration.

Anaemia eventually develops in extremely deficient fish and scoliosis and lordosis do not repair but are walled off by new growth around the afflicted areas of the spine when ascorbic acid is once again added to the ration.

Requirements

In the rainbow trout, reasonable blood and anterior kidney storage levels were obtained with an intake of about 100 mg of vitamin C/kg of dry ration in 10°,

12°, or 15° C water systems. When wound repair experiments were initiated, however, or when fish were exposed to other stress then the requirements doubled or tripled. When severe abdominal or intramuscular wounds were inflicted, young fish needed at least 500 mg of active ascorbate for tissue repair comparable with control fish receiving 1 g or more of ascorbate in the diet/kg of dry diet. Coho salmon appear to need about half of these requirements for adequate tissue levels and for maximum severe wound repair rates. This phenomenon is illustrated in Table showing growth response and tissue repair for rainbow trout and coho salmon. The requirement for ascorbic acid is related to stress, growth rate and size of the animal, as well as to the other nutrients present in the diet.

A compromise value of about 200 mg of ascorbic acid/kg diet for trout and salmon raised in freshwater systems between 10-15°C would ensure reasonable tissue storage levels and furnish some excess for mild stress conditions and for ascorbic acid loss from the diet through oxidation. Large common carp can synthesize some ascorbate and the requirement for this species may be dependent on fish size and the environment in which they are reared.

Table. Growth and Tissue Ascorbate

	Trout	Salmon				
Vitamin C diet treatment	Average weight at 24 weeks	Ascorbate concentrate	Average weight at 24 weeks	Ascorbate concentrate		
		blood	kidney		blood	kidney
mg/100 g	g	μ g/g	μ g/g	g	μ g/g	μ s/s
0	2.4	–	–	5.0	22.3±2.2	89
5	9.6	34.4±2.9	125	6.0	30.5±1.2	132
10	10.6	34.6±1.3	137	5.7	35.8±1.6	265
20	10.1	38.8±3.3	132	6.1	34.2±2.3	183
40	10.2	46.8±6.2	162	6.3	33.7±2.0	225
100	10.8	51.0±4.6	247	6.0	37.8±2.3	321

- Average of five samples for blood ±(S.D.) and two for head kidney tissue
- No fish available for assay

Sources and Protection

Ascorbic acid is widely distributed in nature with citrus fruits, cabbage, liver, and kidney tissue good sources for the vitamin. Fresh insects and fish tissues contain reasonable amounts of the vitamin. Synthetic ascorbic acid is also readily available. Fish food should be protected from oxidizing agents and kept sealed or frozen until used to prevent loss of the vitamin.

Antimetabolites and Inactivation

D-ascorbic acid, the optical isomer of the active form, has no vitamin activity and competes for sites of several enzyme reactions mediated by L-ascorbic acid. 6-deoxy-L-ascorbic acid, dehydroascorbic acid and L-glucoascorbic acid have very little activity.

Clinical Assessment

Ascorbic acid status for experimental animals is normally attempted by tissue ascorbate analysis. Most of the assays previously used measure total ascorbate and not biologically active L-ascorbic acid. Consequently these are fraught with errors and misconception of true vitamin C status. In fish tissues, blood and liver do not adequately reflect the ascorbic acid intake and status, whereas assay of the anterior kidney which contains adrenal tissue provides a fairly representative picture of tissue storage of the vitamin. Stress rapidly reduces the ascorbic acid content of this tissue with concurrent production of adrenal steroids.

Conversely, dietary repletion is reflected by up to four or five-fold storage levels from the deficient state. Examination of fragile support cartilage in the gill filaments under low magnification will detect early hypovitaminosis before clinically acute symptoms become noticeable.

However, the best tissue for routine clinical analysis to assess vitamin C status in trout and salmon appears to be the anterior kidney with samples selected from the junction of the two forward wings. Tissues are blotted free of blood with filter paper, and then assayed for total ascorbate by one of the improved quick methods to determine total ascorbate.

INOSITOL

Muscle ‘sugar’ was discovered by Scherer in 1850 and was characterized as inositol in 1887. Inositol was shown to be an alopecia (a form of hairlessness) preventing factor for mice in 1940. Poor growth and poor food passage in inositol-deficient fish was observed in salmon and carp.

CHEMISTRY

Inositol

Seven optically inactive and two optically active isomers of hexahydroxycyclohexane can exist. One of the optically active forms, myo-inositol, is a white crystalline powder soluble in water and insoluble in alcohol and ether. The material can be synthesized, but is easily isolated from biological material in free or combined forms. The mixed calcium-magnesium salt of the hexophosphate is phytin. Isomers have little biological activity but will compete in chemical reactions. Inositol is a highly stable compound.

POSITIVE FUNCTIONS

Myo-inositol is a structural component in living tissues. It has lipotropic action by preventing accumulation of cholesterol in one type of fatty liver disease and is involved with choline in maintaining normal lipid/metabolism. It is a growth-promoting substance for micro-organisms and prevents an alopecia in mice. It is a reserve carbohydrate in muscle as well as a major component of phosphoglycerides in animal tissues.

DEFICIENCY SYNDROME

Poor growth, increased gastric emptying time, oedema, dark colour, and distended stomachs are symptoms observed in salmon, trout, carp and catfish held for long periods on inositol-deficient test rations. A 'spectacle-eye' condition described for rats has not been observed under the experimental conditions used in fish studies. The major deficiency sign is inefficiency in digestion and food utilization and concomitant poor growth leading to a population of fish with distended abdomens.

SOURCES AND PROTECTION

Myo-inositol occurs ubiquitously in large amounts wherever biological tissue is found. Wheat germ, dried peas, and beans are rich sources. Brain, heart, and glandular tissues are very good sources of biologically active inositol. Citrus fruit pulp and dried yeast also contain inositol. The compound is stable.

ANTIMETABOLITES AND INACTIVATION

Seven optically inactive and one optically active but biologically inactive stereo isomers occur. Since inositol is synthesized in the biologically active form by many microorganisms in the gut, only large quantities of chemically synthesized inactive isomers added to diets would interfere with the metabolism of inositol for growth and normal physiological function. Biologically inactive stereoisomers of myo-inositol do not compete for critical sites in metabolism. Methyl derivatives and mono-, di-, and triphosphoric acid esters occur naturally. Salts of the hexaphosphate or phytin make the bound inositol practically unavailable to the animal.

CLINICAL ASSESSMENT

Assessment in fish has been based on lack of deficiency signs coupled with the most efficient food conversion. Salmon feeding actively in oceans show 1-1.5 mg of inositol/g of fresh liver tissue and young fingerlings raised in freshwater at 10-15°C had 600-700 μ g/g of liver tissue. An alternate, better assessment may be based on a standard muscle section or whole carcass analysis for free or for bound inositol. Projection of inositol intake from normal fish diet ingredients should indicate an excess of this particular vitamin.

CHOLINE

Methylation as a basic metabolic process was postulated by Hoffmeister. Methyl transfer was shown in vivo by Thompson and the interrelationships between choline, methionine, and homo-cystine were shown by du Vigneaud in 1939-42. Trout fed low choline rations developed haemorrhagic kidneys and salmon showed an aversion to food in choline-deficient diet.

CHEMISTRY

Choline

Choline is a very strong organic base and forms many derivatives widely distributed in animal and vegetable tissue. One derivative, acetylcholine, is involved in the transmission of nerve impulses across synapses. Choline is very hygroscopic, very soluble in water, and is stable to heat in acid, but decomposes in alkaline solutions.

Positive Functions

Choline acts as a methyl donor in trans-methylation reactions. It is a lipotropic and antihaemorrhagic factor preventing the development of fatty livers. It is involved in the synthesis of phospholipids and in fat transport. Acetylcholine transmits the excitory state across the ganglionic synapses and neuromuscular junctions. Choline is essential for growth and good food conversion in fish.

Deficiency Syndrome

Deficiency signs include poor growth and poor food conversion accompanied by impaired fat metabolism. Haemorrhagic kidneys and intestines have been reported in trout and increased gastric emptying time has been observed in salmon.

Sources and Protection

Rich sources of choline are wheat germ, beans, brain, and heart tissue. Choline hydrochloride, the commercially available form, may inactivate a - tocopherol and vitamin K when in direct contact with these vitamins and care should be exercised in selecting properly protected (gelatin coated) fat soluble vitamins in fish diet preparation.

Clinical Assessment

Choline status of fish can be estimated from the assay of choline content of the dietary ingredients with the absence of deficiency signs. Maximum liver storage may not be the best criteria to determine choline nutritional status but has been used to assess the tentative requirement listed for the two species of salmon.

P-AMINOBENZOIC ACID

Para-aminobenzoic acid is a white crystalline powder which is water soluble and heat and light stable in aqueous and mild alkaline solution. It is a growth promoting vitamin for micro-organisms which require the vitamin for folic acid synthesis. Large intake has been shown to counteract the antimetabolite effect of sulphonamides in bacterial culture. No positive function or deficiency signs have been observed in fish.

LIPOIC ACID

Lipoic acid is both fat soluble and water soluble. Its active form is the amide derivative known as lipoamide. Lipoic acid functions as a coenzyme in a α - ketoacid decarboxylation. It was discovered independently in several laboratories during the period 1945-50 and shown to be an essential component of multienzyme, the pyruvate dehydrogenase complex. Pyruvate is converted to 'active acetaldehyde' which in turn is picked up by lipoamide. Subsequent oxidation of the aldehyde results in the reduction of lipoamide to a disulphydryl form. The multi-enzyme unit also includes thiamine pyrophosphate, coenzyme A, and flavin adenine dinucleotide. Glandular tissues are good sources of lipoic acid. No requirements have been determined for fish.

RIBOFLAVIN

Riboflavin is a yellow-brown crystalline pigment. The vitamin is very slightly soluble in water but soluble in alkali. It is insoluble in most organic solvents. Riboflavin is stable to oxidizing agents in strong mineral acids and in neutral aqueous solution. It is also stable to dry heat but is irreversibly decomposed on irradiation with ultraviolet rays or visible light, breaking down to lumiflavin. Riboflavin phosphate is the chemically active group of Warburg's 'yellow' enzyme.

POSITIVE FUNCTIONS

Riboflavin functions in the tissues in the form of flavin adenine dinucleotide (FAD) or as flavin mononucleotide (FMN). The flavo-proteins function as enzymes of tissue respiration and are involved in hydrogen transport to catalyze the oxidation of reduced pyridine nucleo-tides (NADH and NADPH).

Thus, they function as coenzymes for many oxidases and redunctases such as cytochrome e reductase, D- and L-amino acid oxidases, xanthine and aldehyde oxidase, succinic dehydrogenase, glucose oxidase and fumaric dehydrogenase. Riboflavin is also involved with pyridoxine in the conversion of tryptophan to nicotinic acid and is most important in the respiration of poorly vascularized tissues such as the cornea of the eye. Riboflavin is involved in the retinal pigment during light adaptation and lack of it causes impaired vision and photophobia in experimental animals, including fish.

DEFICIENCY SYNDROME

Tissue riboflavin is exhausted in young salmonids after 10 to 12 weeks on riboflavin-deficient diets in 10-15 °C water systems. Poor appetite and poor diet efficiency are the first signs, followed by photophobia, mono or bilateral cataract's, corneal vascularization, eye haemorrhage, incoordination, and general anaemia.

Dark pigmentation, coupled with striated constrictions of abdominal wall in salmon have been noted. Skin atrophy has been reported for some fish species and abnormal pigmentation of both skin and iris has been noted. Replacement of riboflavin in the diet reduced the symptoms except when cataracts have developed.

This irreversible condition will continue in monolateral cataracts throughout the life of the fish, whereas bilateral cataracts largely result in eventual starvation and death of the afflicted animal. The first specific signs have consistently appeared in and about the eyes of salmonids, carps and catfish.

REQUIREMENTS

The requirements of fish for riboflavin under experimental conditions in 10-15 °C water supplies. Values for the trout are slightly lower than those reported for salmon. The requirements may vary depending upon the balance of other dietary ingredients, caloric density and environmental conditions under which the fish is raised.

The values established under these standard test conditions should serve as tentative-levels which will satisfy biological demands for normal growth, health, and physiological function.

Most of these studies have been made on very young fish, often initially feeding fry, with the logical assumption that their vitamin needs would be more than that of larger fish having advanced metabolic enzyme systems capable of synthesizing at least some of these vitamins to partially meet the requirements.

SOURCES AND PROTECTION

Riboflavin is widely distributed in plants and in animal glandular tissues. Milk, liver, kidney, heart, yeast, germinated grains, peanuts, soybeans and eggs are rich sources. Keeping feeds from sunlight or intense artificial light is necessary to minimize loss of the vitamin by conversion to lumiflavin.

ANTIMETABOLITES AND INACTIVATION

When the ribose group in the molecule is replaced by other groups, analogues are formed which either have reduced activity or become antimetabolites. Galactoflavin is an antagonist to riboflavin and inhibits growth of rats when the diet contains this compound.

Flavin monosulphate inhibits D-amino acid oxidase and appears to act as a competitor and inhibitor for growth of Lactobacillus casei. The hydroxyethyl analogue is an antagonist for riboflavin function in both rats and bacteria and also shows antifungal activity.

CLINICAL ASSESSMENT

Liver tissue of actively feeding sea salmon contains 6-8 g of riboflavin/g of wet tissue. In a freshwater environment, young fish fed test diets in 15 °C water systems showed liver storage of 3.5-4.0 μ g/g. Riboflavin content of blood plasma does not change significantly in riboflavin deficiency in experimental animals.

However, the erythrocyte riboflavin content has been reported at around 10μ g/100 ml blood for man on low riboflavin intake and approximately twice that level on high intake.

Urinary excretion of riboflavin has been used clinically to indicate dietary status, with 200 μ g/24 it suggesting adequate intake and less than 100 μ g indicating low intake. Excretion of 50 μ g or less daily is strong indication of extended dietary deficiency. Several studies show that excretion of less than 200 μ g of riboflavin/g of creatinine would be indicative of deficiency.

PYRIDOXINE

A new factor which could cure dermatitis in rats was reported by Györgi in 1935 and was named vitamin B_6. The name pyridoxine is also attributed to Györgi. The active material was isolated in 1939 and pyridoxine was subsequently synthesized. First quantitative requirements for fish were described in 1944. Pyridoxine deficiency has since been reported in trout and in salmon.

CHEMISTRY

Pyrodoxine Hydrochloride

Compounds which have vitamin B_6 activity include: pyridoxine, pyridoxal and pyridoxamine. Pyridoxine hydrochloride is readily soluble in water and is, heat stable in either acid or alkaline solution. It acts as a coenzyme in a number of enzyme systems. Pyridoxine is sensitive to ultraviolet light in neutral or alkaline solutions.

Pyridoxamine and pyridoxal in dilute solutions are labile compounds which are rapidly destroyed on exposure to air, heat, or light. Therefore, most vitamin supplementation is in the form of pyridoxine hydrochloride and analysis for pyridoxine activity by microbiological assay of diet ingredients probably measures pyridoxal phosphate and other intermediates as well.

Positive Functions

Pyridoxal phosphate is the coenzyme, codecarboxylase, involved in the decarboxylation of amino acids. It is also the co-factor of the 22 different transaminases present in animal tissues.

As codecarboxylase, pyridoxal phosphate has been shown to participate in the decarboxylation of 5-hydroxytryptophan to produce 5-hydroxytryptamine or serotonin. As a co-factor of the enzyme desulphydrase converts cysteine to pyruvic acid. Pyridoxal phosphate is the co-factor for the synthesis of δ -amino-levulinate, the latter a precursor of haeme. Many neuro hormones require pyridoxal phosphate as a coenzyme in their synthesis.

Pyridoxine is also involved in fat metabolism, especially of the essential fatty acids. It is involved in the synthesis of messenger RNA which determines amino acids sequence in polypeptide synthesis. Pyridoxine plays a most important role in protein metabolism and, as a result, carnivorous fish have stringent requirements for the vitamin in the diet and stores are rapidly exhausted.

Deficiency Syndrome

Since salmonids, ictalurids, and very young cyprinids are carnivorous, with protein requirements for young animals between 40-50 per cent of the ration, pyridoxine stores are rapidly exhausted when fish are held on pyridoxine-deficient rations.

Acute deficiency signs occur in salmon after 14-21 days on a high protein diet of pyridoxine and the entire population dies within 28 days in 12-15 C water. Deficiency is followed by general nervous disorders, and alteration in control of melanophore contraction. Postmortem rigor mortis occurs very rapidly. Rapid and gasping breathing with flexing of the opercules is a common observation and oedema in the peritoneal cavity with colourless serous fluid occurs in fish on some experimental treatments.

Salmon, trout, carp and yellowtail exhibit premortem rigor a few hours before death. Recovery at late stage deficiency is very unlikely unless fish are injected with pyridoxal phosphate. Handling the animals generally induces more damage than vitamin administration will correct.

Recovery among fish still feeding is equally rapid and dramatic upon administration of pyridoxine hydrochloride in the diet. Deficiency signs disappear within a day or two. Erythrocyte and plasma transaminase activities reflect the deficiency state.

Sources and Protection

Good sources of pyridoxine are yeast, whole cereals, egg yolk, liver and glandular tissues. Pyridoxine compounds in phosphorylated form present in agricultural products are fairly stable but are sensitive to ultraviolet radiation. Some pyridoxal phosphate will be lost on exposure to air. Free forms or pyridoxal and pyridoxamine are rapidly destroyed by air, light and heat when moist.

Antimetabolites and Inactivation

Antagonists may compete for reaction sites of the apoenzyme or may react with pyridoxal phosphate to form inactive compounds. Deoxypyridoxine is a potent pyridoxine antagonist because of competition for apoenzyme sites but is a useful agent to accelerate pyridoxine deficiency in experimental animals. This same compound inhibits tyrosine decarboxylase. Methoxypyridoxine is another antagonist. Oxopyrimidine (2-methyl-4-amino-5-hydroxymethyl-pyrimidine) produces liver damage and inhibits glutamate decarboxylase.

Clinical Assessment

Plasma and erythrocyte transaminase activity in most instances reflects pyridoxine status of fish although high tryptophan load in the diet tends to increase the pyridoxine requirement.

Liver storage measured by microbiological assay showed 5-6 V-g of pyridoxine activity/g in fresh sea salmon liver; whereas, fingerling salmon fed a 50 per cent protein diet in freshwater had 2-3 mg/g of wet tissue. Assay of diet for pyridoxine by microbiological methods gives a truer representation of total pyridoxine activity including intermediates.

2

The Central Dogma of Molecular Biology

The discovery that DNA is the prime genetic molecule, carrying all the hereditary information within chromosomes, immediately focused attention on its structure. It was hoped that knowledge of the structure would reveal how DNA carries the genetic messages that are replicated when chromosomes divide to produce two identical copies of themselves.

During the late 1940s and early 1950s, several research groups in the United States and in Europe engaged in serious efforts—both cooperative and rival—to understand how the atoms of DNA are linked together by covalent bonds and how the resulting molecules are arranged in three-dimensional space. Not surprisingly, there initially were fears that DNA might have very complicated and perhaps bizarre structures that differed radically from one gene to another. Great relief, if not general elation, was thus expressed when the fundamental DNA structure was found to be the double helix.

It told us that all genes have roughly the same three-dimensional form and that the differences between two genes reside in the order and number of their four nucleotide building blocks along the complementary strands. Now, some 50 years after the discovery of the double helix, this simple description of the genetic material remains true and has not had to be appreciably altered to accommodate new findings.

Nevertheless, we have come to realise that the structure of DNA is not quite as uniform as was first thought. For example, the chromosome of some small viruses have single-stranded, not double-stranded, molecules. Moreover, the precise orientation of the base pairs varies slightly from base pair to base pair in a manner that is influenced by the local DNA sequence. Some DNA sequences even permit the double helix to twist in the left-handed sense, as opposed to the right-handed sense originally formulated for DNA's general structure. And while some DNA molecules are linear, others are circular.

Still additional complexity comes from the supercoiling (further twisting) of the double helix, often around cores of DNA-binding proteins. Likewise, we now realise that RNA, which at first glance appears to be very similar to DNA, has its own distinctive structural features. It is principally found as a single-stranded molecule. Yet by means of intra-strand base pairing, RNA exhibits extensive double-helical character and is capable of folding into a wealth of diverse tertiary structures. These structures are full of surprises, such as non-classical base pairs, base-backbone interactions, and knot-like configurations. Most remarkable of all, and of profound evolutionary significance, some RNA molecules are enzymes that carry out reactions that are at the core of information transfer from nucleic acid to protein. Clearly, the structures of DNA and RNA are richer and more intricate than was at first appreciated. Indeed, there is no one generic structure for DNA and RNA.

DNA IS COMPOSED OF POLYNUCLEOTIDE CHAINS

The most important feature of DNA is that it is usually composed of two polynucleotide chains twisted around each other in the form of a double helix. The upper part of the figure presents the structure of the double helix shown in a schematic form. Note that if inverted 180°, the double helix looks superficially the same, due to the complementary nature of the two DNA strands. The space-filling model of the double helix, in the lower part of the figure, shows the components of the DNA molecule and their relative positions in the helical structure. The backbone of each strand of the helix is composed of alternating sugar and phosphate residues; the bases project inward but are accessible through the major and minor grooves.

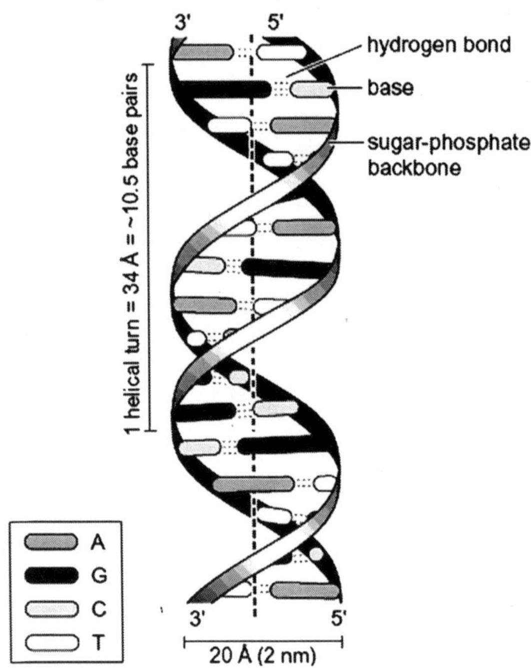

Fig. The Helical Structure of DNA

Let us begin by considering the nature of the nucleotide, the fundamental building block of DNA. The nucleotide consists of a phosphate joined to a sugar, known as 2'-deoxyribose, to which a base is attached. The sugar is called *2'*-deoxyribose because there is no hydroxyl at position 2' (just two hydrogens).

Note that the positions on the ribose are designated with primes to distinguish them from positions on the bases. We can think of how the base is joined to 2'-deoxyribose by imagining the removal of a molecule of water between the hydroxyl on the 1' carbon of the sugar and the base to form a glycosidic bond.

The sugar and base alone are called a nucleoside. Likewise, we can imagine linking the phosphate to 2'-deoxyribose by removing a water molecule from between the phosphate and the hydroxyl on the 5' carbon to make a 5' phosphomonoester.

Adding a phosphate (or more than one phosphate) to a nucleo*s*ide creates a nucleo*t*ide. Thus, by making a glycosidic bond between the base and the sugar, and by making a phosphoester bond between the sugar and the phosphoric acid, we have created a nucleotide. Nucleotides are, in turn, joined to each other in polynucleotide chains through the 3' hydroxyl of 2'-deoxyribose of one nucleotide and the phosphate attached to the 5' hydroxyl of another nucleotide.

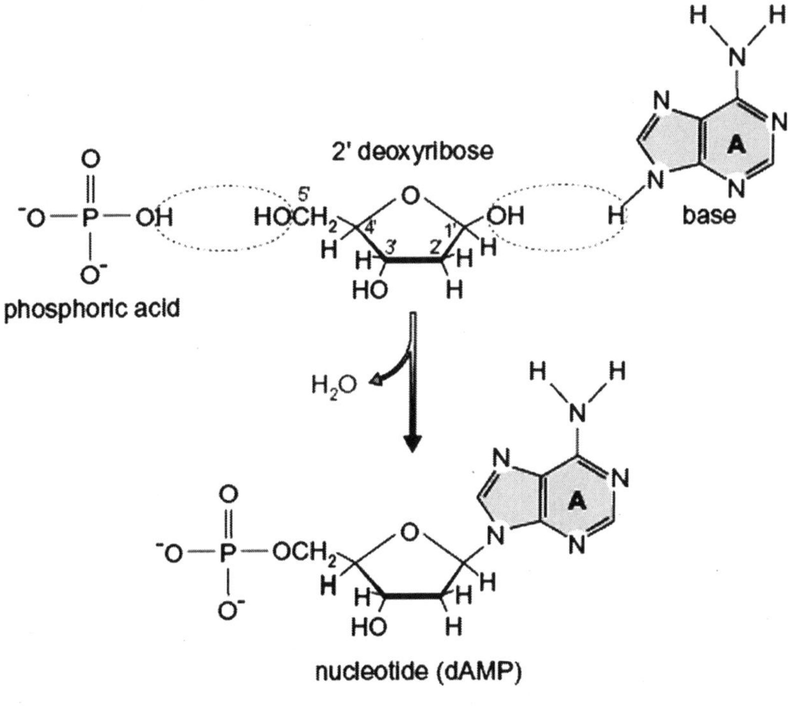

Fig. Formation of Nucleotide by Removal of Water

This is a phosphodiester linkage in which the phosphoryl group between the two nucleotides has one sugar esterified to it through a 3' hydroxyl and a second sugar esterified to it through a 5' hydroxyl. Phosphodiester linkages create the repeating, sugar-phosphate backbone of the polynucleotide chain, which is a regular feature of DNA. In contrast, the order of the bases along the polynucleotide chain is irregular. This irregularity as well as the long length is, as we shall see, the basis for the enormous information content of DNA. The phosphodiester linkages impart an inherent polarity to the DNA chain.

	Base Adenine	Nucleoside Adenosine	Nucleotide Adenosine 5'-phosphate	Deoxynucleotide Deoxyadenosine 5' phosphate
Structure[a]				
M.W.	135.1	267.2	347.2	331.2

This polarity is defined by the asymmetry of the nucleotides and the way they are joined. DNA chains have a free 5' phosphate or 5' hydroxyl at one end and a free 3' phosphate or 3' hydroxyl at the other end. The convention is to write DNA sequences from the 5' end (on the left) to the 3' end, generally with a 5' phosphate and a 3' hydroxyl.

STRANDS OF THE DOUBLE HELIX

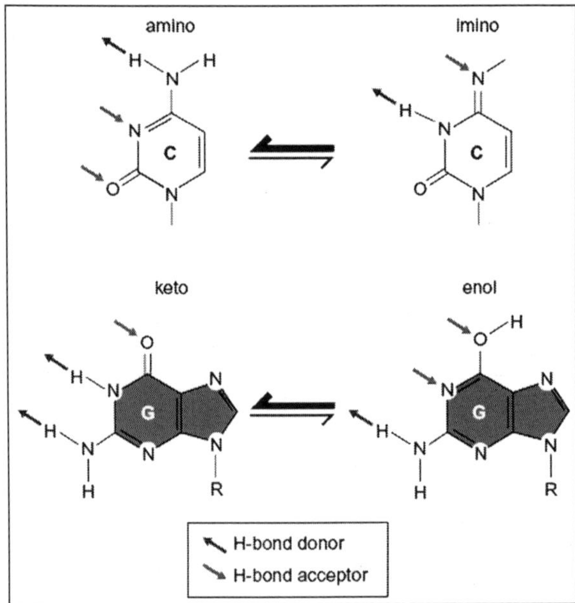

Fig. Base Tautomers

The double helix is composed of two polynucleotide chains that are held together by weak, non-covalent bonds between pairs of bases, as shown in Figure. Adenine on one chain is always paired with thymine on the other chain and, likewise, guanine is always paired with cytosine. The two strands have the same helical geometry but base pairing holds them together with the opposite polarity.

That is, the base at the 5' end of one strand is paired with the base at the 3' end of the other strand. The strands are said to have an anti-parallel orientation. This anti-parallel orientation is a stereochemical consequence of the way that adenine and thymine and guanine and cytosine pair with each together.

CHAINS OF THE DOUBLE HELIX

The pairing between adenine and thymine and between guanine and cytosine results in a complementary relationship between the sequence of bases on the two intertwined chains and gives DNA its self-encoding character. For example, if we have the sequence 5'-ATGTC-3' on one chain, the opposite chain must have the complementary sequence 3'-TACAG-5'. The strictness of the rules for this "Watson-Crick" pairing derives from the complementarity both of shape and of hydrogen bonding properties between adenine and thymine and between guanine and cytosine. Adenine and thymine match up so that a hydrogen bond can form between the exocyclic amino group at on adenine and the carbonyl at in thymine; and likewise, a hydrogen bond can form between of adenine and of thymine.

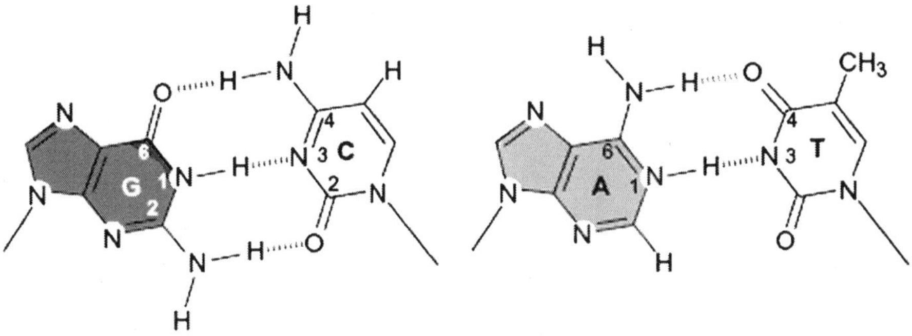

Fig. The Figure shows Hydrogen Bonding between the Bases

A corresponding arrangement can be drawn between a guanine and a cytosine, so that there is both hydrogen bonding and shape complementarity in this base pair as well. A G:C base pair has three hydrogen bonds, because the exocyclic NH_2 at on guanine lies opposite to, and can hydrogen bond with, a carbonyl at on cytosine.

Likewise, a hydrogen bond can form between of guanine and of cytosine and between the carbonyl at of guanine and the exocyclic NH_2 at C_4 of cytosine.

Watson-Crick base pairing requires that the bases are in their preferred tautomeric states. An important feature of the double helix is that the two base pairs have exactly the same geometry; having an A:T base pair or a G:C base pair between the two sugars does not perturb the arrangement of the sugars. Neither does T:A or C:G. In other words, there is an approximately twofold axis of symmetry that relates the two sugars and all four base pairs can be accommodated within the same arrangement without any distortion of the overall structure of the DNA.

HYDROGEN BONDING

The hydrogen bonds between complementary bases are a fundamental feature of the double helix, contributing to the thermodynamic stability of the helix and providing the information content and specificity of base pairing. Hydrogen bonding might not at first glance appear to contribute importantly to the stability of DNA for the following reason. An organic molecule in aqueous solution has all of its hydrogen bonding properties satisfied by water molecules that come on and off very rapidly.

As a result, for every hydrogen bond that is made when a base pair forms, a hydrogen bond with water is broken that was there before the base pair formed. Thus, the net energetic contribution of hydrogen bonds to the stability of the double helix would appear to be modest. However, when polynucleotide strands are separate, water molecules are lined up on the bases. When strands come together in the double helix, the water molecules are displaced from the bases.

This creates disorder and increases entropy, thereby stabilizing the double helix. Hydrogen bonds are not the only force that stabilizes the double helix. A second important contribution comes from stacking interactions between the bases. The bases are flat, relatively waterinsoluble molecules, and they tend to stack above each other roughly perpendicular to the direction of the helical axis. Electron cloud interactions (ð–ð) between bases in the helical stacks contribute significantly to the stability of the double helix. Hydrogen bonding is also important for the specificity of base pairing. Suppose we tried to pair an adenine with a cytosine.

Then we would have a hydrogen bond acceptor of adenine lying opposite a hydrogen bond acceptor of cytosine with no room to put a water molecule in between to satisfy the two acceptors. Likewise, two hydrogen bond donors, the NH_2 groups at of adenine and of cytosine, would lie opposite each other. Thus, an A:C base pair would be unstable because water would have to be stripped off the donor and acceptor groups without restoring the hydrogen bond formed within the base pair.

BASES CAN FLIP OUT FROM THE DOUBLE HELIX

As we have seen, the energetics of the double helix favour the pairing of each base on one polynucleotide strand with the complementary base on the other strand. Certain enzymes that methylate bases or remove damaged bases do so with the base in an extra helical configuration in which it is flipped out from the double helix, enabling the base to sit in the catalytic cavity of the enzyme.

Furthermore, enzymes involved in homologous recombination and DNA repair are believed to scan DNA for homology or lesions by flipping out one base after another. This is not energetically expensive because only one base is flipped out at a time. Clearly, DNA is more flexible than might be assumed at first glance.

DNA IS USUALLY A RIGHT-HANDED DOUBLE HELIX

Applying the handedness rule from physics, we can see that each of the polynucleotide chains in the double helix is right-handed. In your mind's eye, hold your right hand up to the DNA molecule in Figure with your thumb pointing up and along the long axis of the helix and your fingers following the grooves in the helix. Trace along one strand of the helix in the direction in which your thumb is pointing. Notice that you go around the helix in the same direction as your fingers are pointing. This does not work if you use your left hand. Try it! A consequence of the helical nature of DNA is its periodicity. Each base pair is displaced (twisted) from the previous one by about 36°. Thus, in the X-ray crystal structure of DNA it takes a stack of about 10 base pairs to go completely around the helix (360°). That is, the helical periodicity is generally 10 base pairs per turn of the helix.

THE DOUBLE HELIX HAS MINOR AND MAJOR GROOVES

As a result of the double-helical structure of the two chains, the DNA molecule is a long extended polymer with two grooves that are not equal in size to each other. Why are there a minor groove and a major groove? It is a simple consequence of the geometry of the base pair. The angle at which the two sugars protrude from the base pairs (that is, the angle between the glycosidic bonds) is about 120° (for the narrow angle or 240° for the wide angle).

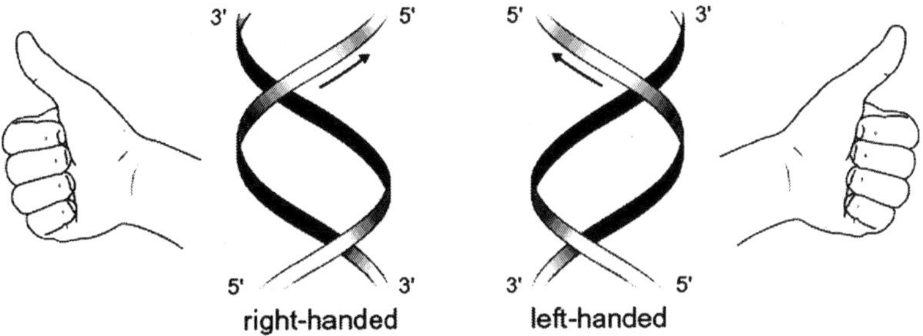

Fig. Left- and Right-Handed Helices

As a result, as more and more base pairs stack on top of each other, the narrow angle between the sugars on one edge of the base pairs generates a minor groove and the large angle on the other edge generates a major groove. (If the sugars pointed away from each other in a straight line, that is, at an angle of 180°, then two grooves would be of equal dimensions and there would be no minor and major grooves.)

THE MAJOR GROOVE IS RICH IN CHEMICAL INFORMATION

The edges of each base pair are exposed in the major and minor grooves, creating a pattern of hydrogen bond donors and acceptors and of van der Waals surfaces that identifies the base pair. The edge of an A:T base pair displays the following chemical groups in the following order in the major groove: a hydrogen bond acceptor (the of adenine), a hydrogen bond donor (the exocyclic amino group on of adenine), a hydrogen bond acceptor (the carbonyl group on of thymine) and a bulky hydrophobic surface (the methyl group on of thymine).

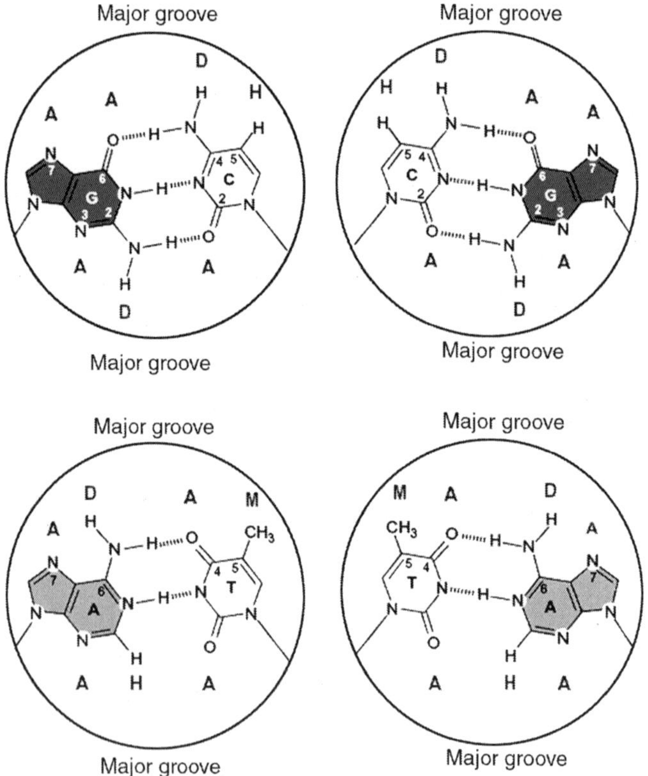

Fig. Chemical Groups Exposed in the Major and Minor Grooves from the Edges of the Base Pairs. The Letters Identify as Hydrogen Bond Acceptors (A), Hydrogen Bond Donors (D), Nonpolar Hydrogens (H), and Methyl groups (M).

Similarly, the edge of a G:C base pair displays the following groups in the major groove: a hydrogen bond acceptor (at of guanine), a hydrogen bond acceptor (the carbonyl on of guanine), a hydrogen bond donor (the exocyclic amino group on of cytosine), a small non-polar hydrogen (the hydrogen at of cytosine). Thus, there are characteristic patterns of hydrogen bonding and of overall shape that are exposed in the major groove that distinguish an A:T base pair from a G:C base pair, and, for that matter, A:T from T:A, and G:C from C:G. We can think of these features as a code in which A represents a hydrogen bond acceptor, D a hydrogen bond donor, M a methyl group, and H a nonpolar hydrogen. In such a code, A D A M in the major groove signifies an A:T base pair, and A A D H stands for a G:C base pair.

Likewise, M A D A stands for a T:A base pair and H D A A is characteristic of a C:G base pair. In all cases, this code of chemical groups in the major groove specifies the identity of the base pair. These patterns are important because they allow proteins to unambiguously recognize DNA sequences without having to open and thereby disrupt the double helix. Indeed, as we shall see, a principal decoding mechanism relies upon the ability of amino acid side chains to protrude into the major groove and to recognize and bind to specific DNA sequences.

The minor groove is not as rich in chemical information and what information is available is less useful for distinguishing between base pairs. The small size of the minor groove is less able to accommodate amino acid side chains. Also, A:T and T:A base pairs and G:C and C:G pairs look similar to one another in the minor groove.

An A:T base pair has a hydrogen bond acceptor (at of adenine), a nonpolar hydrogen (at of adenine) and a hydrogen bond acceptor (the carbonyl on of thymine). Thus, its code is A H A. But this code is the same if read in the opposite direction, and hence an A:T base pair does not look very different from a T:A base pair from the point of view of the hydrogenbonding properties of a protein poking its side chains into the minor groove.

Likewise, a G:C base pair exhibits a hydrogen bond acceptor (at N3 of guanine), a hydrogen bond donor (the exocyclic amino group on guanine), and a hydrogen bond acceptor (the carbonyl on of cytosine), representing the code A D A. Thus, from the point of view of hydrogen bonding, C:G and G:C base pairs do not look very different from each other either. The minor groove does look different when comparing an A:T base pair with a G:C base pair, but G:C and C:G, or A:T and T:A, cannot be easily distinguished.

THE DOUBLE HELIX EXISTS IN MULTIPLE CONFORMATIONS

Early X-ray diffraction studies of DNA, which were carried out using concentrated solutions of DNA that had been drawn out into thin fibers, revealed two kinds of structures, the B and the A forms of DNA. The B form, which is observed at high humidity, most closely corresponds to the average structure of DNA under physiological conditions. It has 10 base pairs per turn, and a wide major groove and a narrow minor groove.

The A form, which is observed under conditions of low humidity, has 11 base pairs per turn. Its major groove is narrower and much deeper than that of the B form, and its minor groove is broader and shallower. The vast majority of the DNA in the cell is in the B form, but DNA does adopt the A structure in certain DNA-protein complexes. Also, as we shall see, the A form is similar to the structure that RNA adopts when double helical.

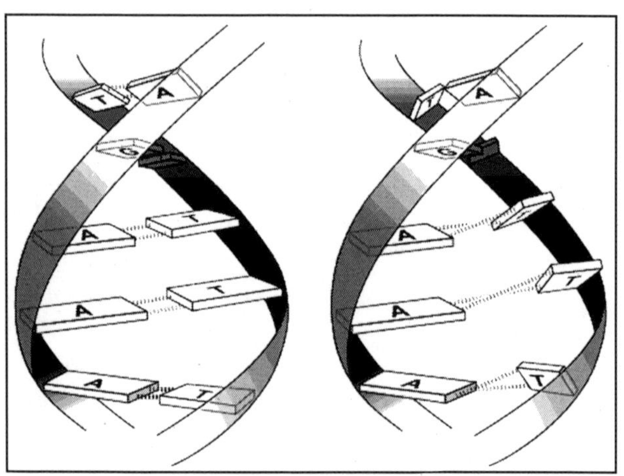

Fig. The Propeller Twist between the Purine and Pyrimidine Base Pairs of a Right-Handed Helix

The B form of DNA represents an ideal structure that deviates in two respects from the DNA in cells. First, DNA in solution, as we have seen, is somewhat more twisted on average than the B form, having on average 10.5 base pairs per turn of the helix. Second, the B form is an average structure whereas real DNA is not perfectly regular. Rather, it exhibits variations in its precise structure from base pair to base pair. This was revealed by comparison of the crystal structures of individual DNAs of different sequences. For example, the two members of each base pair do not always lie exactly in the same plane.

Rather, they can display a "propeller twist" arrangement in which the two flat bases counter rotate relative to each other along the long axis of the base pair, giving the base pair a propeller-like character. Moreover, the precise rotation per base pair is not a constant. As a result, the width of the major and minor grooves varies locally. Thus, DNA molecules are never perfectly regular double helices. Instead, their exact conformation depends on which base pair (A:T, T:A, G:C, or C:G) is present at each position along the double helix and on the identity of neighboring base pairs. Still, the B form is for many purposes a good first approximation of the structure of DNA in cells.

DNA CAN SOMETIMES FORM A LEFT-HANDED HELIX

DNA containing alternative purine and pyrimidine residues can fold into left-handed as well as right-handed helices. To understand how DNA can form a left-handed helix, we need to consider the glycosidic bond that connects the base to the 1' position of 2'-deoxyribose. This bond can be in one of two conformations called *syn* and *anti*.

In right-handed DNA, the glycosidic bond is always in the *anti* conformation. In the left-handed helix, the fundamental repeating unit usually is a purine-pyrimidine dinucleotide, with the glycosidic bond in the *anti* conformation at pyrimidine residues and in the *syn* conformation at purine residues. It is this *syn* conformation at the purine nucleotides that is responsible for the left-handedness of the helix.

The change to the *syn* position in the purine residues to alternating *anti–syn* conformations gives the backbone of left-handed DNA a zigzag look, which distinguishes it from right-handed forms. The rotation that effects the change from *anti* to *syn* also causes the ribose group to undergo a change in its pucker. In solution alternating purine–pyrimidine residues assume the left-handed conformation only in the presence of high concentrations of positively charged ions (e.g., Na') that shield the negatively charged phosphate groups.

At lower salt concentrations, they form typical right-handed conformations. The physiological significance of Z DNA is uncertain and left-handed helices probably account at most for only a small of proportion of a cell's DNA.

DNA STRANDS CAN SEPARATE (DENATURE) AND REASSOCIATE

Because the two strands of the double helix are held together by relatively weak (non-covalent) forces, you might expect that the two strands could come apart easily. Indeed, the original structure for the double helix suggested that DNA replication would occur in just this manner. The complementary strands of double helix can also be made to come apart when a solution of DNA is heated above physiological temperatures (to near 100 °C) or under conditions of high pH, a process known as denaturation. However, this complete separation of DNA strands by denaturation is reversible.

When heated solutions of denatured DNA are slowly cooled, single strands often meet their complementary strands and reform regular double helices. The capacity to renature denatured DNA molecules permits artificial hybrid DNA molecules to be formed by slowly cooling mixtures of denatured DNA from two different sources. Likewise, hybrids can be formed between complementary strands of DNA and RNA.

The ability to form hybrids between two single-stranded nucleic acids (hybridization) is the basis for several indispensable techniques in molecular biology, such as Southern blot hybridization and DNA microarrays. Important insights into the properties of the double helix were obtained from classic experiments carried out in the 1950s in which the denaturation of DNA was studied under a variety of conditions. In these experiments DNA denaturation was monitored by measuring the absorbance of ultraviolet light passed through a solution of DNA. DNA maximally absorbs ultraviolet light at a wavelength of about 260 nm.

It is the bases that are principally responsible for this absorption. When the temperature of a solution of DNA is raised to near the boiling point of water, the optical density (absorbance) at 260 nm markedly increases. The explanation for this increase is that duplex DNA is hypochromic; it absorbs less ultraviolet light by about 40% than do individual DNA chains. The hypochromicity is due to base stacking, which diminishes the capacity of the bases in duplex DNA to absorb ultraviolet light. If we plot the optical density of DNA as a function of temperature, we observe that the increase in absorption occurs abruptly over a relatively narrow temperature range.

The midpoint of this transition is the melting point or Tm. Like ice, DNA melts: it undergoes a transition from a highly ordered double-helical structure to a much less ordered structure of individual strands.

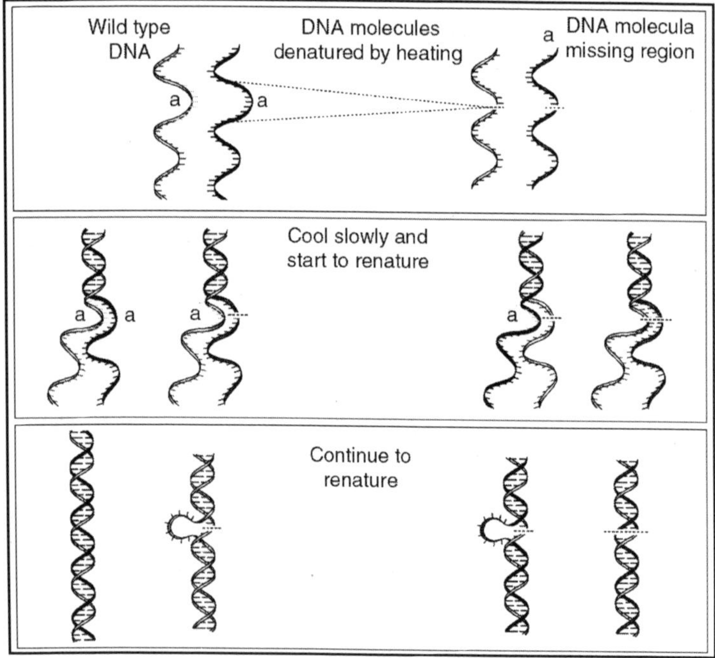

Fig. Reannealing and Hybridization

The sharpness of the increase in absorbance at the melting temperature tells us that the denaturation and renaturation of complementary DNA strands is a highly cooperative, zippering-like process. Renaturation, for example, probably occurs by means of a slow nucleation process in which a relatively small stretch of bases on one strand find and pair with their complement on the complementary strand. The remainder of the two strands then rapidly zipper-up from the nucleation site to reform an extended double helix.

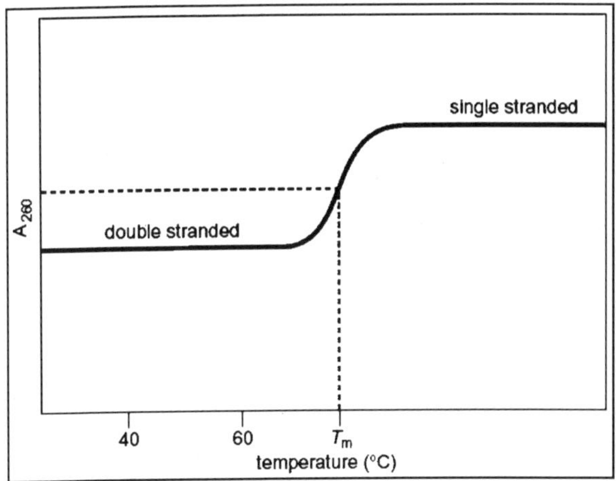

Fig. DNA Denaturation Curve

The melting temperature of DNA is a characteristic of each DNA that is largely determined by the G:C content of the DNA and the ionic strength of the solution. The higher the percent of G:C base pairs in the DNA (and hence the lower the content of A:T base pairs), the higher the melting point. Likewise, the higher the salt concentration of the solution the greater the temperature at which the DNA denatures.

How do we explain this behaviour? G:C base pairs contribute more to the stability of DNA than do A:T base pairs because of the greater number of hydrogen bonds for the former (three in a G:C base pair versus two for A:T) but also importantly because the stacking interactions of G:C base pairs with adjacent base pairs are more favorable than the corresponding interactions of A:T base pairs with their neighboring base pairs. The effect of ionic strength reflects another fundamental feature of the double helix.

The backbones of the two DNA strands contain phosphoryl groups, which carry a negative charge. These negative charges are close enough across the two strands that if not shielded they tend to cause the strands to repel each other, facilitating their separation. At high ionic strength, the negative charges are shielded by cations, thereby stabilizing the helix. Conversely, at low ionic strength the unshielded negative charges render the helix less stable.

CIRCLES CIRCULAR DNA MOLECULES

It was initially believed that all DNA molecules are linear and have two free ends. Indeed, the chromosomes of eukaryotic cells each contain a single (extremely long) DNA molecule. But now we know that some DNAs are circles. For example, the chromosome of the small monkey DNA virus SV40 is a circular, double-helical DNA molecule of about 5,000 base pairs. Also, most (but not all) bacterial chromosomes are circular; *E. coli* has a circular chromosome of about 5 million base pairs.

Additionally, many bacteria have small autonomously replicating genetic elements known as plasmids, which are generally circular DNA molecules. Interestingly, some DNA molecules are sometimes linear and sometimes circular. The most well-known example is that of the bacteriophage ‘, a DNA virus of *E. coli*. The phage ‘ genome is a linear double-stranded molecule in the virion particle. However, when the ‘ genome is injected into an *E. coli* cell during infection, the DNA circularizes. This occurs by base-pairing between single-stranded regions that protrude from the ends of the DNA and that have complementary sequences ("sticky ends").

DNA TOPOLOGY

As DNA is a flexible structure, its exact molecular parameters are a function of both the surrounding ionic environment and the nature of the DNA-binding proteins with which it is complexed. Because their ends are free, linear DNA molecules can freely rotate to accommodate changes in the number of times the two chains of the double helix twist about each other.

But if the two ends are covalently linked to form a circular DNA molecule and if there are no interruptions in the sugar phosphate backbones of the two strands, then the absolute number of times the chains can twist about each other cannot change. Such a covalently closed, circular DNA is said to be topologically constrained. Even the linear DNA molecules of eukaryotic chromosomes are subject to topological constraints due to their entrainment in chromatin and interaction with other cellular components.

Despite these constraints, DNA participates in numerous dynamic processes in the cell. For example, the two strands of the double helix, which are twisted around each other, must rapidly separate in order for DNA to be duplicated and to be transcribed into RNA. Thus, understanding the topology of DNA and how the cell both accommodates and exploits topological constraints during DNA replication, transcription, and other chromosomal transactions is of fundamental importance in molecular biology.

LINKING NUMBER OF COVALENTLY CLOSED, CIRCULAR DNA

Let us consider the topological properties of covalently closed, circular DNA, which is referred to as cccDNA. Because there are no interruptions in either polynucleotide chain, the two strands of cccDNA cannot be separated from each other without the breaking of a covalent bond.

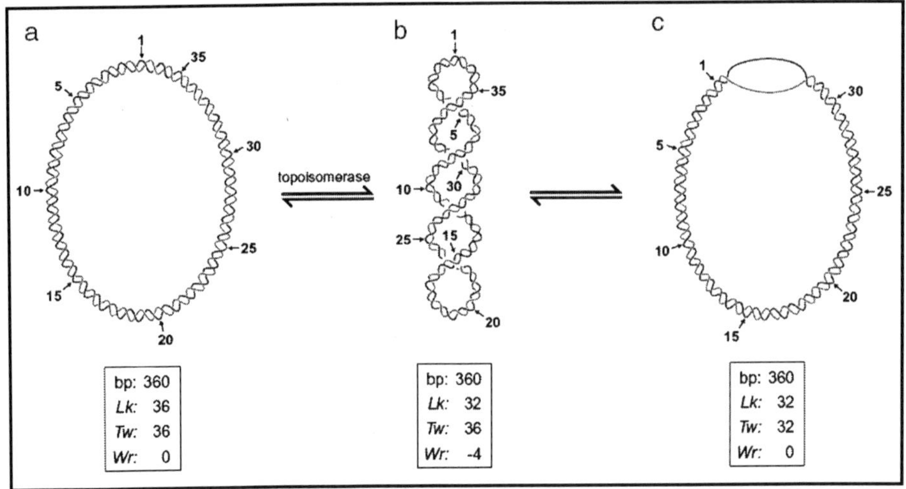

Fig. Topological States of Covalently Closed Circular (ccc) DNA

If we wished to separate the two circular strands without permanently breaking any bonds in the sugar phosphate backbones,we would have to pass one strand through the other strand repeatedly(we will encounter an enzyme that can perform just this feat!).

The number of times one strand would have to be passed through the other strand in order for the two strands to be entirely separated from each other is called the linking number. The linking number, which is always an integer, is an invariant topological property of cccDNA, no matter how much the DNA molecule is distorted.

Linking Number Is Composed of Twist and Writhe

The linking number is the sum of two geometric components called the twist and the writhe. Let us consider twist first. Twist is simply the number of helical turns of one strand about the other, that is, the number of times one strand completely wraps around the other strand. Consider a cccDNA that is lying flat on a plane.

In this flat conformation, the linking number is fully composed of twist. Indeed, the twist can be easily determined by counting the number of times the two strands cross each other. The helical crossovers (twist) in a right-handed helix are defined as positive such that the linking number of DNA will have a positive value. But cccDNA is generally not lying flat on a plane.

Rather, it is usually torsionally stressed such that the long axis of the double helix crosses over itself, often repeatedly, in three-dimensional space. This is called *writhe*. To visualize the distortions caused by torsional stress, think of the coiling of a telephone cord that has been overtwisted. Writhe can take two forms. One form is the interwound or plectonemic writhe, in which the long axis is twisted around itself. The other form of writhe is a toroid or spiral in which the long axis is wound in a cylindrical manner, as often occurs when DNA wraps around protein. The writhing number (*Wr*) is the total number of interwound and/or spiral writhes in cccDNA. For example, the molecule shown in Figure has a writhe of 4 from 4 interwound writhes. Interwound writhe and spiral writhe are topologically equivalent to each other and are readily interconvertible geometric properties of cccDNA.

Also, twist and writhe are interconvertible. A molecule of cccDNA can readily undergo distortions that convert some of its twist to writhe or some of its writhe to twist without the breakage of any covalent bonds.

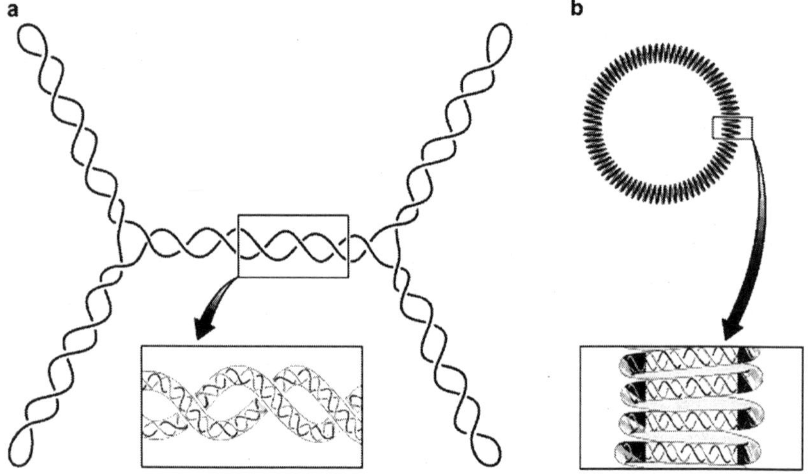

Fig. Two Forms of Writhe of Supercoiled DNA

The only constraint is that the sum of the twist number (*Tw*) and the writhing number (*Wr*) must remain equal to the linking number (*Lk*). This constraint is described by the equation: $Lk = Tw + Wr$.

Lk° Is the Linking Number of Fully Relaxed cccDNA

Consider cccDNA that is free of supercoiling (that is, it is said to be relaxed) and whose twist corresponds to that of the B form of DNA in solution under physiological conditions (about 10.5 base pairs per turn of the helix). The linking number (*Lk*) of such cccDNA under physiological conditions is assigned the symbol $Lk°$. $Lk°$ for such a molecule is the number of base pairs divided by 10.5.

For a cccDNA of 10,500 base pairs, $Lk = +1{,}000$. (The sign is positive because the twists of DNA are right-handed.) One way to see this is to imagine pulling one strand of the 10,500 base pair cccDNA out into a flat circle. If we did this, then the other strand would cross the flat circular strand 1,000 times.

How can we remove supercoils from cccDNA if it is not already relaxed? One procedure is to treat the DNA mildly with the enzyme DNase I, so as to break on average one phosphodiester bond (or a small number of bonds) in each DNA molecule. Once the DNA has been "nicked" in this manner, it is no longer topologically constrained and the strands can rotate freely, allowing writhe to dissipate.

If the nick is then repaired, the resulting cccDNA molecules will be relaxed and will have on average an *Lk* that is equal to $Lk°$. (Due to rotational fluctuation at the time the nick is repaired, some of

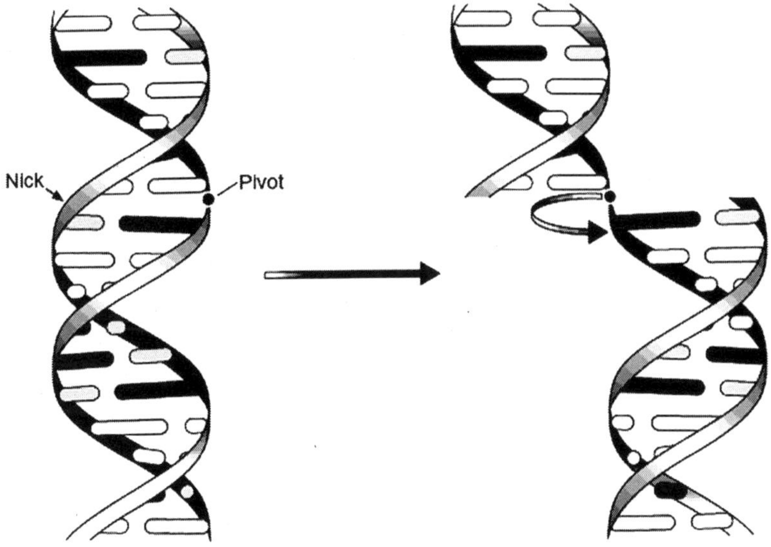

Fig. Relaxing DNA with DNase I

the resulting cccDNAs will have an *Lk* that is somewhat greater than $Lk°$ and others will have an *Lk* that is somewhat lower. Thus, the relaxation procedure will generate a narrow spectrum of topoisomers whose average *Lk* is equal to $Lk°$).

DNA in Cells is Negatively Supercoiled

The extent of supercoiling is measured by the difference between *Lk* and $Lk°$, which is called the linking difference:

$$Lk = Lk - Lk°.$$

If the "*Lk* of a cccDNA is significantly different from zero, then the DNA is torsionally strained and hence it is supercoiled. If $Lk < Lk°$ and "$Lk < 0$, then the DNA is said to be "negatively supercoiled." Conversely, if $Lk < Lk°$ and " $Lk < 0$, then the DNA is "positively supercoiled." For example, the molecule shown in Figure is negatively supercoiled and has a linking difference of -4 because its *Lk* is four less than that for the relaxed form of the molecule shown in Figure. Because " *Lk* and Lk^O are dependent upon the length of the DNA molecule, it is more convenient to refer to a normalized measure of supercoiling. This is the superhelical density, which is assigned the symbol ó and is defined as:

$$ó = "Lk/Lk^{\circ}$$

DNA rings purified both from bacteria and eukaryotes are usually negatively supercoiled, having values of ó of about -0.06. The electron micrograph shown in Figure compares the structures of bacteriophage DNA in its relaxed form with its supercoiled form. What does this mean biologically? Negative supercoils can be thought of as a store of free energy that aids in processes that require strand separation, such as DNA replication and transcription. Because

$$Lk = Tw + Wr,$$

negative supercoils can be converted into untwisting of the double helix. Regions of negatively supercoiled DNA therefore have a tendency to partially unwind. Thus, strand separation can be accomplished more easily in negatively supercoiled DNA than in relaxed DNA. The only organisms that have been found to have positively supercoiled DNA are certain thermophiles, microorganisms that live under conditions of extreme high temperatures, such as in hot springs.

In this case, the positive supercoils can be thought of as a store of free energy that helps keep the DNA from denaturing at the elevated temperatures. In so far as positive supercoils can be converted into more twist (positively supercoiled DNA can be thought of as being overwound), strand separation requires more energy in thermophiles than in organisms whose DNA is negatively supercoiled.

Nucleosomes Introduce Negative Supercoiling in Eukaryotes

DNA in the nucleus of eukaryotic cells is packaged in small particles known as nucleosomes in which the double helix is wrapped almost two times around the outside circumference of a protein core. You will be able to recognize this wrapping as the toroid or spiral form of writhe. Importantly, it occurs in a lefthanded manner.. It turns out that writhe in the form of left-handed spirals is equivalent to negative supercoils. Thus, the packaging of DNA into nucleosomes introduces negative superhelical density.

Topoisomerases Can Relax Supercoiled DNA

As we have seen, the linking number is an invariant property of DNA that is topologically constrained. It can only be changed by introducing interruptions into the sugar-phosphate backbone. A remarkable class of enzymes known as topoisomerases are able to do just that by introducing transient nicks or breaks into the DNA. Topoisomerases are of two broad types.

Type II topoisomerases change the linking number in steps of two. They make transient double-stranded breaks in the DNA, through which they pass a region of uncut duplex DNA before resealing the break. Type II topoisomerases require energy from ATP hydrolysis for their action. Type I topoisomerases, in contrast, change the linking number of DNA in steps of one.

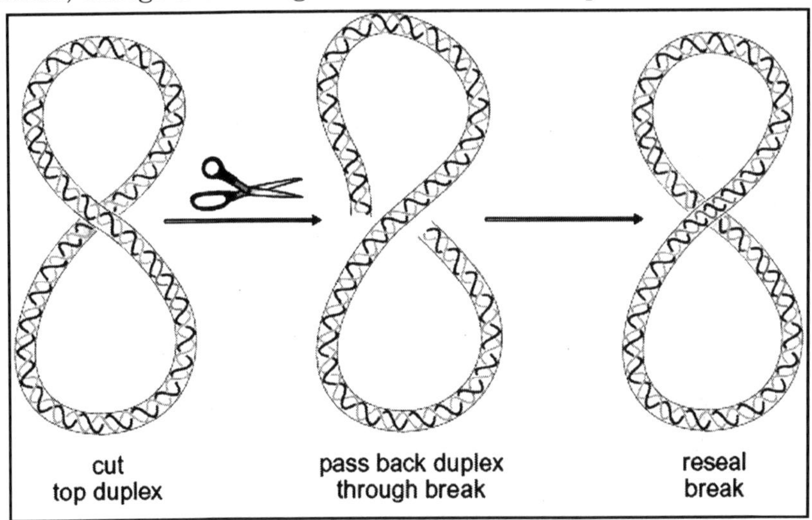

Fig. Schematic for Changing the Linking Number in DNA with Topoisomerase II

They make transient singlestranded breaks in the DNA, allowing one strand to pass through the break in the other before resealing the nick. Type I topoisomerases relax DNA by removing supercoils (dissipating writhe).

They can be compared to the protocol of introducing nicks into cccDNA with DNase and then repairing the nicks, which as we saw can be used to relax cccDNA, except that type I topoisomerases relax DNA in a controlled and concerted manner. In contrast to type II topoisomerases, type I topoisomerases do not require ATP. Both type I and type II topoisomerases work through an intermediate in which the enzyme is covalently attached to one end of the broken DNA.

Prokaryotes Have a Special Topoisomerase that Introduces Supercoils

Both prokaryotes and eukaryotes have type I and type II topoisomerases, which are capable of removing supercoils from DNA. In addition, however, prokaryotes have a special type II topoisomerase known as DNA gyrase that is able to introduce negative supercoils, rather than remove them. DNA gyrase is responsible for the negative supercoiling of chromosomes in prokaryotes, which facilitates unwinding of the DNA duplex during transcription and DNA replication.

DNA Topoisomers Can Be Separated by Electrophoresis

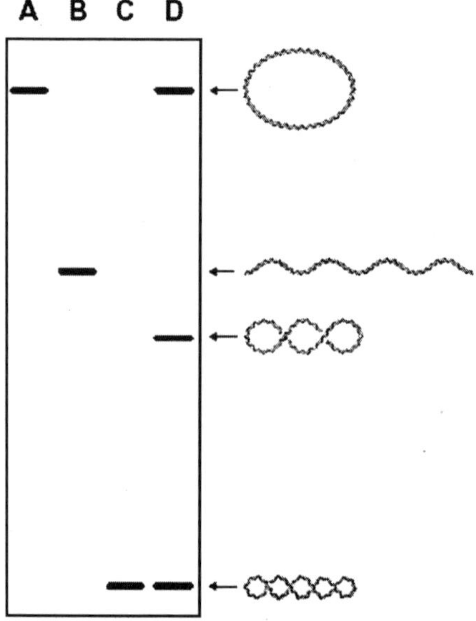

Fig. Schematic of Electrophoretic Separation of DNA Topoisomers

Covalently closed, circular DNA molecules of the same length but of different linking numbers are called DNA topoisomers. Even though topoisomers have the same molecular weight, they can be separated from each other by electrophoresis through a gel of agarose. The basis for this separation is that the greater the writhe the more compact the shape of a cccDNA. Once again, think of how supercoiling a telephone cord causes it to become more compact. The more compact the DNA, the more easily (up to a point) it is able to migrate through the gel matrix.

Thus, a fully relaxed cccDNA migrates more slowly than a highly supercoiled topoisomer of the same circular DNA. Figure shows a ladder of DNA topoisomers resolved by gel electrophoresis. Molecules in adjacent rungs of the ladder differ from each other by a linking number difference of just one. Obviously, electrophoretic mobility is highly sensitive to the topological state of DNA.

Proving That DNA Has a Helical Periodicity of about 10.5 Base

The observation that DNA topoisomers can be separated from each other electrophoretically is the basis for a simple experiment that proves that DNA has a helical periodicity of about 10.5 base pairs per turn in solution. Consider three cccDNAs of sizes 3990, 3995, and 4011 base pairs that were relaxed to completion by treatment with topoisomerase I. When subjected to electrophoresis through agarose, the 3990- and 4011-base-pair DNAs exhibit essentially identical mobilities. Due to thermal fluctuation, topoisomerase treatment actually generates a narrow spectrum of topoisomers, but for simplicity let us consider the mobility of only the most abundant topoisomer (that corresponding to the cccDNA in its most relaxed state). The mobilities of the most abundant topoisomers for the 3990- and 4011-base-pair DNAs are indistinguishable because the 21-base-pair difference between them is negligible compared to the sizes of the rings. The most abundant topoisomer for the 3995-base-pair ring, however, is found to migrate slightly more rapidly than the other two rings even though it is only 5 base pairs larger than the 3990-base-pair ring.

How are we to explain this anomaly? The 3990- and 4011- base-pair rings in their most relaxed states are expected to have linking numbers equal to *Lk*°, that is, 380 in the case of the 3990-base-pair ring (dividing the size by 10.5 base pairs) and 382 in the case of the 4011-base-pair ring. Because *Lk* is equal to *Lk*°, the linking difference (*'Lk* = *Lk* - *Lk*°) in both cases is zero and there is no writhe. But because the linking number must be an integer, the most relaxed state for the 3995-base-pair ring would be either of two topoisomers having linking numbers of 380 or 381. However, *Lk*° for the 3995-base-pair ring is 380.5.

Thus, even in its most relaxed state, a covalently closed circle of 3995 base pairs would necessarily have about half a unit of writhe (its linking difference would be 0.5), and hence it would migrate more rapidly than the 3990- and 4011-base-pair circles. In other words, to explain how rings that differ in length by 21 base pairs (two turns of the helix) have the same mobility whereas a ring that differs in length by only 5 base pairs (about half a helical turn) exhibits a different mobility, we must conclude that DNA in solution has a helical periodicity of about 10.5 base pairs per turn.

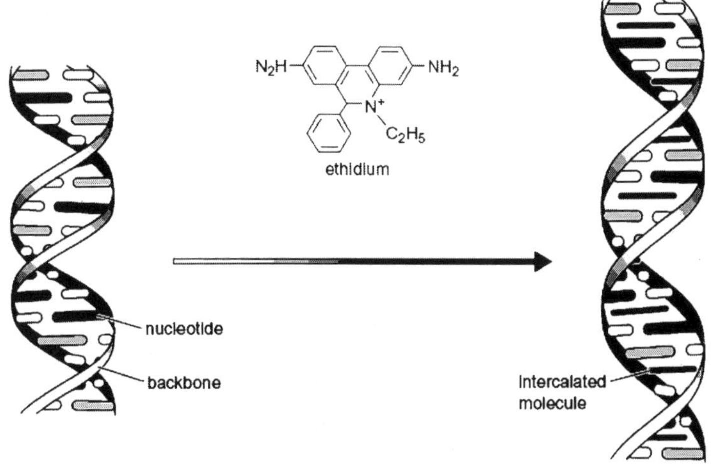

Fig. Intercalation of Ethidium Bromide into DNA

Ethidium Ions Cause DNA to Unwind

Ethidium is a large, flat, multi-ringed cation. Its planar shape enables ethidium to slip (intercalate) between the stacked base pairs of DNA. Because it fluoresces when exposed to ultraviolet light, and because its fluorescence increases dramatically after intercalation, ethidium is used as a stain to visualize DNA. When an ethidium ion intercalates between two base pairs, it causes the DNA to unwind by 26°, reducing the normal rotation per base pair from ~36° to ~10°. In other words, ethidium decreases the twist of DNA.

Imagine the extreme case of a DNA molecule that has an ethidium ion between every base pair. Instead of 10 base pairs per turn it would have 36! When ethidium binds to linear DNA or to a nicked circle, it simply causes the helical pitch to increase. But consider what happens when ethidium binds to covalently closed, circular DNA. The linking number of the cccDNA does not change (no covalent bonds are broken and resealed), but the twist decreases by 26° for each molecule of ethidium that has bound to the DNA.

Because $Lk = Tw + Wr$, this decrease in Tw must be compensated for by a corresponding increase in Wr. If the circular DNA is initially negatively supercoiled (as is normally the case for circular DNAs isolated from cells), then the addition of ethidium will increase Wr. In other words, the addition of ethidium will relax the DNA. If enough ethidium is added, the negative supercoiling will be brought to zero, and if even more ethidium is added, Wr will increase above zero and the DNA will become positively supercoiled.

Because the binding of ethidium increases Wr, its presence greatly affects the migration of cccDNA during gel electrophoresis. In the presence of non-saturating amounts of ethidium, negatively supercoiled circular DNAs are more relaxed and migrate more slowly, whereas relaxed cccDNAs become positively supercoiled and migrate more rapidly.

3

DNA Structure

DNA stands for deoxyribonucleic acid. DNA is pretty unusual in that it is about the only common molecule capable of directing its own synthesis. The processes of mitosis and meiosis were discovered in the 1870s and 1890s. It was observed that, as cells divided, chromosomes moved around in a cell, and people began to wonder what their function was.

It was determined that chromosomes were made of protein and DNA, about which people knew almost nothing. People began to suspect that chromosomes had something to do with genetics, but couldn't explain what/how. When enough evidence was accumulated to confirm that chromosomes did, indeed, have something to do with genetics, most people thought that in some way the protein in the chromosomes served as the genetic material.

People knew that DNA was also in the chromosomes, but because its structure was unknown and people didn't know much about it, few people thought it was the genetic material. Frederick Griffith performed an experiment using pneumonia bacteria and mice. This was one of the first experiments that hinted that DNA was the genetic code material. He used two strains of *Streptococcus pneumoniae*: a "smooth" strain which has a polysaccharide coating around it that makes it look smooth when viewed with a microscope, and a "rough" strain which doesn't have the coating, thus looks rough under the microscope.

When he injected live S strain into mice, the mice contracted pneumonia and died. When he injected live R strain, a strain which typically does not cause illness, into mice, as predicted they did not get sick, but lived. Thinking that perhaps the polysaccharide coating on the bacteria somehow caused the illness and knowing that polysaccharides are not affected by heat, Griffith then used heat to kill some of the S strain bacteria and injected those dead bacteria into mice.

This failed to infect/kill the mice, indicating that the polysaccharide coating was not what caused the disease, but rather, something within the living cell. Since Griffith had used heat to kill the bacteria and heat denatures protein, he next hypothesized that perhaps some protein within the living cells, that was denatured by the heat, caused the disease. He then injected another group of mice with a mixture of heat-killed S and live R, and the mice died! When he did a necropsy on the dead mice, he isolated live S strain bacteria from the corpses.

Griffith concluded that the live R strain bacteria must have absorbed genetic material from the dead S strain bacteria, and since heat denatures protein, the protein in the bacterial chromosomes was not the genetic material. This evidence pointed to DNA as being the genetic material. Transformation is the process whereby one strain of a bacterium absorbs genetic material from another strain of bacteria and "turns into" the type of bacterium whose genetic material it absorbed. Because DNA was so poorly understood, scientists remained skeptical up through the 1940s.

Alfred Hershey and Martha Chase did an experiment which is so significant, it has been nicknamed the "Hershey-Chase Experiment". At that time, people knew that viruses were composed of DNA (or RNA) inside a protein coat/shell called a capsid. It was also known that viruses replicate by taking over the host cell's metabolic functions to make more virus. We are used to thinking and talking about viruses which invade our bodies and make us sick, but there are other, different kinds of viruses that infect other kinds of animals, still other viruses which infect plants, and even some viruses that infect bacteria.

A virus which infects a bacterium is called a bacteriophage because the host bacterium cell is killed as the new virus particles leave the bacterial cell. In order to do all this, the virus must inject whatever is the viral genetic code into the host cell.

Thus, people realized that the viral genetic code material had to be either its DNA or its protein capsid. Hershey and Chase sought an answer to the question, "Is it the viral DNA or viral protein coat (capsid) that is the viral genetic code material which gets injected into a host bacterium cell? To try to answer this question, Hershey and Chase performed an experiment using a bacterium named *Escherichia coli*, or *E. coli* for short (named after a scientist whose last name was Escher) and a virus called that is a bacteriophage that infects *E. coli*. Isolated, like other viruses, is just a crystal of DNA and protein, so it must live inside *E. coli* in order to make more virus like itself.

When the new viruses are ready to leave the host *E. coli* cell (and go infect others), they burst the *E. coli* cell open, killing it (hence the name "bacteriophage"). The results that Hershey and Chase obtained indicated that the viral DNA, not the protein, is its genetic code material.

Hershey and Chase used radioactive chemicals to distinguish between ("label") the protein capsid and the DNA in virus so they could tell which of those molecules entered the *E. coli* cells. Since some amino acids contain sulfur in their side chains, if is grown in *E. coli* with a source of radioactive sulfur, the sulfur will be incorporated into the protein coat making it radioactive. Since DNA has lots of phosphorus in its phosphate ($-PO_4$) groups, if is grown in *E.*

coli with a source of radioactive phosphorus, the phosphorus will be incorporated into the viral DNA, making that radioactive.

Hershey and Chase grew two batches of and *E. coli*: one with radioactive sulfur and one with radioactive phosphorus to get batches of "labeled" with either radioactive S or radioactive P. Then, these radioactive were placed in separate, new batches of *E. coli*, but were left there only 10 minutes. This was to give the time to inject their genetic material into the bacteria, but not reproduce.

In the next step, still in separate batches, the mixtures were agitated in a kitchen blender to knock loose any viral parts not inside the *E. coli* but perhaps stuck on the outer surface. Hopefully, this would differentiate between the protein and DNA portions of the virus. Then, each mixture was spun in a centrifuge to separate the heavy bacteria (with any viral parts that had gone into them) from the liquid solution they were in (including any viral parts that had not entered the bacteria). The centrifuge causes the heavier bacteria to be pulled to the bottom of the tube where they form a pellet, while the light-weight viral "left-overs" stay suspended in the liquid portion called the supernatant.

In the subsequent step, the pellet and supernatant from each tube were separated and tested for the presence of radioactivity. Radioactive sufur was found in the supernatant, indicating that the viral protein did not go into the bacteria. Radioactive phosphorus was found in the bacterial pellet, indicating that viral DNA did go into the bacteria.

Based on these results, Hershey and Chase concluded that DNA must be the genetic code material, not protein as many poeple believed. When their experiment was published and people finally acknowledged that DNA was the genetic material, there was a lot of competition to be the first to discover its chemical structure.

What was known is that DNA contains a nitrogenous base. There are two kinds of these, which include:

Pyrimidine Purine
(6-member ring (that + 5 member ring
of C & N) of C & N)
Cytosine Guanine
Thymine in Adenine
DNAuracil in RNA

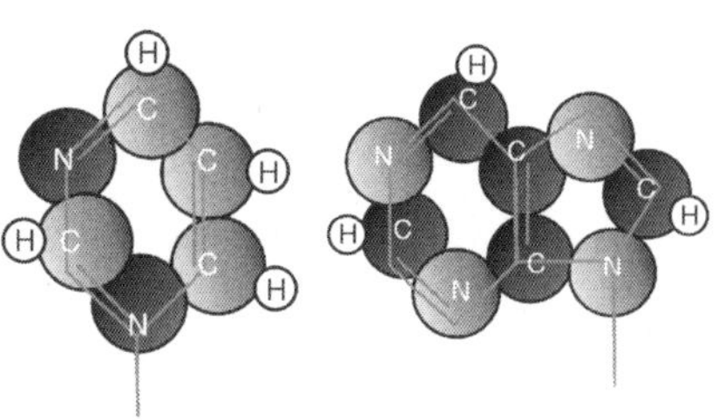

Each nitrogenous base is connected to a molecule of ribose sugar (–1 oxygen in DNA) to form a nucleoside like the *adenosine* in ATP. Each nucleoside is joined to a PO_4 (phosphate group) to form a nucleotide like adenosine monophosphate (which can be turned into ATP by adding phosphate groups). People also knew that nucleotides were somehow linked by dehydration synthesis to form DNA, but the exact structure/arrangement was unknown.

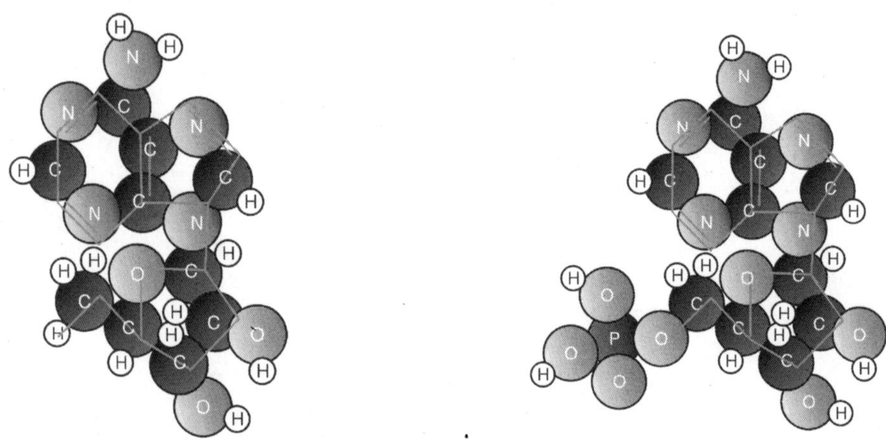

In the early 1950s, Rosalind Franklin, an Englishwoman, was doing research which involved bouncing x-rays off crystals of various substances (a process which is called x-ray crystallography), including DNA, then exposing photographic film to the x-rays. She was studying the scatter patterns made by the x-rays bouncing off the crystals of various substances (Unfortunately, she died of cancer soon afterwards, or she might have been more famous). Other people like Linus Pauling were also attempting to figure out the structure of DNA.

James Watson, a young American scientist was in England working with Francis Crick, another young researcher. Someone else showed them Franklin's photographs of DNA x-ray crystallography, and from her pictures, they were able to determine that the structure of DNA was organized into a double spiral or double helix. Based on Franklin's data, in 1953, Watson and Crick published a paper in which they proposed and described an hypothetical structure for DNA. Subsequent research by many other people has since upheld their hypothesis, and based on subsequent examination of Franklin's lab notes and calculations, she was probably within a couple days of coming to the same conclusion when their paper was published. For their discovery, Watson and Crick received the Nobel prize in 1962. In the intervening time, Rosalind Franklin had died in 1958 of ovarian cancer, probably due in large part to her work with x-rays. Since the Nobel prize is not awarded posthumously, people have often wondered if the Nobel committee would have included Franklin if she had still been alive.

DNA is a double helix. The outer edges are formed of alternating ribose sugar molecules and phosphate groups. The two strands go in opposite directions (1 "up" and 1 "down"). The nitrogenous bases are "inside" like rungs on a ladder. Adenine on one side pairs with thymine (uracil in RNA) on the other by hydrogen bonding, and cytosine pairs with guanine. Note that the C-G pair has three hydrogen bonds while the A-T pair has only two, which keeps them from pairing wrong.

This dictates side-to-side pairing, but says nothing about the order *along* the molecule. Watson and Crick said this variability along the molecule can account for the variety in the genetic code. Their model also accounts for how DNA can replicate itself. They said the molecule "unzips" and new matching bases are added in to create two new molecules. They called this semiconservative replication because each new molecule has one "old" and one "new" strand of DNA.

DNA codes for protein synthesis by first coding for RNA. First, the DNA code is transcribed to RNA code, which is still in the "language" of nitrogenous bases, except that adenine on the DNA pairs with uracil (in place of thymine) on the RNA. The RNA code is then translated to protein code, which is a different "language."

		Second Base					
		U	C	A	G		
First Base	**U**	UUU Phe	UCU Ser	UAU Tyr	UGU Cys	**U**	**Third Base**
		UUC Phe	UCC Ser	UAC Try	UGC Cys	**C**	
		UUA Leu	UCA Ser	UAA Stop	UGA Stop	**A**	
		UUG Leu	UCG Ser	UAG Stop	UGG Trp	**G**	
	C	CUU Leu	CCU Pro	CAU His	CGU Arg	**U**	
		CUC Leu	CCC Pro	CAC His	CGC Arg	**C**	
		CUA Leu	CCA Pro	CAA Gln	CGA Arg	**A**	
		CUG Leu	CCG Pro	CAG Gln	CGG Arg	**G**	
	A	AUU Ile	ACU Thr	AAU Asn	AGU Ser	**U**	
		AUC Ile	ACC Thr	AAC Asn	AGC Ser	**C**	
		AUA Ile	ACA Thr	AAA Lys	AGA Arg	**A**	
		AUG Met or Start	ACG Thr	AAG Lys	AGG Arg	**G**	
	G	GUU Val	GCU Ala	GAU Asp	GGU Gly	**U**	
		GUC Val	GCC Ala	GAC Asp	GGC Gly	**C**	
		GUA Val	GCA Ala	GAA Glu	GGA Gly	**A**	
		GUG Val	GCG Ala	GAG Glu	GGG Gly	**G**	

This process involves ribosomes and two kinds of RNA: mRNA and tRNA. The mRNA codes for the gene in question and is copied off the DNA, while tRNA matches a specific group of nucleotides with a specific amino acid. A "unit" of three nucleotides on the tRNA codes for one amino acid. Each of these "units" is called an anticodon. These match up with corresponding three-nucleotide sequences on the mRNA called codons, and in this manner the amino acids are organized into the correct sequence to build a protein. The ribosome works with the mRNA and tRNA to hook the amino acids together to form a protein.

Here is a list of the mRNA codons and the corresponding amino acids for which they code. Mutations can be caused by a change in the sequence of the nucleotides. Some mutations have more effect than others, depending on where in the code they are and how important that area is to the code. While mutations in some areas of some genes have little effect, sickle cell anemia is caused by a mutation in only one nucleotide. This changes the codon at that location to code for a different amino acid, and that, in turn, significantly changes the shape of the hemoglobin molecules in that person's blood.

When some viruses (especially *Herpes* viruses, including Chicken Pox and Cold Sores) infect us, they insert their DNA into our cells' DNA, and stay resident in our cells for the rest of our lives. These can potentially become active again either making a person sick again (like Shingles in a person who has had Chicken Pox) or just being shed from a person's body (to infect others) without obvious symptoms of illness (like Mononucleosis). Some kinds of cancer may be caused this way. For example, there is some pretty strong evidence linking genital warts (human papillomavirus, HPV) and cervical cancer.

The AIDS virus does things "backwards." This virus contains RNA rather than DNA, yet when it gets into someone's cells, it can do reverse transcription and code from its RNA to make DNA which, then, can code to make more virus.

GENETIC ENGINEERING

We now have the knowledge and ability to transfer genes from one organism to another, which seems to have some benefits associated with it, but may also have many yet-to-be-discovered problems associated with it. Because this is all so new, not enough time has elapsed to allow scientists to study/look for any possible long-term effects of genetically-modified organisms (GMOs).

Many medicines are now made by GMOs. For example, insulin was formerly extracted from the pancreas of animals after they were slaughtered for meat. However, now most insulin is produced by bacteria with the insulin gene spliced onto their chromosome. Using this method of production, drug companies can make more insulin, faster.

In theory, insulin produced in this way should be more "pure" — someone who could not use pig insulin due to a pork allergy may be able to tolerate insulin made in this way (but someone could, potentially, be allergic to some component of the bacteria present in the refined insulin). In the relatively short time that this form of insulin has been available, there is no doubt that it has saved many people's lives — let's hope that some unforseen, long-term, deleterious effects are not discovered later on.

Experimentation is being done to investigate the possible use of genetically-engineered viruses to treat genetic diseases such as cystic fibrosis. In this "treatment," a kind of virus that infects our lungs is used. The genes that enable it to infect our cells are kept while the genes which make us sick are (hopefully) all removed, and the missing human (normal) gene that relieves cystic fibrosis is then inserted into the virus' genome. These viruses are then sprayed into the lungs of a person with CF and allowed to "infect" the cells in that person's lungs.

When the normal gene is inserted into the genetic make-up of the cells lining that person's lungs, those cells function normally and the CF symptoms are alleviated. However, this treatment doesn't last because only that layer of cells is "infected," and when those cells die and are replaced by new cells, the new cells do not contain the genetic code to overcome the CF gene, and the person must inhale more genetically-engineered virus. While this seems to be a promising, life-giving, technique, no data are yet available on long-term effects and safety.

Bacillus thuringiensis (Bt) is a species of bacterium that infects and kills a number of species of caterpillars (Remember that caterpillars turn into butterflies or moths when they grow up.). There is a species of moth whose caterpillar is a "pest" in corn plants, and insects of any kind are more successful when we humans plant huge monocultures of their favourite foods. For a number of years, now, people have realized that in certain situations, by judiciously applying BT, "pest" species of caterpillars can be infected and killed without the use of man-made, chemical insecticides.

More recently, a major US chemical company came up with the idea of creating genetically-engineered corn containing BT genes, which was good news to agri-business firms who plant huge areas of land with monocultures of corn. Because corn is wind-pollinated and therefore makes lots of pollen which, in this case, contains BT genes, many scientists are concerned about the effects of this corn on local butterfly populations.

Some research has indicated that when this pollen settles on nearby caterpillar host plants and is, therefore, consumed by caterpillars, this might cause an increase in mortality (therefore fewer "good" butterflies such as Monarchs). It is assumed that these BT genes in corn should, they think, have no adverse effects on people or cattle who consume this corn, but no long-term testing has been done. Also, this company has convinced the government that this corn should be marketed without any labeling indicating that it is genetically-engineered — they're scared that if you are given a choice, you won't buy/eat their corn if you know it is genetically engineered.

Additionally, to increase their profits, this company has also put "suicide" genes into this corn so that farmers cannot save seed from one year to plant the next year, and have to buy more seed from them, instead. For big agribusinesses, this is of small consequence, but for small, family farmers this is total disaster. To save money, the latter often save seed from one year to plant the next, and even if they don't plant this genetically-engineered corn, if their corn is pollinated by genetically-engineered corn from a neighboring field, it will not produce viable seed.

This same chemical company came up with the idea of "transplanting" what they think is the cold-tolerance gene from a species of cold-water-inhabiting fish into tomatoes, thereby hoping to "invent" cold-tolerant tomatoes, and again, has convinced the government that these tomatoes should not be labeled in any way to indicate that they have fish genes in them, and that you should not have a choice about what you eat.

At the very least, this would be a problem for someone who is a vegetarian and chooses to not eat fish. A more serious consequence, however, would be that a person who is severely allergic to fish could also have an allergic reaction to these tomatoes and end up in the hospital. Again, no tests were done before the government approved these tomatoes and no data are available on the long-term effects on humans (or anything else).

This same chemical company manufactures a widely-used herbicide, and came up with the idea to genetically engineer cotton, soybeans, and other crop plants so they would be immune to the effects of that herbicide. That way, farmers can (have to?) buy their seed from that company, then spray their fields with herbicide also purchased from that company to (hopefully) kill all local plants except the immune crop, simultaneously putting all their money in the corporation's pockets. Preliminary research has shown that these resistant genes have already begun to "jump" into local weeds, thereby making them resistant to that herbicide, and again, no data are available on long-term effects on humans, other animals, or the environment in general.

DISCOVERY OF THE STRUCTURE OF DNA

Most biological experiments are done with samples of living matter- cells, tissues, whole organisms or extracts of these materials. Only a few experiments are done with mathematical or physical models of biological components. Ocassionally, however, a model experiment gives an insight that would be difficult to obtain in any other way. This was true of the discovery of the structure of DNA. A crucial experiment was done by James Watson in 1952 using nothing more complicated than chemical models cut out of cardboard. The experiment is so simple that you can do it yourself and can even show it to your children and friends.

PREVIOUS INFORMATION ABOUT THE STRUCTURE OF DNA

By 1952 conventional laboratory experiments had shown the following:

- DNA was the molecule of heredity: it had been shown that transferring DNA into bacteria could change them genetically. This

made the solving of the DNA structure one of the top priorities in biology
- DNA was known to be composed of phosphate, the 5 carbon sugar deoxyribose and 4 nucleotide bases: adenine, cytosine, guanine and thymine (abbreviated A, C, G & T)
- Chemical structures of all of the components (phosphate, sugar & bases) were known
- The sugar (S) and phosphate (P) were known to be connected together to form a backbone. The bases (B) were known to be stuck out to the side, attached to the sugar of the backbone by one of their N atoms:

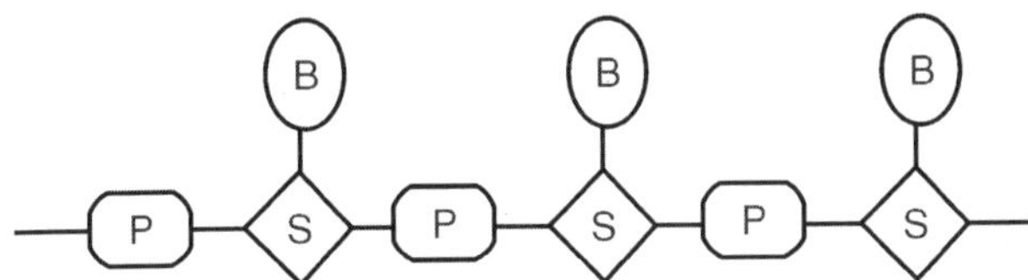

- Rosalind Franklin had pointed out that phosphate has negative charges at cellular pHs. Since these charges would repel each other she insisted that the sugar-phosphate backbone was on the outside of the molecule
- Hydrogen bonds had been shown to be very important in determining the shapes of proteins and it was expected that they would be important in DNA as well
- Rosalind Franklin had taken excellent x-ray diffraction pictures of DNA. The pictures had a distinctive pattern that was known from theory (developed by Francis Crick) to be due to a helical structure (2 or more strands spiraling around each other)
- Erwin Chargaff had been making accurate measurements of the base composition of DNA from many different species. He had observed a very strange relationship: in all cases A = T and C = G

Almost everything needed to solve the structure of DNA is in this list, but although a number of people had this data no one had come up with a reasonable solution (several "unreasonable" solutions had been suggested and even published).

AN EXPERIMENT WITH CARDBOARD MODELS

In late 1952 several laboratories were actively working on the structure of DNA. Linus Pauling at Cal Tech had published an incorrect structure with 3 DNA chains wound in a helix.

Rosalind Franklin and Maurice Wilkins were doing x-ray diffraction experiments at King's College in London. Erwin Chargaff thought that the secret to the structure was in the base composition of DNA and he was doing painstaking chemical analyses at Columbia University.

Finally, at Cambridge University Francis Crick and a young American postdoctoral student, James Watson, were very interested in the DNA structure. They were an unlikely pair because they were doing no laboratory experiments of their own.

Instead they were talking to all of the other participants and building chemical models. After a number of mistakes they ordered a set a metal models of the 4 bases to be made by the university shop.

The models were slow in coming, so one afternoon Watson became impatient and decided to draw the base structures on cardboard and cut them out. He did this and then went out to dinner and the theatre.

The next morning he returned to his models and moved them around in pairs to see how they might fit together. He was starting to think that the DNA had 2 chains and he wanted to see if interactions between the bases might hold them together.

It was immediately apparent that interactions between the bases would involve hydrogen bonds. Within a few minutes the structure of DNA became apparent and Watson had an explanation for Chargaff's results (A = T and C = G) and even a good idea of how DNA replicated itself.

The objective is to move the models around to see how they might fit together. You must pay special attention to possible hydrogen bonds. Hydrogen bonds occur between 2 groups of atoms:

- Hydrogen atoms attached to oxygen and nitrogen
- Hydrogen acceptors: oxygens and nitrogens without attached hydrogens

In the drawings of the A, C, G and T bases below the hydrogens available for H bonds are colored red and the hydrogen acceptors are colored green, as shown in this table:

C	Carbon atoms (don't make good H bonds)
N	Nitrogen atoms attached to deoxyribose sugar (not available for H bonding).
N	Nitrogen atoms already bonded to H (not available for H bonding).
N O	Nitrogen & Oxygen atoms available for H bonding
○	H atoms attached to Carbon (don't make good H bonds)
●	H atoms available for H bonding

Here are the 4 DNA bases with the correct colour-coding:

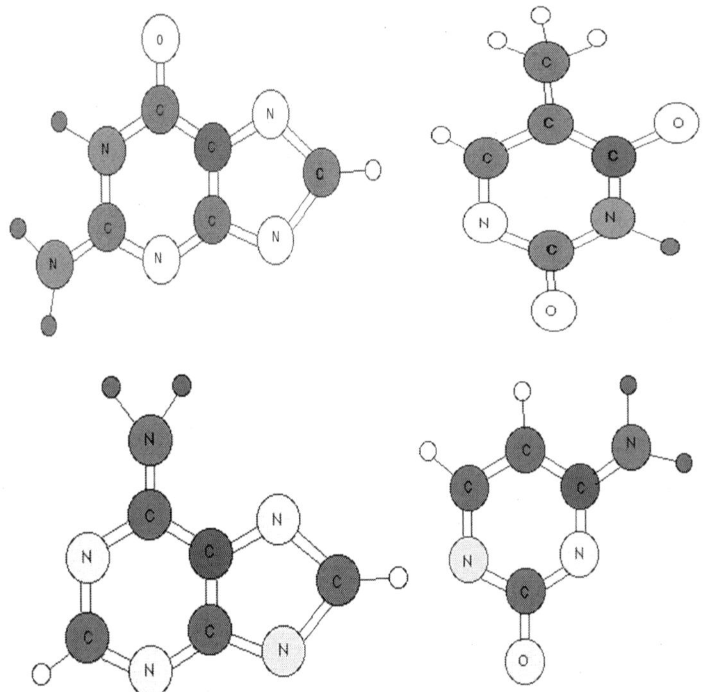

DNA AND MOLECULAR GENETICS

THE PHYSICAL CARRIER OF INHERITANCE

While the period from the early 1900s to World War II has been considered the "golden age" of genetics, scientists still had not determined that DNA, and not protein, was the hereditary material. However, during this time a great many genetic discoveries were made and the link between genetics and evolution was made. Friedrich Meischer in 1869 isolated DNA from fish sperm and the pus of open wounds. Since it came from nuclei, Meischer named this new chemical, nuclein. Subsequently the name was changed to nucleic acid and lastly to deoxyribonucleic acid (DNA).

Robert Feulgen, in 1914, discovered that fuchsin dye stained DNA. DNA was then found in the nucleus of all eukaryotic cells. During the 1920s, biochemist P.A. Levene analyzed the components of the DNA molecule. He found it contained four nitrogenous bases: cytosine, thymine, adenine, and guanine; deoxyribose sugar; and a phosphate group.

He concluded that the basic unit (nucleotide) was composed of a base attached to a sugar and that the phosphate also attached to the sugar. He (unfortunately) also erroneously concluded that the proportions of bases were equal and that there was a tetranucleotide that was the repeating structure of the molecule.

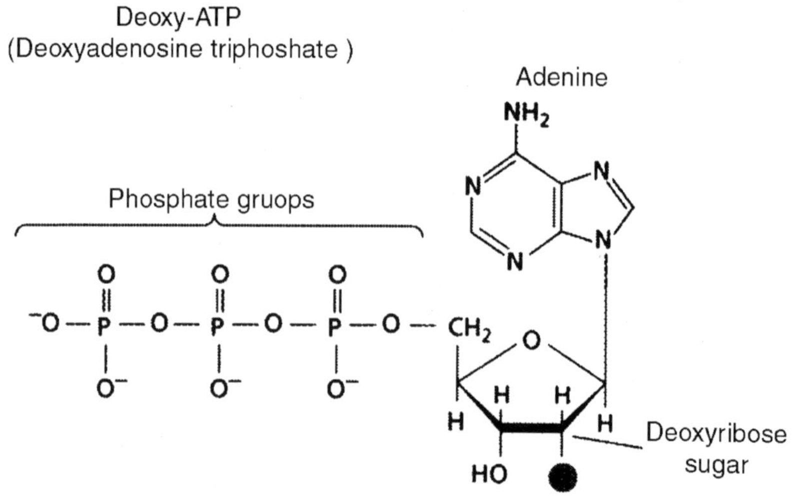

Deoxy-ATP
(Deoxyadenosine triphoshate)

Adenine

NH_2

Phosphate gruops

Deoxyribose sugar

Fig. Molecular Structure of Three Nirogenous Bases

During the early 1900s, the study of genetics began in earnest: the link between Mendel's work and that of cell biologists resulted in the chromosomal theory of inheritance; Garrod proposed the link between genes and "inborn errors of metabolism"; and the question was formed: what is a gene?

The answer came from the study of a deadly infectious disease: *pneumonia*. During the 1920s Frederick Griffith studied the difference between a disease-causing strain of the pneumonia causing bacteria (*Streptococcus peumoniae*) and a strain that did not cause pneumonia.

The pneumonia-causing strain (the S strain) was surrounded by a capsule. The other strain (the R strain) did not have a capsule and also did not cause pneumonia. Frederick Griffith was able to induce a nonpathogenic strain of the bacterium *Streptococcus pneumoniae* to become pathogenic. Griffith referred to a transforming factor that caused the non-pathogenic bacteria to become pathogenic. Griffith injected the different strains of bacteria into mice. The S strain killed the mice; the R strain did not. He further noted that if heat killed S strain was injected into a mouse, it did not cause pneumonia.

When he combined heat-killed S with Live R and injected the mixture into a mouse (remember neither alone will kill the mouse) that the mouse developed pneumonia and died. Bacteria recovered from the mouse had a capsule and killed other mice when injected into them!

Hypotheses:

- The dead S strain had been reanimated/resurrected.
- The Live R had been transformed into Live S by some "transforming factor".

Further experiments led Griffith to conclude that number 2 was correct.

In 1944, Oswald Avery, Colin MacLeod, and Maclyn McCarty revisited Griffith's experiment and concluded the transforming factor was DNA. Their evidence was strong but not totally conclusive. The then-current favourite for the hereditary material was protein; DNA was not considered by many scientists to be a strong candidate.

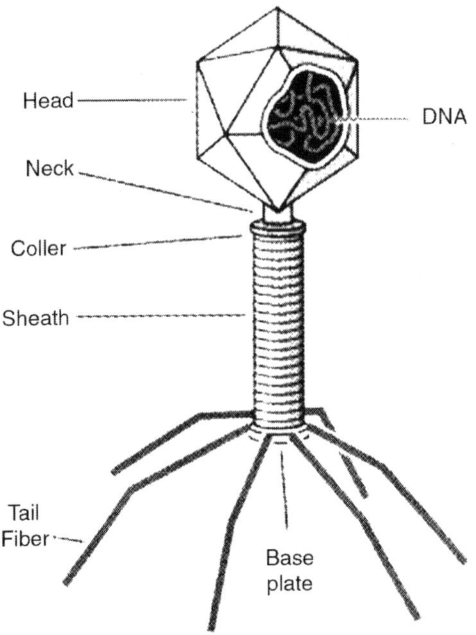

Fig. Structure of a Bacteriophage Virus

The breakthrough in the quest to determine the hereditary material came from the work of Max Delbruck and Salvador Luria in the 1940s. Bacteriophage are a type of virus that attacks bacteria, the viruses that Delbruck and Luria worked with were those attacking *Escherichia coli*, a bacterium found in human intestines. Bacteriophages consist of protein coats covering DNA. Bacteriophages infect a cell by injecting DNA into the host cell.

This viral DNA then "disappears" while taking over the bacterial machinery and beginning to make new virus instead of new bacteria. After 25 minutes the host cell bursts, rcleasing hundreds of new bacteriophage.

Phages have DNA and protein, making them ideal to resolve the nature of the hereditary material. In 1952, Alfred D. Hershey and Martha Chase conducted a series of experiments to determine whether protein or DNA was the hereditary material. By labeling the DNA and protein with different (and mutually exclusive) radioisotopes, they would be able to determine which chemical (DNA or protein) was getting into the bacteria.

Such material must be the hereditary material (Griffith's transforming agent). Since DNA contains Phosphorous (P) but no Sulfur (S), they tagged the DNA with radioactive Phosphorous-32. Conversely, protein lacks P but does have S, thus it could be tagged with radioactive Sulfur-35. Hershey and Chase found that the radioactive S remained outside the cell while the radioactive P was found inside the cell, indicating that DNA was the physical carrier of heredity.

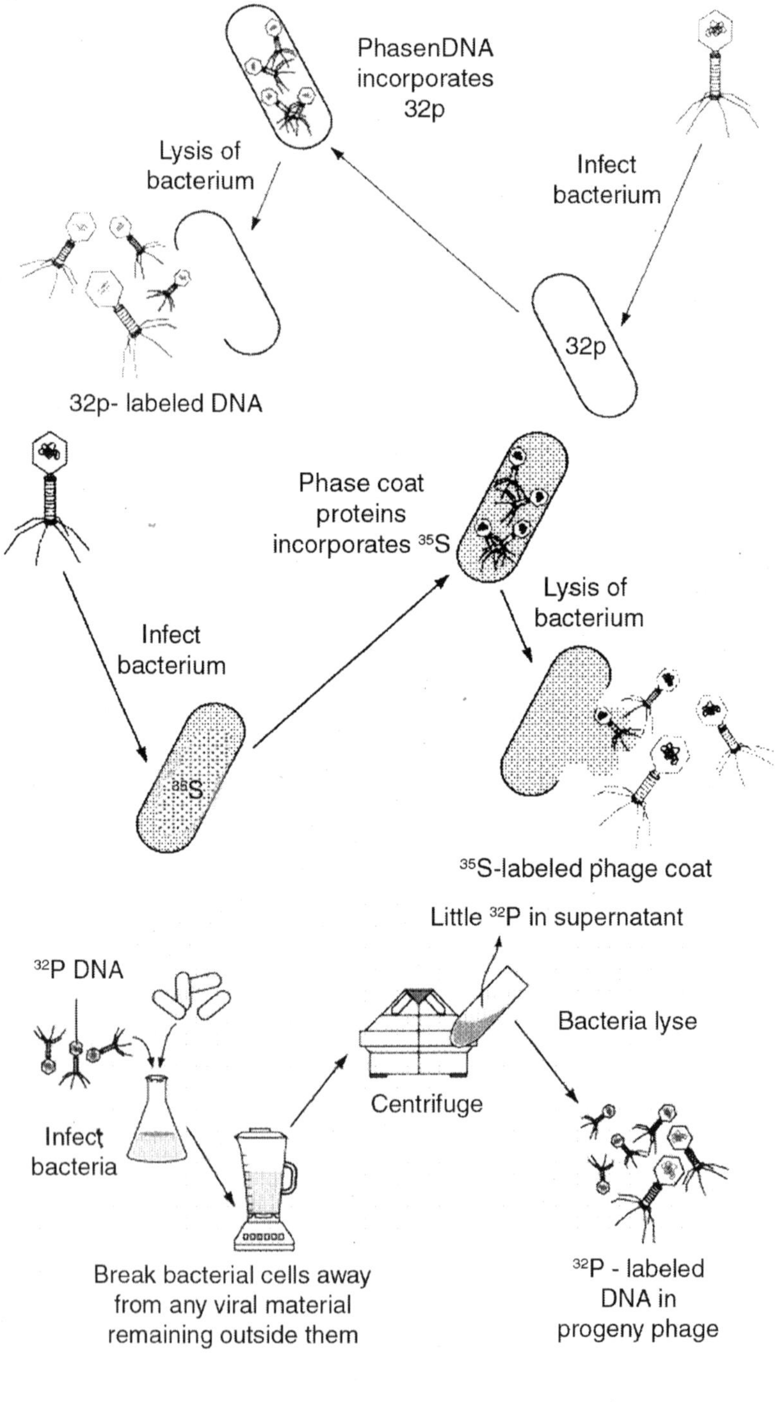

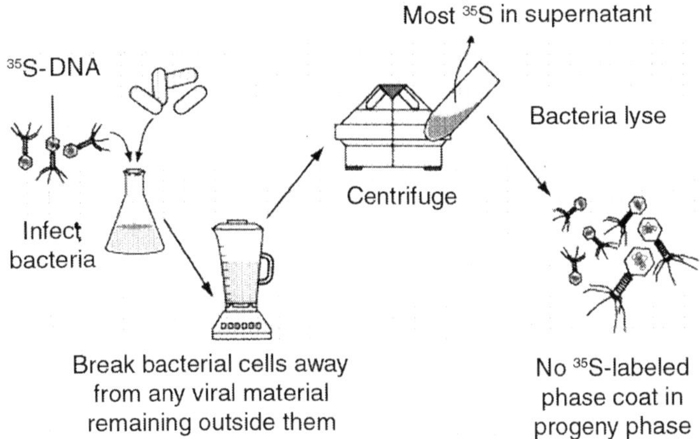

Fig. Diagrams Illlustrating the Hershey and Chase Experiment that Supported DNA as the Hereditary Material while it also showed Protein was NOT the Hereditary Material

THE STRUCTURE OF DNA

Erwin Chargaff analyzed the nitrogenous bases in many different forms of life, concluding that the amount of purines does not always equal the amount of pyrimidines (as proposed by Levene). DNA had been proven as the genetic material by the Hershey-Chase experiments, but how DNA served as genes was not yet certain.

DNA must carry information from parent cell to daughter cell. It must contain information for replicating itself. It must be chemically stable, relatively unchanging. However, it must be capable of mutational change. Without mutations there would be no process of evolution.

Many scientists were interested in deciphering the structure of DNA, among them were Francis Crick, James Watson, Rosalind Franklin, and Maurice Wilkens. Watson and Crick gathered all available data in an attempt to develop a model of DNA structure.

Franklin took X-ray diffraction photomicrographs of crystalline DNA extract, the key to the puzzle. The data known at the time was that DNA was a long molecule, proteins were helically coiled (as determined by the work of Linus Pauling), Chargaff's base data, and the x-ray diffraction data of Franklin and Wilkens.

DNA is a double helix, with bases to the centre (like rungs on a ladder) and sugar-phosphate units along the sides of the helix (like the sides of a twisted ladder). The strands are complementary (deduced by Watson and Crick from Chargaff's data, A pairs with T and C pairs with G, the pairs held together by hydrogen bonds).

Notice that a double-ringed purine is always bonded to a single ring pyrimidine. Purines are Adenine (A) and Guanine (G). We have encountered Adenosine triphosphate (ATP) before, although in that case the sugar was ribose, whereas in DNA it is deoxyribose. Pyrimidines are Cytosine (C) and Thymine (T). The bases are complementary, with A on one side of the molecule you only get T on the other side, similarly with G and C. If we know the base sequence of one strand we know its complement.

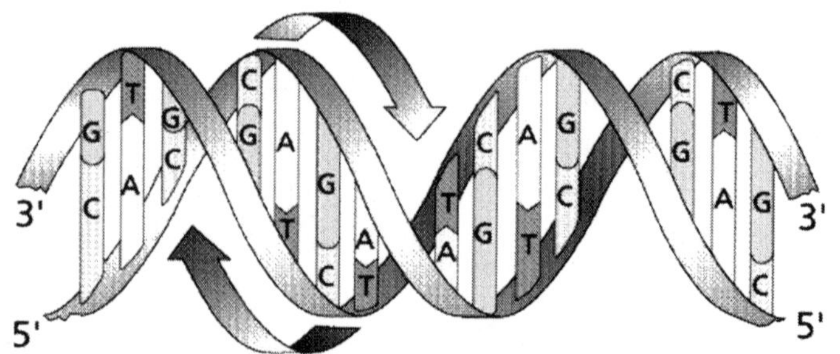

Fig. The Ribbon Model of DNA

DNA REPLICATION

DNA was proven as the hereditary material and Watson et al. had deciphered its structure. What remained was to determine how DNA copied its information and how that was expressed in the phenotype. Matthew Meselson and Franklin W. Stahl designed an experiment to determine the method of DNA replication. Three models of replication were considered likely.

Conservative replication

Would somehow produce an entirely new DNA strand during replication.

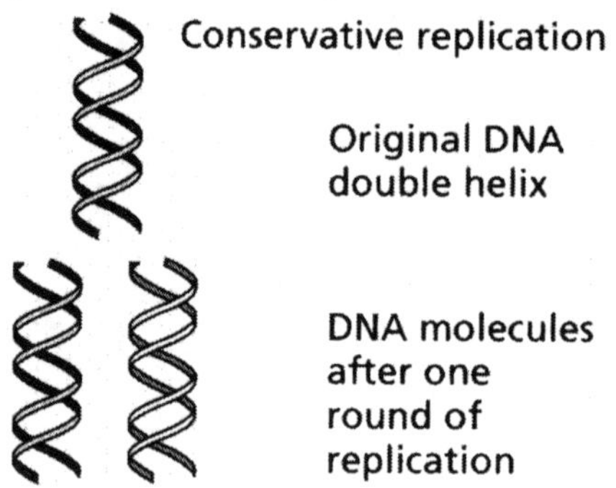

Fig. Conservative Model of DNA Replication

Semiconservative Replication

Would produce two DNA molecules, each of which was composed of one-half of the parental DNA along with an entirely new complementary strand. In other words the new DNA would consist of one new and one old strand of DNA. The existing strands would serve as complementary templates for the new strand.

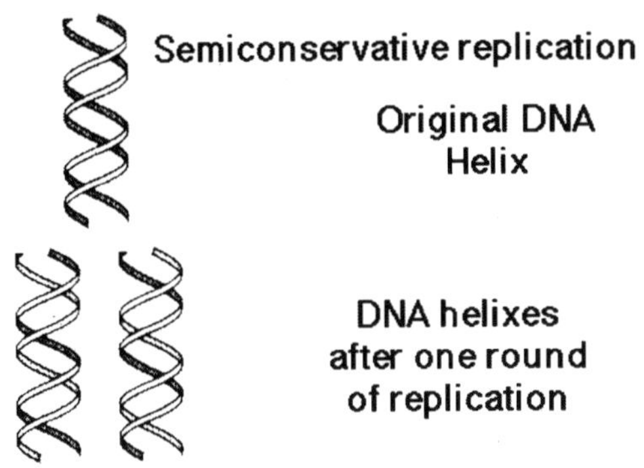

Fig. The Semiconservative Model of DNA Structure

Dispersive replication

Involved the breaking of the parental strands during replication, and somehow, a reassembly of molecules that were a mix of old and new fragments on each strand of DNA.

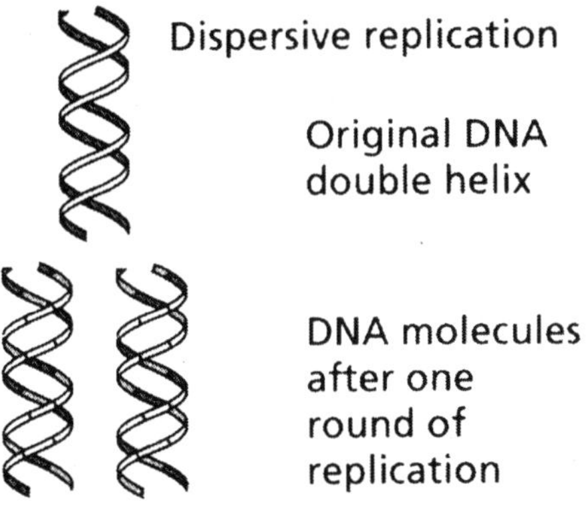

Fig. The Dispersive Replication Model of DNA Replication

The Meselson-Stahl experiment involved the growth of *E. coli* bacteria on a growth medium containing heavy nitrogen (Nitrogen-15 as opposed to the more common, but lighter molecular weight isotope, Nitrogen-14). The first generation of bacteria was grown on a medium where the sole source of N was Nitrogen-15. The bacteria were then transferred to a medium with light (Nitrogen-14) medium. Watson and Crick had predicted that DNA replication was semi-conservative. If it was, then the DNA produced by bacteria grown on light medium would be intermediate between heavy and light.

It was. DNA replication involves a great many building blocks, enzymes and a great deal of ATP energy (remember that after the S phase of the cell cycle cells have a G phase to regenerate energy for cell division). Only occurring in a cell once per (cell) generation, DNA replication in humans occurs at a rate of 50 nucleotides per second, 500/second in prokaryotes. Nucleotides have to be assembled and available in the nucleus, along with energy to make bonds between nucleotides.

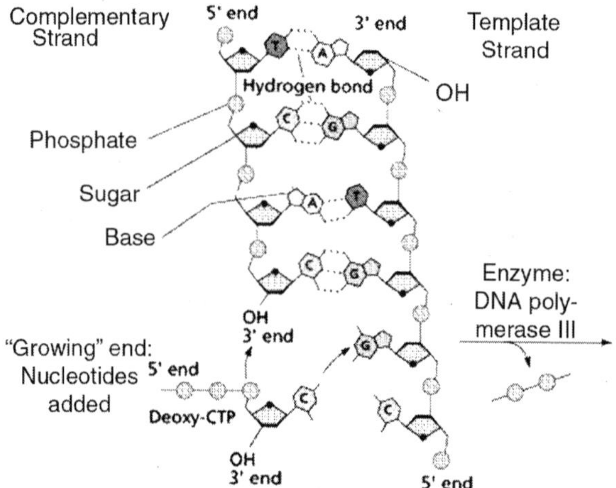

DNA polymerases unzip the helix by breaking the H-bonds between bases. Once the polymerases have opened the molecule, an area known as the replication bubble forms (always initiated at a certain set of nucleotides, the origin of replication). New nucleotides are placed in the fork and link to the corresponding parental nucleotide already there (A with T, C with G). Prokaryotes open a single replication bubble, while eukaryotes have multiple bubbles. The entire length of the DNA molecule is replicated as the bubbles meet.

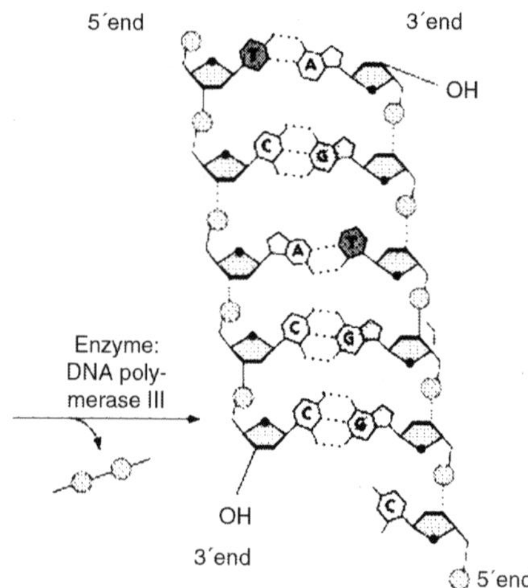

Fig. The Roles of DNA Polymerases in

Since the DNA strands are antiparallel, and replication proceeds in thje 5' to 3' direction on EACH strand, one strand will form a continuous copy, while the other will form a series of short Okazaki fragments.

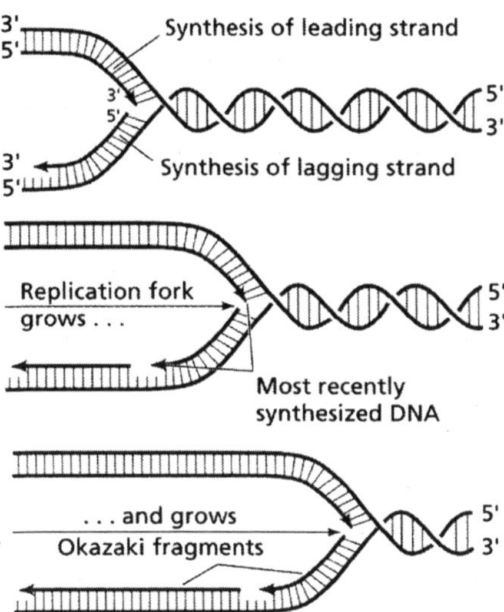

ENZYME STRUCTURE PROVIDES CLUES TO DNA

Before a cell can begin to divide or differentiate, the genetic information within the cell's DNA must be copied, or "transcribed," onto complementary strands of RNA. RNA polymerase II (pol II) is an enzyme that, by itself, can unwind the DNA double helix, synthesize RNA, and proofread the result. When combined with other molecules that regulate and control the transcription process, pol II is the key to successful interpretion of an organism's genetic code.

However, the size, complexity, scarcity, and fragility of pol II complexes have made analysis of these macromolecules by x-ray crystallography a formidable challenge. A team of structural biologists has met this challenge using data obtained from both the Stanford Synchrotron Radiation Laboratory and the Macromolecular Crystallography Facility at the ALS. The resultant high-resolution model of a 10-subunit pol II complex suggests roles for each of the subunits and will allow researchers to begin unraveling the intricacies of DNA transcription and its role in gene expression. In this work, the researchers studied the pol II enzyme from the yeast *Saccharomyces cerevisiae*, which is likely to be an excellent model for the human enzyme in light of its highly similar gene sequences.

It is also the best-characterized form of the pol II enzyme, having been the subject of many biochemical and low-resolution structural studies in the past. To obtain a high-resolution structure, the research team drew on its considerable expertise in the preparation of protein crystals: two-dimensional crystals of pol II (minus two small subunits found to impede crystal growth) were used as seeds for growing three-dimensional crystals. These crystals, when produced in an inert atmosphere to prevent oxidation, enabled the collection of data to 3.5-angstrom resolution. The addition of a final soaking procedure to produce uniform crystals, combined with high-brightness x-ray sources, resulted in a resolution of 3.0 angstroms.

The current results bring into focus the somewhat fuzzy features previously observed in or inferred from earlier experiments. More importantly, the structural details suggest possible explanations for some of the unusual characteristics of this enzyme, which include a high processivity (the ability to synthesize very long strands of RNA) and the tendency to work in periodic spurts separated by pauses.

While it is known that additional proteins (transcription factors) play a role in controlling the activity of pol II, scientists have yet to understand how such proteins interact with pol II binding sites to perform their various functions. The pol II model reported here establishes the positions of the various subunits and provides detailed information about the DNA/RNA binding domains.

The data reveal two main subunits (Rpb1 and Rpb2) separated by a deep cleft where DNA can enter the complex. At the end of the cleft is the active site, where the DNA can be unwound for a short distance (the "transcription bubble") and a DNA/RNA hybrid can be produced. Two prominent grooves lead away from the active site, either of which could accommodate the exiting RNA transcript. An opening below the active site may allow the entry of nucleotides (for manufacturing RNA) and transcription factors (for regulating the process).

The same opening may provide room for the leading end of the RNA strand during "backtracking" maneuvers, which are important for proofreading and for traversing obstacles such as DNA damage.

Other notable features that might help account for the great stability of this transcribing complex include a pair of "jaws" that appear to grip the DNA strands as they enter the complex and, closer to the active site, a clamp on the DNA that could possibly be locked in the closed position by the presence of RNA.

The high-resolution pol II structure reported here is a landmark achievement, pulling together threads from numerous diverse research efforts into a cohesive whole. Further study should yield many new insights into the detailed mechanisms of pol II and its transcription factors. Construction of an atomic model is already well underway.

UNRAVELING DNA

Encoded into the double-helical strands of DNA, the human genome is the complete set of instructions required to make a human being. While the mapping of the human genome (roughly three billion components) is certainly a Herculean accomplishment, it is only the first step toward realizing the full potential of genomic medicine in the diagnosis, monitoring, and treatment of disease. Beyond knowing what the genetic blueprint says, scientists must understand how that blueprint gets interpreted, or "expressed" as an individual with unique traits. The pol II enzyme is the catalyst for a major step in this process.

As a pol II molecule slides along a DNA molecule, it "unzips" the strands of the DNA double helix, synthesizes a complementary strand of RNA (which will carry the genetic information to where it is needed), and verifies that no mistakes have occurred.

This process is regulated by transcription factors—separate molecules that bind to pol II and determine which genes are expressed, at what stage of development, and in which tissue.

Done correctly, this process results in healthy cell growth and differentiation; otherwise, aberrations such as cancer can be the result. Thus, details of the structure of pol II, including information about its binding sites and how they interact with transcription factors, will provide valuable insight into the detailed mechanisms underlying the flow of genetic information from DNA to RNA to protein, which is necessary for life and health.

HYDROGEN BONDING IN DNA

Without hydrogen bonds there could be no life because they hold the double helix of DNA together, and this they do by charge attractions. A hydrogen bonded to oxygen, or nitrogen, becomes slightly positively charged which enables it to attract a centre of negative charge on another molecule, such as another oxygen or nitrogen atom. The hydrogen bond is then written, e.g., O-H...N, with the dotted line signifying the hydrogen bond. There are also O-H...O, N-H...O and N-H...N bonds, the last being among the weakest.

The secondary effects they have on structures, molecular vibrations, etc., can be used to infer hydrogen bonding, but there is no primary way of observing them because of their inherent weakness. NMR appears to be the least useful technique because neither of the common isotopes, oxygen-16 or nitrogen-14, has a magnetic nucleus. However, nitrogen-15 has a magnetic moment, and by replacing ^{14}N by N, Grzesiek has opened up a new area of investigating these enigmatic bonds.

Paper reports for the first time the direct observation by NMR of an N-H...N hydrogen bond between nucleic acids enriched with ^{15}N, by measuring the coupling of the nitrogen atoms. Grzesiek, working with Andrew Dingley of the Heinrich-Heine University in Düsseldorf, has been able to do this and show that the coupling is surprisingly large. Normally atomic nuclei only couple with each other if they are linked by normal chemical bonds, and in theory hydrogen bonds have neither the strength nor stability for this to occur.

The German researchers studied an ^{15}N enriched sample of the T1 domain of the potato spindle tuber viroid and were able to prove that N-H...N hydrogen bonding was present between the base pairs, uridine...adenosine and guanosine...cytidine, with couplings of approximately 7 Hz. How could they be certain that the signals they were observing are due to N-H...N hydrogen bonds? The answer was to use triple resonance techniques to examine base pairs that hydrogen bond only via O-H...N hydrogen bonds, and show that the signal they had previously observed was absent.

Grzesiek's second paper on hydrogen bonds, coauthored by Florence Cordier, Heinrich-Heine University, extends the work in an even more remarkable way by measuring the NMR coupling between nitrogens and carbons in the backbone hydrogen bonds of the human protein ubiquitin. The carbons are part of a carbonyl (C=O) group, so are one removed from the hydrogen bond, i.e., N-H...O=C. This time they used material enriched with ^{15}N and ^{13}C (normal ^{12}C has no nuclear magnet), and there too was the evidence for these hydrogen bonds, albeit with an interaction an order of magnitude weaker (at -0.25 to -0.9 Hz) than the N-H...N coupling. Nevertheless, the couplings correlate with the strength of the hydrogen bond, being stronger in the stronger bonds. These findings were confirmed by paper #9, which is from the researchers of Ad Bax's group based at NIH, Bethesda, Maryland.

Grzesiek's third paper, was done in conjunction with Dingley and researchers at UCLA. Together they studied not only the hydrogen bonding of Watson-Crick base pairs but also of Hoogsten base pairs within a DNA triplex consisting of one purine and two pyrimidine strands. Four different base pairs were identified, their various couplings distinguished–including those of the weaker interactions at the "frayed ends" of the DNA chains–and relationships with other hydrogen bonding parameters, such as bond length, were established. In addition they were able to show that density functional computer simulations by computer could reproduce these findings exactly.

DNA REPLICATION

The process of copying a double-stranded DNA molecule to form two double-stranded molecules DNA replication. The process of DNA replication is a fundamental process used by all living organisms as it is the basis for biological inheritance. As each DNA strand holds the same genetic information, both strands can serve as templates for the reproduction of the opposite strand. The template strand is preserved in its entirety and the new strand is assembled from nucleotides. This process is called "semiconservative replication".

The resulting double-stranded DNA molecules are identical; proofreading and error-checking mechanisms exist to ensure near perfect fidelity. In a cell, DNA replication must happen before cell division can occur. DNA synthesis begins at specific locations in the genome, called "origins", where the two strands of DNA are separated. RNA primers attach to single stranded DNA and the enzyme DNA polymerase extends the primers to form new strands of DNA, adding nucleotides matched to the template strand. The unwinding of

DNA and synthesis of new strands forms a replication fork. In addition to DNA polymerase, a number of other proteins are associated with the fork and assist in the initiation and continuation of DNA synthesis. DNA replication can also be performed artificially, using the same enzymes used within the cell. DNA polymerases and artificial DNA primers are used to initiate DNA synthesis at known sequences in a template molecule. The polymerase chain reaction (PCR), a common laboratory technique, employs artificial synthesis in a cyclic manner to rapidly and specifically amplify a target DNA fragment from a pool of DNA.

ORIGINS OF REPLICATION

The first step in DNA replication is the separation of the two DNA strands that make up the helix that is to be copied. DNA Helicase untwists the helix at locations called replication origins. The replication origin forms a Y shape, and is called a replication fork. The replication fork moves down the DNA strand, usually from an internal location to the strand's end. The result is that every replication fork has a twin replication fork, moving in the opposite direction from that same internal location to the strand's opposite end. Single-stranded binding proteins (SSB) work with helicase to keep the parental DNA helix unwound. It works by coating the unwound strands with rigid subunits of SSB that keep the strands from snapping back together in a helix. The SSB subunits coat the single-strands of DNA in a way as not to cover the bases, allowing the DNA to remain available for base-pairing with the newly synthesized daughter strands.

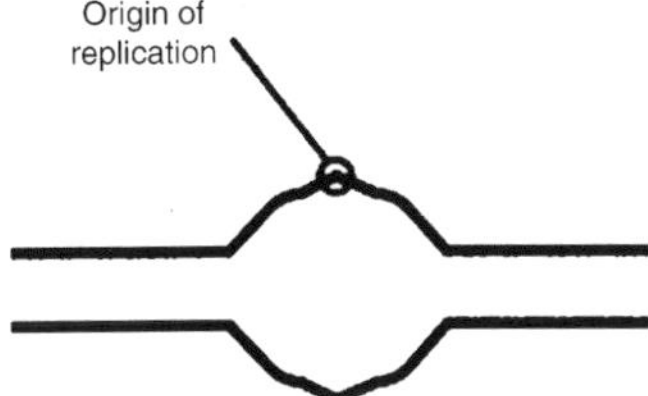

Fig. Replication Fork

As you can see in figure when the two parent strands of DNA are separated to begin replication, one strand is oriented in the 5' to 3' direction while the other strand is oriented in the 3' to 5' direction. DNA replication, however, is inflexible: the enzyme that carries out the replication, DNA polymerase, only functions in the 5' to 3' direction.

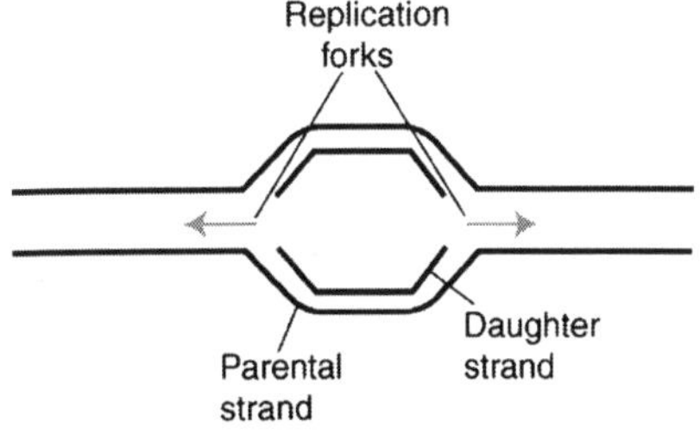

Fig. Replication Fork

This characteristic of DNA polymerase means that the daughter strands synthesize through different methods, one adding nucleotides one by one in the direction of the replication fork, the other able to add nucleotides only in chunks. The first strand, which replicates nucleotides one by one is called the leading strand; the other strand, which replicates in chunks, is called the lagging strand.

DNA REPLICATION IS SEMI-CONSERVATIVE

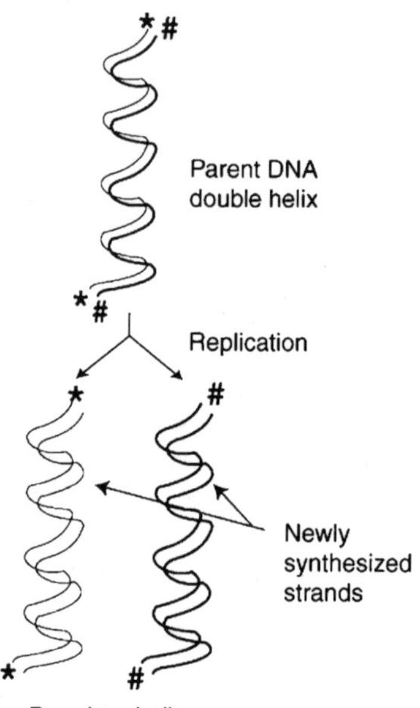

Fig. The Semi-Conservative Nature of DNA Replication

DNA replication of one helix of DNA results in two identical helices. If the original DNA helix is called the "parental" DNA, the two resulting helices can be called "daughter" helices. Each of these two daughter helices is a nearly exact copy of the parental helix (it is not 100% the same due to mutations). DNA creates "daughters" by using the parental strands of DNA as a template or guide.

Each newly synthesized strand of DNA (daughter strand) is made by the addition of a nucleotide that is complementary to the parent strand of DNA. In this way, DNA replication is semi-conservative, meaning that one parent strand is always passed on to the daughter helix of DNA.

THE LAGGING STRAND

Whereas the DNA polymerase on the leading strand can simply follow the replication fork, because DNA polymerase must move in the 5' to 3' direction, on the lagging strand the enzyme must move away from the fork. But if the enzyme moves away from the fork, and the fork is uncovering new DNA that needs to be replicated, then how can the lagging strand be replicated at all? The problem posed by this question is answered through an ingenious method.

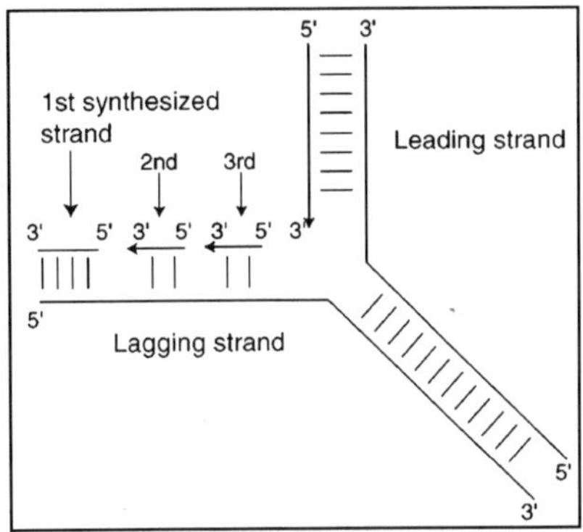

Fig. Leading and Lagging Strands

The lagging strand replicates in small segments, called Okazaki fragments. These fragments are stretches of 100 to 200 nucleotides in humans (1000 to 2000 in bacteria) that are synthesized in the 5' to 3' direction away from the replication fork. Yet while each individual segment is replicated away from the replication fork, each subsequent Okazaki fragment is replicated more closely to the receding replication fork than the fragment before. These fragments are then stitched together by DNA ligase, creating a continuous strand. This type of replication is called discontinuous

As you can see in the figure above, the first synthesized Okazaki fragment on the lagging strand is the furthest away from the replication fork, which is itself receding to the right. Each subsequent Okazaki fragment starts at the replication fork and continues until it meets the previous fragment. The two fragments are then stitched together by DNA ligase.

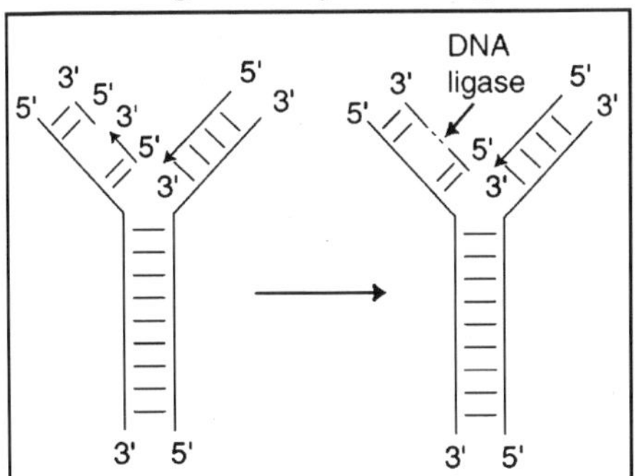

Fig. Patching Up Okazaki Fragments

In figure above, we can also see how replication on the lagging strand remains slightly behind that on the leading strand. Because synthesis on the lagging strand takes place in a "backstitching" mechanism, its replication is slightly delayed in relation to synthesis on the leading strand. The lagging strand must wait for a patch of the parent helix to open up a short distance in front of the newly synthesized strand before it can begin its synthesis back to the end of the daughter strand.

This "Lag" time does not occur in the leading strand because it synthesizes the new strand by following right behind as the helix unwinds at the replication fork. Another complication to replication on the lagging strand is the initiation of replication. Whereas the RNA primer on the leading strand only has to trigger the initiation of the strand once, on the lagging strand each individual Okazaki fragment must be triggered.

On the lagging strand, then, an enzyme called primase that moves with the replication fork synthesizes numerous RNA primers, each of which triggers the growth of an Okazaki fragment. The RNA primers are eventually removed leaving gaps that are filled by the replication machinery.

THE LEADING STRAND

Since DNA replication moves along the parent strand in the 5' to 3' direction, replication can occur very easily on the leading strand. As seen in figure, the nucleotides are added in the 5' to 3' direction. Triggered by RNA primase, which adds the first nucleotide to the nascent chain, the DNA polymerase simply sits near the replication fork, moving as the fork does, adding nucleotides one after the other, preserving the proper anti-parallel orientation. This sort of replication, since it involves one nucleotide being placed right after another in a series, is called continuous.

DNA STRUCTURE

DNA usually exists in a double-stranded structure, with both strands coiled together to form the characteristic double-helix. Each single strand of DNA is a chain of four types of nucleotide: adenine, cytosine, guanine, and thymine. A nucleotide consists of a phosphate and a deoxyribose sugar forming the backbone of the DNA double helix plus a base that points inwards. Nucleotides are matched between strands through hydrogen bonds to form base pairs.

Adenine pairs with thymine and cytosine pairs with guanine. The physical pairing of bases in DNA means that the information contained within each strand is redundant. The nucleotides on a single strand can be used to reconstruct nucleotides on a newly synthesized partner strand. DNA strands have a directionality, and the different ends of a single strand are called the "3' end" and the "5' end" (these refer to the carbon atom in ribose that the next phosphate in the chain attaches to).

In addition to being complementary, the two strands of DNA are antiparallel: they are orientated in opposite directions. This directionality has consequences in DNA synthesis, because DNA polymerase can only synthesize DNA in one direction by adding nucleotides to the 3' end of a DNA strand.

DNA POLYMERASE

DNA polymerases are a family of enzymes critical for all forms of DNA replication. A DNA polymerase synthesizes a new strand of DNA by extending the 3' end of an existing nucleotide chain, adding new nucleotides matched to the template strand one at a time. Some DNA polymerases may also have some proofreading ability, removing nucleotides from the end of a strand in order to remove any mismatched bases. DNA polymerases are generally extremely accurate, making less than one error for every million nucleotides added. The energy for the process of DNA polymerization comes from the two additional phosphates attached to each of the unincorporated nucleotides.

These free nucleotides, also known as nucleoside triphosphates, contain a total of three phosphates. When a nucleotide is being added to a growing DNA strand, two of the phosphates are removed and the energy produced is used to attach the remaining phosphate to the growing chain. The energetics of this process may also explain the directionality of synthesis - if DNA were synthesized in the 3' to 5' direction, the energy for the process would come from the 5' end of the growing strand rather than from free nucleotides.

During proofreading, if the 5' nucleotide needed to be removed this triphosphate end would be lost, losing the energy source required to add a new nucleotide to the end. DNA polymerase can only extend an existing DNA strand paired with a template strand, it cannot begin the synthesis of a new strand. To do this a short fragment of DNA or RNA, called a primer, must be created and paired with the template strand before DNA polymerase can synthesize new DNA.

DNA REPLICATION WITHIN THE CELL

Origins of Replication

For a cell to divide, it must first replicate its DNA. This process is initiated at particular points within the DNA, known as “origins”, which are targeted by proteins that separate the two strands and initiate DNA synthesis. Origins contain DNA sequences recognized by replication initiator proteins (eg. dnaA in *E coli'* and the Origin Recognition Complex in yeast).

These initiator proteins recruit other proteins to separate the two strands and initiate replication forks. Initiator proteins recruit other proteins to separate the DNA strands at the origin, forming a bubble. Origins tend to be “AT-rich” (rich in adenine and thymine bases) to assist this process because A-T base pairs have two hydrogen bonds (rather than the three formed in a C-G pair)—strands rich in these nucleotides are generally easier to separate. Once strands are separated, RNA primers are created on the template strands and DNA polymerase extends these to create newly synthesized DNA.

As DNA synthesis continues, the original DNA strands continue to unwind on each side of the bubble, forming replication forks. In bacteria, which have a single origin of replication on their circular chromosome, this process eventually creates a "theta structure". In contrast, eukaryotes have longer linear chromosomes and initiate replication at multiple origins within these.

THE REPLICATION FORK

The replication fork is a structure which forms when DNA is being replicated. It is created through the action of helicase, which breaks the hydrogen bonds holding the two DNA strands together. The resulting structure has two branching "prongs", each one made up of a single strand of DNA.

Leading Strand Synthesis

In DNA replication, the leading strand is defined as the new DNA strand at the replication fork that is synthesized in the 5"!3' direction in a continuous manner. When the enzyme helicase unwinds DNA, two single stranded regions of DNA (the "replication fork") form. On the leading strand DNA polymerase III is able to synthesize DNA using the free 3' OH group donated by a single RNA primer and continuous synthesis occurs in the direction in which the replication fork is moving.

Lagging Strand Synthesis

The lagging strand is the DNA strand at the opposite side of the replication fork from the leading strand, running in the 3' to 5' direction. Because DNA polymerase cannot synthesize in the 3"!5' direction, the lagging strand is synthesized in short segments known as Okazaki fragments. Along the lagging strand's template, primase builds RNA primers in short bursts.

DNA polymerases are then able to use the free 3' OH groups on the RNA primers to synthesize DNA in the 5"!3' direction. The RNA fragments are then removed (different mechanisms are used in eukaryotes and prokaryotes) and new deoxyribonucleotides are added to fill the gaps where the RNA was present. DNA ligase then joins the deoxyribonucleotides together, completing the synthesis of the lagging strand.

DYNAMICS AT THE REPLICATION FORK

As helicase unwinds DNA at the replication fork, the DNA ahead is forced to rotate. This process results in a build-up of twists in the DNA ahead. This build-up would form a resistance that would eventually halt the progress of the replication fork. DNA topoisomerases are enzymes that solve these physical problems in the coiling of DNA. Topoisomerase I cuts a single backbone on the DNA, enabling the strands to swivel around each other to remove the build-up of twists. Topoisomerase II cuts both backbones, enabling one double-stranded DNA to pass through another, thereby removing knots and entanglements that can form within and between DNA molecules.

Bare single-stranded DNA has a tendency to fold back upon itself and form secondary structures; these structures can interfere with the movement of DNA polymerase. To prevent this, single-strand binding proteins bind to the DNA until a second strand is synthesized, preventing secondary structure formation.

Clamp proteins form a sliding clamp around DNA, helping the DNA polymerase maintain contact with its template and thereby assisting with processivity. The inner face of the clamp enables DNA to be threaded through it. Once the polymerase reaches the end of the template or detects double stranded DNA, the sliding clamp undergoes a conformational change which releases the DNA polymerase. Clamp-loading proteins are used to initially load the clamp, recognizing the junction between template and RNA primers.

Eukaryotes

Within eukaryotes, DNA replication is controlled within the context of the cell cycle. As the cell grows and divides, it progresses through stages in the cell cycle; DNA replication occurs during the S phase (Synthesis phase). The progress of the eukaryotic cell through the cycle is controlled by cell cycle checkpoints. Progression through checkpoints is controlled through complex interactions between various proteins, including cyclins and cyclin-dependent kinases.

The G1/S checkpoint (or restriction checkpoint) regulates whether eukaryotic cells enter the process of DNA replication and subsequent division. Cells which do not proceed through this checkpoint are quiescent in the "G0" stage and do not replicate their DNA.

Replication of chloroplast and mitochondrial genomes occurs independent of the cell cycle, through the process of D-loop replication.

Bacteria

Most bacteria do not go through a well-defined cell cycle and instead continuously copy their DNA; during rapid growth this can result in multiple rounds of replication occurring concurrently. Within *E coli*, the most well-characterized bacteria, regulation of DNA replication can be achieved through several mechanisms, including: the hemimethylation and sequestering of the origin sequence, the ratio of ATP to ADP, and the levels of protein DnaA. These all control the process of initiator proteins binding to the origin sequences.

Because *E coli* methylates GATC DNA sequences, DNA synthesis results in hemimethylated sequences. This hemimethylated DNA is recognized by a protein (SeqA) which binds and sequesters the origin sequence; in addition, dnaA (required for initiation of replication) binds less well to hemimethylated DNA. As a result, newly replicated origins are prevented from immediately initiating another round of DNA replication. ATP builds up when the cell is in a rich medium, triggering DNA replication once the cell has reached a specific size. ATP competes with ADP to bind to DnaA, and the DNA-ATP complex is able to initiate replication. A certain number of DnaA proteins are also required for DNA replication — each time the origin is copied the number of binding sites for DnaA doubles, requiring the synthesis of more DnaA to enable another initiation of replication.

TERMINATION OF REPLICATION

Because bacteria have circular chromosomes, termination of replication occurs when the two replication forks meet each other on the opposite end of the parental chromosome. *E coli* regulate this process through the use of termination sequences which, when bound by the Tus protein, enable only one direction of replication fork to pass through. As a result, the replication forks are constrained to always meet within the termination region of the chromosome.

Eukaryotes initiate DNA replication at multiple points in the chromosome, so replication forks meet and terminate at many points in the chromosome; these are not known to be regulated in any particular manner. Because eukaryotes have linear chromosomes, DNA replication often fails to synthesize to the very end of the chromosomes (telomeres), resulting in telomere shortening.

This is a normal process in somatic cells — cells are only able to divide a certain number of times before the DNA loss prevents further division. (This is known as the Hayflick limit.) Within the germ cell line, which passes DNA to the next generation, the enzyme telomerase extends the repetitive sequences of the telomere region to prevent degradation. Telomerase can become mistakenly active in somatic cells, sometimes leading to cancer formation.

ROLLING CIRCLE REPLICATION

Another method of copying DNA, sometimes used *in vivo* by bacteria and viruses, is the process of rolling circle replication. In this form of replication, a single replication fork progresses around a circular molecule to form multiple linear copies of the DNA sequence. In cells, this process can be used to rapidly synthesize multiple copies of plasmids or viral genomes.

In the cell, rolling circle replication is initiated by an initiator protein encoded by the plasmid or virus DNA. This protein is able to nick one strand of the double-stranded, circular DNA molecule at a site called the double-strand origin (DSO) and remains bound to the 5' phosphate end of the nicked strand. The free 3' hydroxyl end is released and can serve as a primer for DNA synthesis. Using the unnicked strand as a template, replication proceeds around the circular DNA molecule, displacing the nicked strand as single-stranded DNA.

Continued DNA synthesis produces multiple single-stranded linear copies of the original DNA in a continuous head-to-tail series. *In vivo* these linear copies are subsequently converted to double-stranded circular molecules.

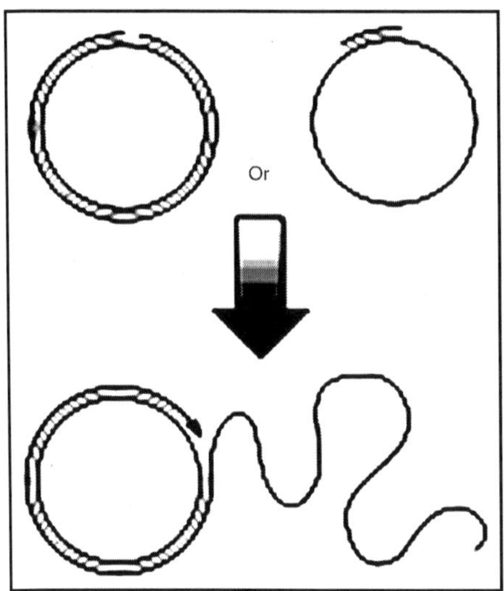

Fig. Rolling Circle Replication

Rolling circle replication can also be performed *in vitro* and has found wide uses in academic research and biotechnology, often used for amplification of DNA from very small amounts of starting material. Replication can be initiated by nicking a double-stranded circular DNA molecule or by hybridizing a primer to a single-stranded circle of DNA. The use of a reverse primer (or random primers) produces hyperbranched rolling circle amplification, resulting in exponential rather than linear growth of the DNA molecule.

POLYMERASE CHAIN REACTION

In vitro, researchers commonly replicate DNA using the polymerase chain reaction (PCR). PCR uses a pair of primers to span a target region in template DNA, polymerizing partner strands in each direction. This process can be repeated through multiple cycles through the use of a thermostable polymerase. At the start of each cycle, the mixture of template and primers is heated, separating the newly synthesized molecule and template. Then, as the mixture cools, both of these become templates for new primers to anneal to, and the polymerase extends from these. As a result the number of copies of the target region doubles each round, growing exponentially.

PROKARYOTIC DNA REPLICATION

Duplication of DNA is one of the fundamental properties of "molecule of life". Duplication or replication takes place once in a cell cycle. The duration and initiation point differs from one system to another. When conditions are favorable cell cytoplasmic mass increases and when cytoplasmic mass reaches a ratio to that of the cell size to 2L, bacterial cell division is triggered.

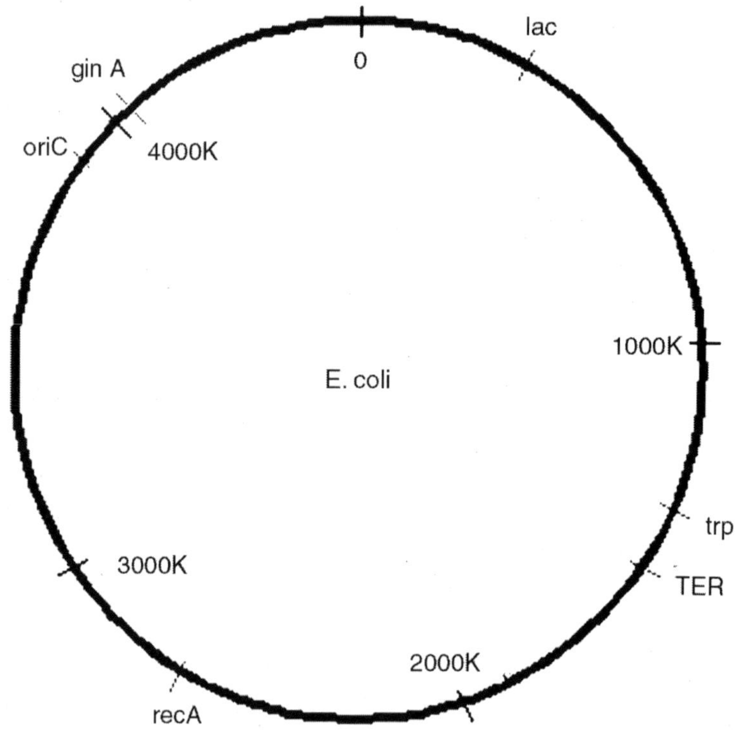

Fig. The Basic Circular Model of the Chromosomes with Map Positions such as ori C and TER Regions

To complete its cell cycle it requires hardly 30 -40 minutes. Several genes involved in different steps of cell division have been characterized. Par excellent system to understand molecular process of prokaryotic cell division is E.coli. The size of the E.coli chromosome is $4.6x10^6$ base pairs, but it is compacted by the binding of histone like proteins.

First the nucleoid i.e the chromosome frees from the membrane and unwinds. Such DNA initiates replication and completes in a matter of 15-20 minutes. Then the cytoplasm starts dividing almost in the centre of the cell by central septal ring formation. It is at this time the daughter DNA molecules are partitioned and segregated to the two compartments. The central periseptal ring that started 1-2 minutes after DNA replication now furrows inwards and divides the cytoplasm in the middle region. Then mid-wall splits in the middle and daughter cell are set free, hence the Schizomycetes to bacteria (E.coli)

Bacterial cell division requires the participation of the following genes and gene products. Eukaryotic cell division is more complex and highly regulated. The paradigm for it is yeast, for it is a unicellular system, grows fast and requires hardly 60 minutes for one cell cycle. One can obtain different kinds of mutation like conditional and null or auxotrophic kinds. In culture cells, the duration of one cell cycle is about 12 hrs. Onion root tip cells divide once in 12 hrs.

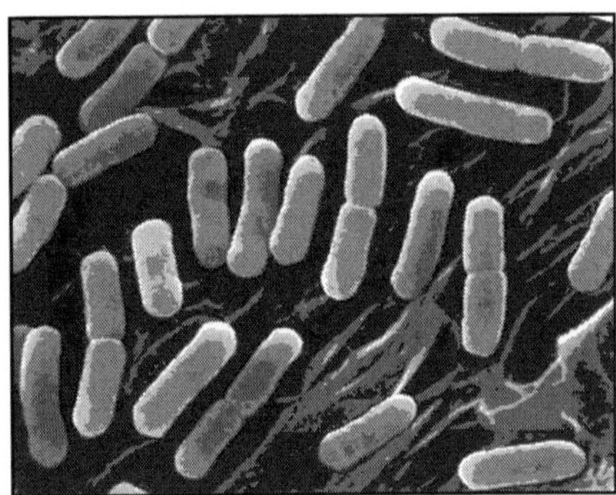

Fig. Bacterial Cells – some Divided and some Yet to be Divided

Whereas embryonic cells in its early stages require just 30 minutes for one cell cycle and they go through 6 to 8 such cell divisions very fast. Replication of DNA in cell division is an important phase. The replication is very accurate and it is nucleotide by nucleotide and there is no provision for making any mistakes, even if mistakes are made they are immediately corrected or repaired, otherwise consequences are serious and deleterious. In general DNA replication is precise, exact and regulated and involves initiation, elongation and termination steps, but mechanisms and rate of replication vary from one system to the other.

REPLICATION IS SEMI CONSERVATIVE

Meselson and Stahl showed that DNA replicates in semi conservative mode. There were some doubts about the mode whether it is conservative, dispersive or semi conservative. But Meselson and Stahl used nonradioactive heavy Nitrogen isotope (15N), as the source of Nitrogen. When cells were grown on such source, 15^N incorporated nitrogen bases get incorporated into DNA, thus it becomes heavier than the DNA that is grown in normal (14N) nitrogen media. When two such DNAs, one grown in the presence of 15N source and the other in 14N (normal), are extracted separately and stained with ethidium Bromide, then subjected to equilibrium density centrifugation at ultra speed of 42000rpm for about 24 –36 hrs, the DNA, separates as two distinct bands, the one that is heavier (15N) is the lower band than the normal (14N) upper DNA.

If cells grown in heavy isotopes are shifted to normal Nitrogen source and grown for one or two cell generation and if DNA is isolated from it and if one combine this with the first two DNA sources and subject them to equilibrium density ultra centrifugation (using cesium chloride or cesium sulphate), one can observe three bands, the top DNA band is from cells grown in 14N without isotope, the second band is the hybrid band with one strand 15^N and the other with normal 14N-Nitrogen and the third band is with heavy nitrogen where both the strands are labeled with 15N.

This experiment virtually and unambiguously proved that DNA replicates in semi conservative mode. Taylor using root tips, labeled with 32^P radioisotopes, of Vicia faba demonstrated semi conservative mode of replication at chromosomal level. Semi conservative means that the two daughter molecules produced at the end of replication, each of the molecules retains one parental strand and the other one is the newly made one. So it is semi conservative. The mode of replication is absolutely semi conservative irrespective the kind of DNA, whether it is ds DNA or ssDNA, whether it is linear or circular, and the basic mechanism is more or less similar; each of them uses their own specialized mechanisms.

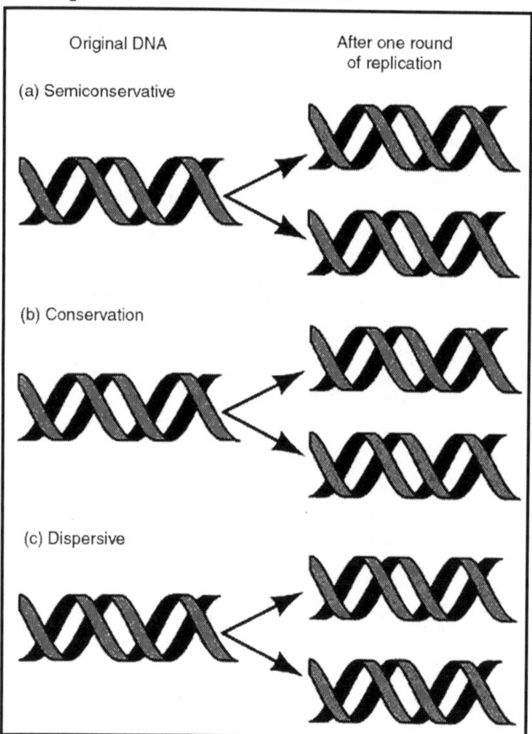

Experimentally semiconservative mode of replication has been shown by density gradient centrifugation of non-radioactive isotopes such as 15N and 14N sources.

DNA REPLICATION

DNA Replication (also known as DNA synthesis) is a process where the double stranded Deoxyribonucleic Acid (DNA) is copied. Replication is an important first step in cell division, as cells must duplicate their entire genetic constitution before they can divide into two daughter cells. DNA replication is also a critical requirement for DNA repair.

REQUIREMENTS OF DNA REPLICATION

Replication requires three important components in order for this process to work.

Template Strand

The DNA serves as a template to guide the incoming nucleotides. The template strand attaches to the RNA primer strand in order to replicate the strand of DNA. The template just guides the nucleotides to the place that they need to go in. This is another component that is needed in order for DNA replication to even take place. If the DNA only had the template strand and none of the rest of the things then the strand could not replicate. The template starnd is very thin. Through an enzyme called primase the DNA template strand is synthesized one nucleotide at a time.

DNA Polymerase

There are many DNA polymerases within a cell, but only one of which is used in the replication of DNA. In order for DNA to replicate, it must have a primer, which is a small strand of RNA. The DNA template strand synthesizes one nucleotide at a time by the use of primase. The adding of nucleotides to the 3' end, which is continued until that section of DNA is replicated. The other DNA polymerases that are within the cell are used in other ways besides replication. DNA polymerase III is one certain polymerase, which aids in replication of what is known as the "lagging strand". DNA ligase is the final enzyme, which combines the lagging strand

Free 3' Hydroxyl

The free 3' hydroxyl is the starter strand called a primer which is required in order for DNA to be replicated. The primer strand is not very big, it is shorter then a single strand of RNA. The primer is the complementary strand to the template strand of DNA. This is another one of the things that are needed in order to make the process of DNA replication even possible. The DNA polymerase is added to the 3' end of the primer, which keeps growing until the section is finally complete. After a while the RNA primer is degraded and disappears.

Types of Replication

During the discovery of DNA replication they also discovered that there are three modes of replication that a DNA strand can take. The three types of DNA replication are semiconservative, conservative, and dispersive. These three types of replication were tested on in order to see the possible patterns that would result in the complementary base pairing.

Semiconservative Replication

Through this type of replication the parent strand serves as a template for the new strand. The two offsprings would have one of the parent strand and one new strand.

Conservative Replication

Through this type of replication the double helix serves as a template. This although does not contribute to the new double helix.

Dispersive Replication

In this type of replication the fragments from the parent DNA molecule. The DNA serves as a template for the assembly of the new molecule. The new double helix contains old and new parts of the DNA strands.

DNA Repairing

After the DNA is replicated, there are three DNA repair mechanisms. They are to be used when they feel necessary. This will lessen the chance of human mutation. The rate of error in which the DNA polymerases makes a mistake is very high. This is why the DNA has mechanisms to cause fewer mutations. The three mechanisms are:

Proofreading

Every time a nucleotide is introduced to an already growing chain, the DNA polymerase checks the connection. If the pair is mismatched, it will be removed and redone. This lowers the overall rate of mutations.

Mismatch Pair

After the replication takes place, another set of proteins check to make sure that there are no more mismatched base pairs and also which strand is the wrong one.

Excision Pair

There can still be damage while the cell is living. Because of this, there are enzymes constantly checking the cell. If they detect a problem, the enzyme cuts the strand and also cuts away the opposite base. DNA polymerase and ligase then fix this base sequence.

LABORATORY TECHNIQUE

Scientists use two techniques in a laboratory that involve DNA replication. They use these techniques to observe genes and genomes. One of the techniques can make multiple copies of DNA from only a short piece. Another technique allows scientists to determine the base sequence of DNA. The two techniques are:

Polymerase Chain Reaction (PCR)

Through this process, a short strand of DNA is copied repetitively. There are three steps that are repeated in the process. The DNA strands are denatured or heated in order to separate the strand. Next, they add a primer, which was artificially made. They also add DNA polymerase and the four deoxyribon-ucleotide triphosphates. These three are needed in order to replicate DNA. The final step is then the DNA polymerase catalyzes the production of a complementary strand.

DNA Sequencing

This allows scientists to determines the base sequence of a DNA. The process is the DNA is first denatured and a single strand is placed in a test tube. They then add DNA polymerase, primer, the four dNTP's, and small amounts of ddNTP's. The DNA replicates within the tube and after the DNA fragments are denatured. The fragments then go through electrophoresis, which sorts the lengths of DNA fragments. The fragments then pass through a laser beam which excites the fluorescent tags. This information is fed into a computer at which tells the sequence.

Mendelson- Stahl Experiment

Through this experiment Meselson- Stahl convinced the scientists that the correct model for DNA replication was the semiconservative replication. They used density labeling in order to distinguish the old and new strands of DNA. In their experiment they used "heavy" isotopes of nitrogen. This nitrogen is a nonradioactive isotope that made the molecule more dense. The chemically identical molecules containing an isotope known as ^{14}N. they needed a way to distinguish the different DNA densities.

Because of this they needed to come up with something to measure the densities of solutions. Meselson, Stahl, and Jerome Vanguard came up with centrifuge, which is a procedure that involves the spinning of solutions at high speed, which causes the particles to separate and form gradient according to the different densities of the solution. The first part of their experiment they grew cultures of Escherichia coli. One of the cultures was grown in ^{15}N, which made the DNA "heavy".

The other culture was placed in a medium using ^{14}N rather than ^{15}N, which made all the DNA "light". In order to see if the use of centrifuge worked they combined the two cultures and when centrifuged it formed two DNA bands. This showed them the different densities with in the two cultures. The next part of their experiment they grew E. Coli in the ^{15}N medium and then transferred the bacteria from that medium to the ^{14}N. Through this they came up with the conclusion that through E. Coli DNA replicates every 20 minutes. They then took DNA samples from each generation through, which they found that the density gradient was different each generation.

Their observations could only be explained with the semiconservative model of DNA replication. Their results showed that when the DNNA first underwent replication, it was in 15N, which caused the DNA to be heavy. Because in the semiconservative model of DNA replication, the one strand acts as a template for a second strand which the DNA was in a ^{14}N DNA strand and a ^{15}N and they were of intermediate density.

ENZYMOLOGY OF DNA REPLICATION

One group of enzymes involved in DNA replication which have been studied in great detail are the DNA dependent DNA polymerases. At least three are known which catalyse the synthesis of a complementary structure using DNA as a template. The first of these. DNA polymerase I, was isolated and characterized by Arthur Kornberg for which he was awarded the Nobel Prize in 1959. Till the early sixties, it was believed that this was the only enzyme with the ability to catalyse DNA synthesis.

Later in the sixties, Roy Curtiss and others showed that bacterial mutants lacking this enzyme still had the ability to replicate DNA and since then, two other DNA polymerases (DNA polymerase II and DNA polymerase III) which can copy DNA have been recognized. In addition to these two enzymes, it is now known that a ligase (joining enzyme), and a primase (to prime the synthesis), in addition to a topoisomerase and other replication proteins are also involved.

DNA polymerase I, was thought to be the major enzyme that joins together de-oxynucleotides in vivo but now it is realized that DNA polymerase III is the main polymerizing enzyme while DNA polymerase I is a repair enzyme and fills In the gaps between the small fragments. The role of DNA polymerase II is not yet clear. All these three known polymerases extend the DNA molecules from the 5 end to the 3 end only. This led to the idea that the replication of double stranded DNA occurs in short pieces discontinuously and these short segments are later joined with the aid of a ligase to form a continuous strand.

Also, the DNA polymerases can add nucleotides only to a perfectly base paired nucleotide sequence and they cannot initiate new DNA synthesis. It now appears that the primary reaction is carried out by a DNA dependant RNA polymerase, which synthesizes short degradable RNA primers. It is also believed that in addition to the RNA polymerases, specific "primases", which differ from the RNA polymerases, synthesize the primary RNA which is then elongated by the DNA polymerase such a primase has been identified in the bacterium E. coli and also in the bacteriophages T4 and T7 Because of the difficulties in understanding the complex process of replication of bacterial double stranded DNA, investigators have used the single stranded DNA bacteriophage Øx174 as a test system to understand the exact mechanism of DNA replication.

In this virus, a complementary strand (-) is first synthesised on the parental single strand (+ strand) to yield a double helical structure. The new strand (-strand) then serves as the template for the synthesis of more of + strands which are then incorporated into the viral particles. The process appears to be straight forward but is not as simple as stated. Discontinuous synthesis appears to occur even during this type of replication.

Evidence for discontinuous replication of DNA in bacteria first came from Okazaki in 1967, who was able to isolate short pieces of' DNA during replication. It now appears that not all replicating DNA molecules rely on RNA priming and Okazaki fragments to solve the problem posed by the properties of DNA polymerase. The involvement of t-RNA primers in the synthesis of DNA copies of tumor viruses has also been reported. Recently an explanation to describe the mechanism by which single stranded DNA in certain animal viruses replicates has been given.

It is found that in single stranded DNA, "inverted repeats" (hair pin like structures) at each end of the molecule act as primers for initiation of replication. The secondary structure of the DNA is a double stranded helix. For replication to occur semi conservatively, the structure must undergo unwinding. Over the years, a number of DNA binding proteins, (DNA destabilizing proteins and untwisting enzymes) have been found both in procaryotic and eucaryotic cells and are in-volved in opening up of the double stranded structure to allow replication.

The latest addition to the list are the topoisomerases including the gyrase first isolated from E. coli. This new enzyme catalyses the introduction of negative super coils into double helical DNA in an ATP dependant reaction and appears to be essential for in vivo replication of DNA both in bacteria and bacteriophages. Antibiotics such as nalidixic acid and novobiocin inhibit this enzyme and there by prevent DNA replication at or beyond the replication point.

It is believed to relieve the positive supercoiling strains which builds up during replication and aid in unwinding of the double helix. One other point of interest yet to be solved with regard to DNA replication is the origin of replication and the number of replicating points per DNA molecule. Although a number of origins of replica-tion have been sequenced the mechanism of control is not yet clear. It is however, generally agreed that a DNA molecule may have more than one replicating points from where replication can be initiated.

The base sequence at the origin is not identical but has similarities. A variety of models have been described in literature to explain the mechanism of vivo DNA replication. Unfortunately, none of these provide a complete answer to the problem. The one that has received much attention an d is close to acceptance is the rolling circle model. According to this model, replication starts with a specific cut in one strand of the parental duplex molecule.

This generates a terminal nucleotide with a free 3' OH group while the other end has a phosphate group at the 5' end. As replication proceeds, the 5' end of the open strand is rolled out as a free tail of increasing length. The replicating structure is called a rolling circle since the unraveling of the free single strand is accompanied by a rotation of the double helical template about its axis. The 5' end tail serves as a template for the synthesis of small DNA fragments which are eventually joined together by the DNA ligase. Such growing tails have a double stranded character soon after their formation. Elongation of such tails sometimes goes on to produce tails many times the length of the original circle. It is believed that the tails are cut by specific endonucleases and this is followed by circularization by pairing between the sticky ends.

One of the prevailing fashions in bioscience these days is the application of genomics to eukaryotic gene expression. Largely eclipsed are the approaches of a few decades ago in which enzymes derived from microorganisms blazed the trail to much of our current understanding of macromolecular biosynthesis and gene regulation. In this Commentary I will give an anecdotal account of the lessons I learned from my attempts to resolve and reconstitute biological events in DNA replication and reflect on how these lessons may still apply to solving the current problems of growth and development and the aberrations of disease.

The beginning of the 20th century saw the birth of modern biochemistry with the demonstration that alcoholic fermentation could be observed in the juice of yeast cells. This led to the discovery of the dozen enzymes that convert sucrose to alcohol and ultimately to the reconstitution of alcoholic fermentation at a refined molecular level. Along with these early biochemical studies with yeast came the discovery that virtually the same enzymes and pathway were responsible in mammalian cells for the conversion of glycogen to lactic acid, which provides ATP energy for muscle contraction.

This astonishing fact, along with many other such examples in metabolic and biosynthetic pathways, made it clear that mechanisms and molecules have been preserved in bacteria, fungi, plants, and animals, essentially intact through billions of years of Darwinian evolution. I regard this insight as one of the great revelations of the 20th century.

RELY ON ENZYMOLOGY TO CLARIFY BIOLOGIC QUESTIONS

Based on the conviction that all reactions in the cell are catalyzed and directed by enzymes, the first commandment commands that enzymology can be relied on to clarify a biologic question. Chemists once bridled at this. But time and again, spontaneous reactions, such as the melting of DNA and the folding of proteins, are found to be driven and directed by enzymes; in the case of DNA, its melting in a cell is catalyzed by several different helicases. The first and crucial step is to find a way to observe the phenomenon of interest in a cell-free system.

Should that succeed, then one should be able to reduce the event to its molecular components by enzyme fractionation. This confidence is derived from the fact that, as mentioned, alcoholic fermentation, which had eluded understanding for centuries, was clarified by fractionation of a cell-free yeast extract, as was glycolysis by fractionation of muscle extracts and in the same vein luminescence in the extracts of a firefly and replication of DNA in microbial cell lysates.

Fractionation procedures for these extracts revealed the molecular mechanisms and machines for the catalysis and regulation of many complex reactions and pathways, as recounted here for DNA replication.

With a cell-free system in hand that recreates a biologic event, the biochemist should be able to perform the process as well as the cell does it. Even better! After all, the cell is under great constraints to provide a consensus medium that supports thousands of diverse reactions, only some of which operate under optimal conditions. By contrast, the biochemist enjoys the freedom to saturate each enzyme with its substrate, trap the products, and provide the optimal pH and salt and metal ion concentrations.

The biochemist can thus be creative and effective in analyzing the molecular basis of a reaction or pathway. The refinement of methods to purify proteins by chromatography and to establish their homogeneity by gel electrophoresis, when combined with the power of reverse genetics and genomics, has made the isolation and large-scale preparation of enzymes relatively easy compared to what it was years ago.

Despite this, discrete events in the proliferation, differentiation, and adaptations of cells and organisms are almost always analyzed by genetic means.

Striking phenotypes are produced by mutations and transfections, but the alterations in enzymes and pathways are generally only inferred. Rarely are they verified by the isolation of proteins with demonstrable functions. To many cell biologists and developmental biologists, the need to examine an event in a cell-free system does not come up on their radar screen.

TRUST THE UNIVERSALITY OF BIOCHEMISTRY AND THE POWER OF MICROBIOLOGY

The universality of biochemistry from microbes to humans in basic metabolic and biosynthetic pathways has led to the silly quip: "What's true for *E. coli* is true for elephants, and what's not true for *E. coli* is not true." My faith in this universality encouraged me to focus on how prokaryotes, particularly

Escherichia coli, replicate their own genomes and those of their phages and plasmids. Microbial generation times, unlike those of eukaryotes, are measured in minutes rather than hours and days. This is where the light on replication shines brightest. I made the choice to work with prokaryotes with the confidence that these systems would be reliable prototypes for how the so-called higher organisms replicate their DNA.

In recent years, exciting advances have been made in discovering and characterizing the eukaryotic replication enzymes: the helicases, topoisomerases, polymerases, primases, ligases, and other components of the chromosomal replicases.

The variations from prokaryotic enzymes are fascinating. Yet virtually all these enzymes and mechanisms were already familiar and adhere to the basic themes discovered earlier in the prokaryotic systems.

DO NOT BELIEVE SOMETHING BECAUSE YOU CAN EXPLAIN IT

In 1950, having found the enzymes that incorporate nucleotides into coenzymes and curious about how they might become part of nucleic acids, I needed first to determine how the purine and pyrimidine bases became substrates for assembly into these polymers. In the course of exploring the biosynthesis of nucleotides, I learned how to use labeled bases and how to tag each of the phosphates of the nucleoside diphosphates (NDPs) and triphosphates (NTPs). In 1954, we observed an activity in an extract of *E. coli* that incorporated the label of [á-^{32}P]ATP into an acid-insoluble form that we presumed to be RNA.

While making progress in purifying this activity, we learned of a discovery in the laboratory of Severo Ochoa in New York. While observing an exchange of orthophosphate with ADP in extracts of *Azotobacter vinelandii*, they discovered an enzyme that converted the ADP and other NDPs into an RNA-like polymer. Acting on this information, we substituted ADP for ATP and found that our activity was far greater. Clearly ADP was the preferred substrate over ATP. The *E. coli* enzyme we then purified was the same polynucleotide phosphorylase that the Ochoa laboratory had first identified in *Azotobacter*. As was learned later, the role of the phosphorylase was to degrade RNA rather than effect its synthesis. Had we persisted with ATP as substrate, we would surely have found RNA polymerase, the true synthetic enzyme, a year earlier than we did DNA polymerase and several years before it was discovered in 1961 by the late Sam Weiss.

CLEAN THINKING ON DIRTY ENZYMES

The late Efraim Racker enunciated this commandment, and I have been one of its ardent disciples. A dramatic example is the discovery of DNA replication. I first observed DNA synthesis in an *E. coli* extract in 1955, when I found that 50 counts out of a million of thymidine were incorporated into an acid-insoluble form. Those few counts above background seemed real because they were susceptible to DNase.

Could we possibly figure out what was going on in so crude a system, let alone in an intact cell? We identified and purified the first DNA polymerase, but not before our fractionation procedures disclosed a variety of novel enzymes that acted initially on the DNA, the [^{14}C]thymidine, and the ATP in our incubation mixtures. First, the [^{14}C]thymidine we added had to be phosphorylated by ATP via a new enzyme, thymidine kinase, to become thymidylate.

The added calf thymus DNA proved to be a substrate for DNases that produced the four hitherto unknown deoxynucleoside 52 -monophosphates. These were then phosphorylated by four distinct nucleotide kinases to the corresponding diphosphates, which were in turn phosphorylated by nucleoside diphosphate kinase to the respective and previously unknown dNTPs.

The DNA we had added served three additional functions beyond being a source of the four building blocks. It was a template to direct the precise order of nucleotide assembly, a source of primer termini for chain elongation, and a pool to protect the tiny amount of synthesized DNA from degradation by the nucleases that are abundant in cell extracts. The *E. coli* extract was thus the source of seven new enzymes in addition to the enzyme we named DNA polymerase. With that, template and primer were introduced into the language of all polymerase actions. We learned all this from fractionating the *E. coli* extract into its multiple activities and finally putting the purified enzymes and their products back together. These studies taught us the basic features of how DNA polymerases act and, incidentally, that the strands of duplex DNA are oriented in opposite directions, not known at the time. Cell extracts are by their nature "dirty enzymes"; intact cells and organisms are "dirtier" still. F. G. Hopkins, a prescient pioneer in the biochemical basis of nutrition, said it best back in 1931: "(The biochemist's word) may not be the last in the description of life, but without his help the last word will never be said."

And so it has been with many cellular events, most recently the fully reconstituted transcription by the 48-subunit yeast RNA polymerase II initiation complex and the sorting of proteins to specific subcellular compartments by fractionated vesicles and enzymes. Purification of an enzyme to homogeneity now opens the door to reverse genetics, and the enzymes themselves still provide unique reagents, as commandments IX and X will describe. Sometimes, an apparently pure enzyme may be found on further purification to harbor a contaminant of great importance.

As one example, an extra step in the purification of DNA polymerase I rendered the enzyme inactive. The reason: the template-primer used in our routine assays was DNA activated by being nicked many times by a DNase. We were not aware at the time that these nicks had been enlarged upon by an exonuclease in our polymerase preparation to create a stretch of exposed template needed by the polymerase. The exonuclease activity we had fractionated away, which we then purified and named exonuclease III, proved to be a crucial reagent in the discovery of recombinant DNA.

In 1967, after 10 years of trying and failing to prove that our enzymatically synthesized DNA was biologically active, we finally succeeded. What made the difference was the use of a single-stranded, circular DNA of a bacterial virus as template and the discovery of DNA ligase that could circularize the linear product. With X174 DNA, we could make quantities of the circularly closed, infectious viral DNA. The sequence of 5,386 nucleotides was correct, and there was no need for novel nucleotides or other components, as had been conjectured.

We also pointed out that this in vitro system afforded the means to introduce novel nucleotides for site-directed mutagenesis. The appearance of our paper announcing the test tube synthesis of infectious DNA generated a huge crush of media attention, a congratulatory phone call from President Lyndon Johnson and headlines worldwide, all based on the belief that we had synthesized a big, hairy virus and "created life in the test tube." I had to explain to the assembled reporters that it was not I who assembled the long DNA chain of the virus but rather it was the awesome enzyme, DNA polymerase, that I had identified and isolated from *E. coli* cells.

And further, I had to make it clear that it was these bacteria in a culture flask that imbibed the viral DNA to make the infectious virus particles. As for "creation of life in the test tube," some might dispute that a virus is even a "living" creature.

DO NOT WASTE CLEAN ENZYMES ON DIRTY SUBSTRATES

Three years after the hoopla about the synthesis of a viral DNA by our DNA polymerase, serious questions remained. How is a DNA chain started? How is the accumulating genetic evidence for additional polymerases and other factors needed for replication explained? A cartoon at the time showed the apparatus at a replication fork discreetly obscured by a fig leaf, and there were polemical attacks in *Nature New Biology* that dismissed our DNA polymerase as merely a repair enzyme with little relevance to replication, in essence a "red herring." Over the years, we had tried a variety of DNA samples to demonstrate the start of a DNA chain.

The results were negative or equivocal. Then it dawned on me that we were violating the fifth commandment. We were using a pure DNA polymerase on a dirty DNA substrate: frayed, gapped, fragmented, denatured and heterogeneous. When we finally switched to the intact, single-stranded, circular

DNA of a small bacteriophage, we discovered how a DNA chain is started: priming with RNA. Single-stranded phages provided not only the DNA substrate with which we could discover the RNA priming of new chains, but also the enzyme systems responsible for the priming and subsequent replication. The filamentous phage M13 depends on the host RNA polymerase to make a short transcript of an origin region, whereas the icosahedral phage X174 appropriates a complex primosome, used by the host to prime the start of chains on the lagging strand at the replication fork.

Whereas the conversion of the single M13 viral strand to the duplex replicative form was readily resolved and reconstituted, the conversion of the X174 single-stranded circle was far more complex and required the discovery of 15 new proteins, which constitute the apparatus at the host chromosomal replicating fork. With these many proteins in hand we could attempt to discover how replication was initiated at the origin of the intact *E. coli* chromosome.

CORRECT FOR EXTRACT DILUTION WITH MOLECULAR CROWDING

Cells are gels. Half of the cell dry weight is made up of proteins packed in highly organized communities. That some of their functions, individually and collectively, can be observed despite great dilution (20-fold or more) is a fortunate break for biochemistry. But there is an absolute need in some cases to restore the crowded molecular state, as we learned from our attempts to observe initiation of replication at the origin of an intact chromosome. We were given a 5-kb plasmid containing the origin of the 4,000-kb *E. coli* chromosome that is replicated in the cell with the physiological and genetic features of the host chromosome, in effect a minichromosome.

When Seichi Yasuda came from Japan with this *oriC* plasmid, I thought we would soon resolve and reconstitute its replication much as we had done with phage X174. But it took 10 man-years of utter frustration before we finally succeeded in making a cell-free system work. Success in achieving *oriC* plasmid replication in a cell-free state depended on two strange maneuvers. One was to include a high concentration of polyethylene glycol (PEG) (10% [wt/vol]), 10,000 Da) in the incubation mixture. As is true of such hydrophilic polymers, the PEG gel occupies most of the aqueous volume and excludes a small volume into which large molecules are crowded.

This concentration is essential when several proteins are needed in the consecutive steps of a pathway. The other maneuver repeated an earlier experience in which we proceeded to fractionate an inactive lysate with ammonium sulfate. Progressive additions of the salt yielded precipitates, in one of which the active proteins were present and concentrated when dissolved in a small volume. Just as important, the supernatant fraction we discarded contained a potent inhibitor, a nuclease that relaxed the plasmid DNA from its essential supercoiled state. Along with a purified, origin-binding DnaA protein, we provided primosomal, replication, and ligase proteins with other factors to obtain rapid, origin-specific, extensive replication of the *oriC* minichromosome. With these many proteins added in sufficient amounts, PEG was no longer needed. The mechanisms we discovered in *oriC* replication were found to apply to the replication of many microbial plasmids and phages and to some eukaryotic viruses and episomes.

RESPECT THE PERSONALITY OF DNA

For years, DNA was regarded as a rigid rod devoid of personality and plasticity. Only upon heating did DNA change shape, melting into a random coil of its single strands. Then we came to realise that the shape of DNA is dynamic in ways essential for its multiple functions.

Chromosome organization, replication, transcription, recombination, and repair have revealed that DNA can bend, twist, and writhe, can be knotted, catenated, and supercoiled (positive and negative), can be in A, B, and Z helical forms, and can breathe. Especially noteworthy is breathing, the transient thermodynamic-driven opening (melting) of the duplex that facilitates the binding of specific proteins such as the helicase responsible for priming and the onset of replication.

Certain DNA sequences are also predisposed to a more extensive form of melting ("heavy breathing") that creates a relatively large opening for transcription. The resulting RNA-DNA duplex (R-loop) can activate an inert origin of replication by altering its structure, even hundreds of base pairs away, which facilitates its opening by origin-binding and replication proteins. Negative supercoiling supplies the energy for the breathing and other features that direct the shape and movements of DNA at the *oriC* origin of replication. These DNA responses have led to an appreciation of the role of transcriptional activation of replication origins near primers in large chromosomes, both prokaryotic and eukaryotic.

USE REVERSE GENETICS AND GENOMICS

Direct genetics, in which a randomly mutated gene can ultimately be linked to a deficiency in a single enzyme, was a landmark discovery in biologic science. This approach served well by providing *E. coli* mutants defective in replication, some in initiation of a chromosome (e.g., *dnaA*) and others in elongation (e.g., *dnaB*, *dnaC*, *dnaE*, and *dnaG*). But randomly generated mutants do not readily disclose the products of their genes nor their particular functions.

Nevertheless, these replication mutants were crucial in validating our assays because DNA synthesis was absent in the extracts of mutant cells and restored when extracts or purified fractions from wild-type cells were added. Reverse genetics and genomics have now made the enzymologic approach even more powerful. Unlike direct genetics, enzymology starts with a defined function, after which finding the responsible genes has become relatively easy.

With even a picomole of a purified enzyme or a band on a gel, a peptide sequence can be determined and the encoding gene identified, cloned, and overexpressed; genomics facilitates the process by providing the complete genome sequences of *E. coli*, yeast, and many other microbes. Profound insights into the physiologic role of an enzyme or pathway emerge from the behaviour of cells with a null, point, or truncated mutation of a gene or modulated levels of its overexpression. The ease with which large quantities of pure enzymes can be produced by overexpression of a cloned gene has made their use as reagents even more attractive (commandment X).

"DNA shuffling", a new technique, has made enzyme reagents compelling. By creating a very large number of random rearrangements of a gene or genome, a particular gene product can be selected for a desired property (e.g., heat resistance) with wide applications in industry and biomedical science.

ENZYMES AS UNIQUE REAGENTS

Biochemistry is replete with examples in which enzymes have been employed as analytic and preparative reagents. From basic research to industrial processes, proteases, amylases, phospholipases, kinases, and phosphatases, etc., have been crucial in operations that were beyond the capabilities of available chemical technology.

I will mention just a few examples of applications to DNA and its replication and one from my recent research on inorganic polyphosphate (poly P). The key discovery in 1944 that identified DNA as the genetic substance was based on the destruction by crystalline pancreatic DNase of the factor that transformed one strain of *Pneumococcus* sp. to another.

It was the action of this DNase again, as mentioned in commandment IV, which in 1955 made me believe that the few counts of [^{14}C]thymidine incorporated by an *E. coli* extract into an acid-insoluble form signaled the synthesis of DNA. Many more examples can be cited in which an enzyme reagent was decisive: the circularization of linear DNAs by ligases, the creation of "sticky" tails by specific exonucleases used to prepare the first recombinant DNAs, the innumerable uses of restriction nucleases, and on and on.

Enzyme reagents have been decisive in my approach to determine the functions of poly P, an inorganic polymer of hundreds of phosphate residues linked by "high-energy" anhydride bonds. Likely present on prebiotic earth, poly P is now found in every living cell, but for lack of any known functions, was earlier regarded as a "molecular fossil."

True to the first commandment, I have sought and isolated enzymes that make and act upon poly P. With these enzymes we developed assays that are definitive, facile, and sensitive in place of those that are ambiguous, laborious, and insensitive. Together with the use of reverse genetics, we have learned that many microbes need poly P to adapt to adverse conditions and to survive in the stationary phase.

The kinase that makes poly P from ATP is highly conserved in some of the major pathogenic bacteria, and mutants lacking the kinase are defective in motility and virulence. Thus, this enzyme, absent from eukaryotes, may prove to be an attractive target for antimicrobial drugs.

ENZYMOLOGY OF THE REPAIR OF FREE RADICALS-INDUCED DNA DAMAGE

Two main pathways appeared during evolution: the release of oxygen by photosynthesis and the aerobic respiration. These highly efficient metabolic systems produce useful, although dangerous molecules for the cells: free radicals (NO· ¼) and reactive oxygen species (ROS) such as $O_2{\cdot}^-$, H_2O_2 and OH·. These molecules are constantly formed in cells by the cellular metabolism and by spontaneous chemical degradation of some biomolecules.

Free radicals are implicated in a wide range of cellular processes but an excess of such molecules could have deleterious effects on living cells leading to cell injury or cell death. Oxidative damage of main cellular components such as DNA, lipids, carbohydrates and proteins has been implemented in the ageing phenomena, ischemia, cancer, autoimmune diseases and neural cell death.

Activation of oncogenes like c-*myc* and ras can also induce ROS by alteration of specific metabolic pathways. Organisms are equally exposed to exogenous factors such as ionising radiation and chemical cancerogens, which can also generate ROS. Evidence has accumulated that lack of protection against free radicals and lack of repair of oxidative damage in biological macromolecules have a significant role in mutagenesis and consequently on carcinogenesis.

Thus maintenance of the cellular equilibrium between prooxidant (ROS) and antioxidant species should be tightly regulated in cells. Specialised systems have evolved, involving cellular antioxidants and DNA repair, to protect cells against ROS-induced injury. Indeed, the defence system, such as DNA repair, is highly conserved from bacteria to human. It is generally assumed that most oxidative DNA damages - base damage, sugar damage and abasic sites - are dealt with by BER. A number of recent reviews report the main features of the BER pathway.

The goal of DNA glycosylases is to locate fast and efficiently the aberrant base amongst a huge excess of normal ones. Very little is known about how these proteins achieve this goal. The comparison of the crystal structures of a number of DNA glycosylases revealed structural homologies leading to the concept of a superfamily of BER DNA glycosylases, the helix-hairpin-helix (HhH) superfamily, having similar HhH fold and a Gly/Pro-rich stretch with nearby Asp (GPD) motifs, although very little sequence similarity. This HhH motif plays an important role in the flipping out of the modified base. The rate of repair measured for the excision of modified bases is not always optimal and should be improved by the identification and the use of accessory proteins. The recent identification of new DNA polymerases able to replicate efficiently and accurately miscoding and modified bases have to be taken into account in the understanding of BER.

FREE RADICAL SPECIES AND OXIDATIVE DAMAGE OF DNA

DNA has a limited chemical stability (intrinsic or induced by exogenous agents) and is one of the most biologically important targets of ROS. Maintenance of its integrity is a major goal for cells. About 100 different kinds of base and sugar damage have been identified. Free-radicals can damage nucleobases and sugar units in DNA either directly, or indirectly.

Hydroxyl radicals, which are the most active species, predominantly, react with the C_8 of purines forming 7,8-dihydro-8-oxo-2'-deoxyguanosine (8-oxoG) and imidazol ring-opened products such as 2,6-diamino-4-hydroxy-5-formamidopyrimidine (Fapy); with C_5-C_6 double bond of pyrimidines forming glycol and pyrimidines hydrates and with C8-C5' of purines forming 8,5'-cyclopurine deoxynucleosides. The abstraction of a hydrogen atom from deoxyribose at C1' and C4' generates DNA-strand breaks with 3'-phosphoglycolate ester and 3'-phosphate.

Indirectly, ROS can generate reactive aldehydes, as a product of membrane lipids peroxidation, which react with DNA bases forming the exocyclic adducts 1,N^6-ethenoadenine, 1,N^2-ethenoguanine, N^2,3-ethenoguanine, and 3,N^4-ethenocytosine and pyrimidopurinone such as M_1G. Damaged bases such as 8-oxoguanine; 5-hydroxy-2'-deoxycitidine, hypoxanthine, ethenoadducts and pyrimido-purinone have miscoding properties and, if not repaired, lead to mutation upon replication.

Others such as oxidised deoxyribose, formamidop-yrimidine, fragmented thymine and thymine glycol cause replication block and therefore are believed to have a strong cytotoxic effect. Ionising radiation induces mutation and chromosomal aberrations in cells, which are believed to lead to cancer and loss of neural function in humans. Humans are daily exposed to low doses of radiation during air travel or from radon in homes and areas of low-level contamination.

Energy from ionising radiation, such as X-rays and gamma rays, is transmitted in the water surrounding the DNA molecule in such a way that between 2 to 5 radical pairs are generated within a radius of 1 to 4 nm. Clustered multiple damaged sites induced by ionising radiation, have been observed, most of them being modified bases rather than DNA strand breaks. The complexity of radiation-induced clustered DNA-damage depends on the ionising density of LET (Linear Energy Transfer) radiation. Ionising radiation was hypothesised to produce clustered damage, and clusters spanning several kilobase pairs to a few base pairs were modeled. Irradiation of DNA in non-radioquenching solution with gamma rays was shown to induce 1 double strand break to ~0.5 Nth protein-recognised oxidised pyrimidine cluster: 1.5 Nfo protein-recognised abasic cluster: 2 Fpg protein-recognised oxidised purine clusters. Use of the polyamines putrescine to measure abasic clusters refractory to Nfo protein cleavage reveals twice as many abasic clusters in -irradiated DNA as does Nfo, suggesting that even more clusters are produced than these values indicate. The dose-response relations in a log-log plot are straight lines with slopes near 1, showing that clusters are formed by a single radiation hit.

All DNA lesions are not equally frequent components of clusters, with about 15% of oxidised pyrimidines, oxidised purines and abasic sites in clusters, but only about 8% of the strand breaks are in DSBs. Since double strand break yields are ~100 times higher in the absence of radical scavengers fewer clusters are expected in DNA irradiated in the presence of a radical scavenger such as Tris. Indeed the absolute levels of bistranded cluster levels are significantly reduced in radioscavenging solution, and, surprisingly, the ratios of specific cluster types also changed strikingly. Cluster levels in human cells exposed to 50 kVp X-rays shows that X-rays induce 1 DSB: 0.75 Nfo-abasic cluster: 1 Fpg-oxidized purine: 0.9 Nth-oxidized pyrimidine cluster. In cells, non-DSB clusters are at least ~70% of the complex damages. Clusters are postulated to be critical because they may be more difficult to repair than random lesions. In contrast to randomly located oxidised bases, clustered modified bases are within half a turn of the double helix, i.e. 5 nucleotides and some of them on the two strands, therefore the excision repair pathways would have difficulty in removing clustered lesions.

It is anticipated that the excision/incision of oxidatively damaged bases, including AP sites, on both strands will, if not tightly regulated, either inhibit certain steps of repair or produce double strand breaks and thus be lethal for the cells. Studies of purified enzymes acting on oligonucleotides with defined lesions at specific relative spacing on opposing strands indicate that clusters may comprise non-repairable, highly repair-resistant and pre-mutagenic damage, and that lesion spacing and polarity are important. However, the precise repair mechanisms for the clustered lesions are so far very poorly understood. *E. coli* generate high levels of DSBs during repair after irradiation, and the repair-generated DSBs likely result from abortive cluster repair. Although rodent cells exposed to high doses (500 Gy) also increase DSB levels, their origin and whether they are generated at low doses was not clear.

BASE EXCISION REPAIR

Oxygen radicals generate mostly non-bulky DNA lesions, most of them are substrates for BER enzymes. Only few oxidative DNA lesions such as cyclopurine and pyrimidopurinone are substrates for nucleotide excision repair (NER) enzymes. Key enzymes of the BER pathway are DNA-glycosylases. They remove damaged and mispaired bases from DNA by cleavage of the *N*-glycosylic bond between the abnormal base and deoxyribose, leaving an abasic site in DNA. Most DNA glycosylases are highly specific and can excise various types of modified bases. There are two types of DNA glycosylases: mono- and bifunctional.

The mono-functional ones cleave the *N*-glycosylic bond releasing the modified base and generating an AP site as a final product which in turn is recognised by an AP endonuclease. The bifunctional ones cleave the *N*-glycosylic bond, liberate the modified base and in a concerted manner cleave the phosphodiester bond 3' to the resulting AP site by a or - elimination mechanism (-lyase activity) generating a single-strand break with 3'-phosphate and 3'-phosphoglycolate (PGA) extremities respectively.

Then the DNA backbone next to the abasic site or 3'-phosphate/PGA terminus is cleaved by an AP-endonuclease allowing a DNA polymerase to fill the gap before DNA ligase reseals the DNA. Experiments with *E. coli* mutants deficient in BER pathway demonstrated the crucial role of DNA glycosylases and AP-endonucleases in protecting cells from mutagenic and cytotoxic effects of free radicals. In human cells BER proceeds via two alternative pathways, either 'short-patch' DNA polymerase -dependent pathway which involves the replacement of a single nucleotide or 'long patch' PCNA-dependent pathway which involves the replacement of up to six nucleotides.

The former pathway was reconstituted *in vitro* with the purified human proteins uracil-DNA glycosylase (UDG), AP endonuclease 1 (APE1), DNA polymerase (pol), the scaffold protein XRCC1 and DNA ligase (I or III). Long-patch pathway can be achieved with AP endonuclease, Pol and PCNA. The antibodies directed against PCNA totally suppress repair patches longer than one nucleotide. Reconstitution *in vitro* of long-patch repair showed that DNase IV/FEN 1 is required in addition to proteins implicated in short-patch pathway and that PCNA greatly stimulates the reaction. Either DNA polymerase/ or DNA polymerase have been shown to perform the synthesis step in this sub-pathway. Recently, it has been demonstrated that removal of 8-oxoG could be achieved when using only hOGG1, Ape1, pol and DNA ligase I.

BASE EXCISION REPAIR IN PROKARYOTES

Mechanistic studies of both classes of DNA glycosylases led to formulation of a model of unified catalytic mechanism. Monofunctional DNA glycosylases cleave the glycosidic bond by a hydrolytic mechanism, activating a water molecule to attack the C1' of the damaged base, resulting in AP site as a final product. In contrast, the DNA glycosylases/AP lyases employ a nucleophilic group in the enzyme to perform attack at C1'.

In all DNA glycosylases/AP lyases so far characterised, an amino group has been implicated as a nucleophile. As a result of nucleophilic attack by amino group, a covalent imino intermediate (Schiff base) is formed between the C1' of the lesion and the enzyme. By abstracting the C2'-*pro-S* proton, the enzyme initiates electron rearrangement that leads to the release of the 3'-phosphate. The Schiff base is then hydrolysed, releasing the enzyme and leaving a nick with an,-unsaturated aldehyde at the 3'-end and a phosphate at the 5'-end of the DNA. Upon completion of -elimination, some enzymes such as the Fpg protein abstract proton at the C4' and promote a -elimination.

The resulting structure is a single-base gap with phosphates at both 3'- and 5'-ends. Sodium borohydride and cyanoborohydride have been used to show that the reaction involves a Schiff base intermediate. When the reduction reaction occurs in the glycosylase-DNA complex, the enzyme becomes irreversibly crosslinked to DNA. *The Fpg protein (formamidopyrimidine-DNA glycosylase):* Among the DNA glycosylases, the *E. coli* Fpg protein (formamidopyrimidine-DNA glycosylase/MutM protein) has been one of the most extensively studied. The gene coding for the Fpg protein was cloned and the physical and enzymatic properties of the protein established. It is a globular monomer of 30.2 kDa of 269 amino acids. *In vitro*, the Fpg protein excises a broad spectrum of modified purines, particularly 2,6-diamino - 4 - hydroxy - 5N - methylformamidopyrimidine (Fapy) and 7,8-dihydro-8-oxoguanine (8-oxoG) residues.

Moreover, the Fpg protein is also able to excise various pyrimidine oxidation products such as 5-hydroxycytosine and 5-hydroxyuracil, the ring fragmentation product of thymine (RT), thymine glycol and 5,6-dihydrothymine. The Fpg protein is a bifunctional DNA glycosylase, endowed of an AP lyase activity that incises DNA at abasic sites by a --elimination mechanism and an activity excising 5'-terminal deoxyribose phosphate (dRPase). *In vivo*, the Fpg protein has an antimutator effect preventing G/CT/A spontaneous transversion. It acts in concert with the MutY protein. The *fpg mutY* double mutant CC104 has an extreme mutator phenotype that can be reversed by plasmids carrying the fpg gene.

Interestingly, expression of the bacterial fpg gene in mammalian cells reduces the mutagenicity of -rays but has no effect on survival. The Fpg protein in its COOH-terminus contains the zinc-finger motif (Cys-X_2-Cys-X_{16}-Cys-X_2-Cys-X_2-COOH) that is mandatory for Fpg binding to DNA and to its enzymatic activities.

The active site of Fpg protein is located within the first 73 amino acid residues of the amino terminus. The targeted mutagenesis of conserved amino acid residues was used to elucidate the mechanism of enzymatic catalysis. It was found that conserved residues lysine 57 (K57G), lysine 155, glutamates 2, 5, 131 and 173 and proline 2 (P2G) dramatically reduce the cleavage of oxidised bases such as 8-oxoG. Recently, major progress has been achieved in solving the three-dimensional structure of the Fpg protein from various origins.

Two Fpg proteins have been isolated from the highly radioresistant bacteria *Deinococcus radiodurans*. Both excise Fapy and 8-oxoG residues and present a -lyase activity. In addition one excises thymine glycols. The genes have been cloned and the detailed specificities established using pure proteins. *The Nth protein (Endonuclease III):* The Nth protein was initially identified as an endonuclease that specifically cleaves DNA damaged by X-rays, UV light and free radicals. It is a DNA glycosylase with a broad substrate specificity, excising ring-saturated, ring-opened, and ring-fragmented pyrimidines, such as thymine glycol, 5,6-dihydrothymine, 5-hydroxy-6-hydrothymine,5,6-dihydrouracil, alloxan, 5-hydroxy-6-hydrouracil, uracil glycol, 5-hydroxy-2'-deoxy-cytidine, 5-hydroxy-2'deoxyridine, -ureidoisobutiric acid and also -R-hydroxy--ureidoisobutiric acid, a fragmentation product of the 5R-thymidine C5-hydrate.

Unexpectedly, it has been demonstrated that Nth is implicated in the repair of 8-oxoG, since it removes 8-oxoG from 8-oxoG/G mispairs and the triple mutant *fpg nth nei* displays increased spontaneous G/CC/G transversions as compared to single and double mutants in these three genes. This leads to the proposal of a new model in which Nth and Nei repairs 8-oxoG in nascent and transcriptionally active DNA. The nicking activity at abasic sites of Nth is due to its AP lyase function. The phosphodiester bond cleavage occurs via -elimination (the protein does not cleave at reduced AP sites) generating 5' ends bearing 5'-phosphate and 3' ends with 2,3-unsaturated abasic residue 4-hydroxy-2-pentanal. This terminus requires further processing to be used for DNA repair synthesis. The gene encoding for the Nth protein, the *nth* gene, was cloned and protein purified to homogeneity. The three dimensional structure of Nth was also established. The Nth protein is a monomeric protein of 23.4 kDa (211 aa).

It is an elongated protein with a cleft separating two similar size domains: a continuous domain formed by six--helices and a second domain formed by three C-terminal -helices and the N-terminal helix. The C terminal loop contains an iron-sulphur centre (4Fe-4S) that is anchored to the protein by a Cys-X6-Cys-X2-Cys-X5-Cys sequence. This cluster has a structural function in positioning basic residues for DNA-binding and it is also present in the MutY protein. *E. coli nth* mutants deficient in the Nth protein does not show apparent phenotype. They are not sensitive to X-rays, H_2O_2 or other agents that produce ring saturation and fragmentation products of pyrimidines in DNA. However the double mutant *nth nei* exhibits a strong spontaneous mutator phenotype and it is hypersensitive to ionising radiation and H_2O_2.

The Nei protein

The capacity of *E. coli nth* mutants defective in most of the activity to excise thymine glycol, allowed the identification of a new activity excising this oxidised base, that was named endonuclease VIII or Nei protein. It has been purified from *E. coli* extract lacking the Nth protein. Similarly to the Nth protein, Nei removes the oxidised forms of thymine such as thymine glycol, dihydrothymine, -ureidoisobutyric acid, and urea residues when present in DNA and cleaves phosphodiester bond at AP sites.

The cloned *nei* gene encodes a 263 amino acid polypeptide with a striking similarity to the Fpg protein. The catalytic site's residues are highly conserved in these two proteins, however, their substrate specificities are different. Characterisation of apparently homogenous Nei protein has shown that it can equally remove the oxidised forms of cytosine such as 5-hydroxycytosine and 5-hydroxyuracil. More recently, genetic evidence for Nei implication in the prevention of spontaneous GT transversions was confirmed since Nei was found to cleave 8-oxoG efficiently when opposite to A and G.

Furthermore, extended the substrate specificity of the Nei protein and have demonstrated that Nei and Nth significantly differ from each other in terms of kinetic although they share common substrate specificity. The crystal structure at 1,25 Å of the Nei covalent intermediate complex with DNA largely confirms that its structure is similar to that of Fpg.

MONOFUNCTIONAL DNA GLYCOSYLASES

AlkA protein

The main substrates for the *E. coli* AlkA and Tag I proteins are alkylated bases. The Tag I protein is a major DNA glycosylase removing 3-methylpurines residues. However, the AlkA protein in contrast to Tag I is also involved in repair of oxidative DNA damage. Hypoxanthine (HX) is generated in DNA by spontaneous deamination of adenine and also by the free radical nitric oxide. HX residues in DNA are mutagenic since they can pair with cytosine, generating AT to GC transitions after DNA replication.

Hypoxantine-DNA glycosylase releases HX from DNA containing dIMP. Cloning the gene coding for the HX-DNA glycosylase in *E. coli*, it was shown that the 3-methyladenine-DNA glycosylase coded by the alkA gene carries the HX-DNA glycosylase activity. ANPG, APDG and MAG proteins, the human, rat, and yeast functional homologues respectively of AlkA protein excise HX residues when present in DNA, the mammalian enzymes being most efficient. The AlkA protein also catalyses the excision of ethenobases N^2,3-ethenoguanine and 1,N^6-ethenoadenine.

The 5-formyluracil (5-foU) residue, an oxidised thymine lesion is a major DNA damage induced by ionising radiation. It induces A/T to G/C transition and is repaired in *E. coli* by the AlkA protein. In addition, AlkA protein excise RT residues when present in DNA with an apparent K_m=170 nM. Thus the AlkA protein has very broad substrate specificity. The three dimensional structure of AlkA reveals a compact globular protein with a prominent hydrophobic cleft on its surface and three equal-sized domains.

Uracil DNA glycosylase

Base excision repair was first established for uracil. Four sub-families of uracil-DNA glycosylase (UDG) have been identified. The best studied family of UDGs is *E. coli* Ung protein. Ung is specific for uracil and it is present in a wide range of living organisms from bacteria, lower to higher eukaryotes, DNA viruses and human. So far UDG has not been implicated in the repair of oxidative DNA damage. The second family corresponds to the mismatch-specific uracil-DNA glycosylase (MUG) identified in some eukaryotes and in several prokaryotes. MUG excises thymine from G/T mismatches in DNA and it is also active on mispaired uracil in G/U pair. The crystal structure of Ung and Mug proteins show that they are structurally similar, despite low sequence homology.

Another sub-family has been characterised from thermophilic archea and several bacteria. These double strand UDGs (dsUDG/DUG) are able to remove uracil from double strand DNA either in U/A and U/G context. The fourth sub-family, initially identified in vertebrates is represented by the human single-strand mismatch-specific uracil-DNA glycosylase (hSMUG1). However hSMUG1 is more active on double-stranded than on single-stranded DNA, it prefers uracil opposite A and G.

The 5-hydroxymethyluracil residue is a product of oxidation and subsequent deamination of 5-methylcytosine. Teebor's group identified the mammalian 5-hydroxy-methyluracil DNA N-glycosylase activity as SMUG1. A fifth sub-family has been characterised recently and is represented by *pa*-UDGb isolated from *Pyrobaculum aerophilum*. *pa*-UDGb has broad substrate specificity. It can remove uracil and also hypoxanthine when present in DNA.

Among the members of the UDG family several enzymes are implicated in the repair of exocyclic DNA adducts which are highly mutagenic oxidative DNA lesions. The *E. coli* MUG protein is able to remove 3,N^4-ethenocytosine, 8-(hydroxymethyl)-3,N^4-ethenocytosine and 1,N^2-ethenoguanine when present in DNA.

MutY

The *E. coli mutY* gene encodes a DNA glycosylase of 39 kDa, which excise A opposite G and C. MutY also removes A opposite to 8-oxoG and 7,8-dihydro-8-oxoadenine (8-oxoA). The protein shows significant sequence homology to Nth and is also an iron-sulphur protein which contains the conserved (4Fe-4S) cluster/HhH domain. The protein was extensively characterised using various approaches including site-directed mutagenesis. However, there is still controversy regarding the AP lyase activity of the MutY protein. Proteolytic cleavage of MutY generates two fragments of 26 kDa and 13 kDa. The 26 kDa fragment corresponds to the catalytic core domain, it retains normal DNA binding and adenine-DNA glycosylase activity but lacks its activity against A/ 8-oxoG.

The specificity for 8-oxoG residues is due to the C-terminal domain of MutY. It is thought that MutY plays an important role in the maintenance of genome integrity following oxidative DNA damage. The presence of AP endonucleases (Xth and Nfo) greatly enhances the excision rate of A when opposite to G by MutY. These data suggest that the MutY-DNA complex interacts with Nfo and Xth proteins *in vivo*. The crystal structure of MutY catalytic core domain (cMutY) has been solved. cMutY is an all- protein that retains the canonical bilobal architecture of Endonuclease III.

The compression of intra-strand phosphate distance by HhH and pseudo HhH domains permit the DNA bending and the nucleotide flipping. The flipped nucleotide is recognised by specified residues of the active side pocket located between the two domains. Crystal structures and mutagenesis results define that MutY cleaves the N-glycosylic bond through a hydrolytic mechanism requiring Asp138, with uncoupled, inefficient AP lyase activity due to Lys142.

AP ENDONUCLEASES: XTH AND NFO

Abasic sites (AP sites) occur in DNA through spontaneous depurination/ depyrimidination, by the action of ROS and as secondary lesions after excision of modified bases by DNA glycosylases. AP sites have miscoding properties since the replication machinery incorporates preferentially adenine opposite to AP site. This observation was later named 'A rule'. The occurrence of an enzyme recognising AP sites in *E. coli* was first described in the seminal work of Verly.

Two types of enzymes recognise and excise DNA at AP sites: AP endonucleases such as Nfo and Xth proteins act on this lesion by a hydrolytic mechanism whereas enzymes carrying a -lyase activity such as Fpg, Nei and Nth proteins incise the AP site via a - and elimination mechanism respectively.

THE NFO PROTEIN (ENDONUCLEASE IV)

The Nfo protein is an EDTA-resistant AP endonuclease that represents only 5% of the total AP endonucleasse activity in *E. coli* wild type. It is inducible by oxidative stress and is under the control of the soxRS system. The gene coding in *E. coli* for Endonuclease IV, *nfo*, has been cloned, and the protein characterised. It is a monomer of 30 kDa having several activities: AP endonuclease, 3'-phosphatase and 3-phosphoglycoaldehyde diesterase activities.

It has been suggested that these 3' repair activities clean the ends, a prerequisite for DNA synthesis. *E. coli nfo* mutants deficient in the product of the *nfo* gene are extremely sensitive to the lethal effects of oxidative agents such as bleomycin and t-butyl-hydroperoxide. In addition, this mutation enhances the sensitivity of *xth* mutants to H_2O_2, and alkylating agents. The double mutants *xth nfo* are also sensitive to -radiation. We will discuss below the role of the Nfo protein and its yeast homologue Apn1 in the nucleotide incision repair pathway (NIR). The high-resolution structure of Nfo protein has been solved. It shows that Nfo is an protein arranged as a $_{88}$-barrel. This structure, well suited for the binding to large molecules such as DNA, contains three Zn^{2+} ions at the active site which are critical for the activity.

The triple Zn centre is ligated to protein by conserved residues that cluster at the centre of the crescent-shaped deep pocket. The Nfo protein detects AP site by insertion of its side chains into the DNA minor groove, it flips the target AP site and the opposite nucleotide out of the DNA base stack to produce a 90° bend in the DNA.

The Xth Protein

The Xth protein, coded for by the xth gene, originally identified as a 3'5' exonuclease active on double-stranded DNA and a 3'-phosphatase, is the major AP endonuclease of *E. coli*: over 80% of the total AP endonuclease activity in wild type. This protein is also active on single stranded DNA. Beside its exonuclease and 3'-phosphatase activities it has 3' repair diesterase and ribonuclease H activities. The 3' termini generated by Xth are normal nucleotides with 3' hydroxyl group that are effective primers for DNA polymerases.

The *xth* mutants are extremely sensitive to H_2O_2, and to oxidative DNA damage generated by near-UV light. The crystal structure of the Xth protein has been solved, and it has been assumed that the base opposite to the abasic site plays an important role for the recognition of the lesion.

Mechanism of action and three-dimensional structure of bifunctional DNA glycosylases: the Fpg case: It was hypothesised that DNA glycosylase/AP lyases use a mechanism involving the nucleophilic attack on the C1' prime of the modified deoxynucleoside targeted for excision, thus displacing the aberrant base and forming a transient intermediate (Schiff base) with the C1'-deoxyribose moiety. The amino acid sequence of the Fpg protein begins with a Met-1 which is processed, thus Pro-2 being the N-terminal.

The N-terminal proline (P2) residue of the Fpg protein, a highly conserved residue, was shown to be linked with DNA containing 8-oxoG residues, suggesting that proline is the nucleophile initiating the excision of 8-oxoG. Site-directed mutagenesis demonstrated the mandatory role of the N-terminal proline residue in the 8-oxoG-DNA glycosylase activity of the Fpg protein *in vitro* and *in vivo*, as well as in its AP lyase activity upon pre-formed AP sites but less of a role in Fapy-DNA glycosylase activity.

The role of the conserved lysine 57 and 155 (K57 and K155) residues upon the various catalytic activities of the Fpg protein was examined by targeted mutagenesis. The lysine 57glycine (FpgK57G) mutant protein had a dramatically reduced (55-fold) DNA glycosylase activity for the excision of 8-oxoG residues. The FpgK57G protein was poorly effective in the formation of Schiff base complex with 8-oxoG/C DNA. However, the mutant could partially restore the ability to prevent spontaneously induced G/CT/A transversions in *E. coli* BH990 (*fpg, mutY*) cells.

The DNA glycosylase activity of FpgK57G using FapyGua residues as the substrate was comparable to that of the wild type enzyme. These results suggested that K57 could participate either in a direct interaction with the C_8 oxygen of the 8-oxo purines or in the opening of the furanose ring, in the first step of the model proposed for the mechanism of action of the Nth protein.

Effect of mutations of conserved glutamic and aspartic acid residues to glutamines and asparagines, have been studied. While the Asp to Asn mutants had no effect on the incision activity on 8-oxoG DNA, several of the substitutions at glutamates reduced Fpg activity on the 8-oxoguanosine DNA, with the E3Q and E174Q mutants being essentially devoid of activity. Unexpectedly, the AP lyase activity of all of the glutamic acid mutants was slightly reduced as compared to the wild-type enzyme. Furthermore, it has been shown that lysine 57 but not proline 2 is crucial for catalysis of oxidatively damaged pyrimidines.

Sugahara and co-workers determined the crystal structure of Fpg homologous protein from an extreme thermophile, *Thermus thermophilus* HB8 (HB8-Fpg) at 1.9 Å resolution. It reveals that the Fpg protein molecule is composed of two distinct domains connected by a flexible hinge.

More recently, Castaing and colleagues resolved the structure of a non-covalent complex between the *Lactococcus lactis* Fpg and a 1,3-propanediol (Pr) abasic site analogue-containing DNA. They have shown that Fpg pushes out the Pr site from the DNA double helix, recognising the cytosine opposite the lesion and inducing a 60 degree bend of the DNA.

Finally, the structure of a trapped catalytic intermediate of *E. coli* Fpg protein has been determined at 2.1 Å resolution. In agreement with the structure of HB8-Fpg the *E. coli* Fpg is a bilobal protein with a wide negatively charged DNA-binding groove. Highly conserved residues Lys-57, His-71, Asn-169 and Arg-259 are involved in binding the phosphodiester backbone of DNA, which is sharply kinked at the lesion site.

Conserved residues Met-74, Arg-110 and Phe-111 are inserted into DNA helix to fill the void in DNA after nucleotide eversion. A deep hydrophobic pocket in the active site is positioned to accommodate the damaged base.

BASE EXCISION REPAIR IN EUKARYOTES

Yeast

Monofunctional DNA glycosylase: spMYH: The *E. coli* MutY homolog was identified in *Schizosaccharomyces pombe* by a homology search in genome databases. The spMYH gene encodes for a 461 amino acids protein which is highly homologous to MutY and human MYH. spMYH possesses the HhH motif and the [4Fe-4S] DNA binding cluster domain found in Nth/MutY.

It has a DNA glycosylase and probably an AP lyase activity. SpMYH incises A opposite G (A/G), A/8-oxoG, 2-aminopurine/G and A/2-aminopurine containing DNA. Cells deleted for the spMYH gene (*spMyH*) have a mutator phenotype and are more sensitive to oxidative agents, such as hydrogen peroxide, than wild type cells.

These data show that spMYH is implicated in the avoidance of 8-oxoG induced mutations and it has an important role in the protection against oxidative stress. Interestingly, similar to human MYH (hMYH), spMYH also interacts directly with PCNA. This interaction is fundamental for the biological function of spMYH in the control of mutation avoidance since the mutation rate of *spMyH* expressing hMYH is reduced but it stays unchanged when *spMyH* cells express a mutant hMYH protein unable to interact with PCNA.

Bifunctional DNA glycosylases S. cerevisiae *yOGG1, Ntg1 and Ntg2: Saccharomyces cerevisiae* 8-oxoguanine-DNA glycosylase (yOGG1) is the yeast counterpart of the *E. coli* Fpg protein. It was identified by functional suppression of the *E. coli fpg mutY* mutator phenotype, leading to the concept that eukaryotes use the DNA glycosylase pathway to repair oxidised purines in DNA. Comparison of amino acid sequences of the yOGG1 protein and *E. coli* Fpg protein does not reveal any obvious homology. The yOGG1 protein has associated DNA glycosylase/AP lyase activity and cleaves abasic sites via - elimination. In contrast to the bacterial enzyme, the yOGG1 protein repairs 8-oxoG only when paired with pyrimidines.

The yOGG1 also removes Fapy guanine-derived residues from DNA but less efficiently than 8-oxoG and -elimination was not observed. The major difference between Fpg and yOGG1 in terms of substrate specificity is that Fpg also removes adenine-derived Fapy lesions and oxidatively damaged pyrimidines. The yOGG1 is not an essential gene, as its disruption had no effect on the viability of a *ogg1* haploid mutant but *ogg1* cells display a mutator phenotype characterised by an augmentation of G/CT/A transversions. Like *fpg* cells, *ogg1* haploid mutants do not show particular sensitivity to H_2O_2. These data suggest that yOGG1 is the functional homologue of *E. coli* Fpg.

Based on the fact that the helix-hairpin-helix (HhH) DNA-binding motif was found in many DNA binding proteins, searches in databases for sequences containing this motif led to identification of the NTG1 gene on chromosome I of *S. cerevisiae*. The NTG1 gene encodes a 45 kDa protein, Ntg1p, which is homologous to the *E. coli* endonuclease III (Nth) but lacks the (4Fe-4S) cluster DNA binding domain found in the other members of this family.

Ntg1p is a DNA glycosylase/AP lyase which removes thymine glycols and unexpectedly releases formamidopy-rimidine residues from DNA with a high efficiency comparable to that of the *E. coli* Fpg protein. Ntg1p also excises other lesions generated by oxidative stress such as 5,6-dihydrouracil, and 5-hydroxy-purines and 8-oxoG only when paired with guanine. Targeted disruption of the NTG1 gene results in viable cells. The Ntg1 protein localises both in the nucleus and in the mitochondria and is induced by cell exposure to DNA-damaging agents purified and characterised a second *S. cerevisiae* Ogg-like protein, called Ogg2, which preferentially acts on 8-oxoG/G base pairs. Later it turned out that Ogg2 is identical to Ntg1.

NTG2 was identified during analysis of *S. cerevisiae* genome as a second gene homolog to *E. coli* Nth in yeast. The NTG2 gene is localised on chromosome XV and codes for a 43.7 kDa protein, Ntg2p, which contains (4Fe-4S) cluster DNA binding domain also present in the *E. coli* Nth protein. The substrate specificity of Ntg2p is similar but not identical to Ntg1p; it cannot incise 8-oxoG mispaired with any of the four DNA bases. Recent results show that Ntg2p, but not Ntg1p, is implicated in the repair of certain degradation products of 8-oxoG such as 8-hydroxydeoxy-guanosine (8-OH-dG).

The targeted disruption of NTG2 results in viable cells but, similarly to *ntg1*⁻ cells, greatly increases the rate of spontaneous and H_2O_2-induced mutations. Ntg2p is a nuclear enzyme constitutively expressed in cells. Interestingly, Ntg2p interacts with the DNA mismatch repair protein Mlh1p. Ntg1p and Ntg2p are both required for repair of spontaneous and induced oxidative DNA damage.

S. cerevisiae *AP endonucleases Apn1 and Apn2:* Homologs of *E. coli* Xth and Nfo have been identified in eukaryotes. The yeast APN1 gene encodes AP endonuclease I (Apn1) homologous to *E. coli* Nfo protein. This 41.4 kDa protein is the major AP endonuclease in yeast, accounting for >90% of total AP endonuclease activity. Apn1 has AP endonuclease, 3'-diesterase and 3'-phosphatase activities.

Like the *E. coli* homolog, Apn1 is a metalloenzyme excising 3'-phosphoglycoaldehyde, 3'-phosphoryl groups, and 3'-, unsaturated aldehydes. Yeast mutant lacking Apn1 (*apn1*) is viable but it is hypersensitive to both oxidative (H_2O_2 and t-butylhydroperoxide) and alkylating (methyl- and ethylmethane sulfonate) agents, it has 6- to 12-fold higher rate of spontaneous mutation than wild-type. This mutator phenotype is mainly characterized by a 60-fold increase in A/T to G/C transversion rate. The second yeast AP endonuclease (Apn2/ETH1) was identified by blast homology search with the sequences of *E. coli* Xth and human Ape1 in the *S. cerevisiae* genome.

Apn2 is a 520 amino acid protein, its expression (at the mRNA level) is induced by DNA damage. Apn2 has AP endonuclease, 3'-phosphodiesterase and 3'5' exonuclease activities, it can remove 3'-phosphate and 3' phosphoglycolate termini produced by H_2O_2. Interestingly, the 3' phosphodie-sterase and the 3'5' exonuclease activities of Apn2 are 30-40-fold more active than its AP endonuclease activity. Apn2 could be an important factor in the repair of oxidative DNA damage. Yeast lacking apn2 (*apn2*) alone are viable and do not show a particular phenotype.

However *apn1 apn2* cells are remarkably sensitive to DNA damaging agents such as H_2O_2, MMS and phleomycin D1 and they present an increase of spontaneous mutation rate. Apn2 contains a carboxy-terminal domain which is absent in the other members of the family (Xth/Ape1). Deletion of this domain does not affect the enzymatic activity of Apn2 *in vitro* but the truncated protein cannot remove AP sites *in vivo*. These data strongly suggest that this domain is indispensible for protein-protein interaction and that Apn2 protein functions *in vivo* as a part of multiprotein complex.

Interestingly, it has been shown that the human homolog of Apn2 protein, Ape2, interacts with PCNA.

MAMMALS (HUMAN, RODENT)

Bifunctional mammalian DNA glycosylases OGG1, NTH1 and NEH1
hOGG1: Several groups reported the molecular cloning of the cDNA of the human and murine 8-oxoguanine DNA glycosylase (OGG1). The amino acid sequence of hOGG1 showed 33% identity and 54% similarity to *S. cerevisiae* OGG1 and conserved HhH/PVD motif which is present in endoIII/MutY/AlkA superfamily of DNA glycosylases. The human OGG1 gene, located on chromosome 3, codes for seven alternatively spliced forms of mRNA.

These transcripts were classified as type 1 and 2 depending on their last exon. The main type 1 and 2 transcripts encode, respectively, a 36 kDa nuclear and a 40 kDa mitochondrial polypeptide. These two proteins differ by their C-terminus.

Both forms possess the same catalytic activity. Purified hOGG1 protein, similarly to *E. coli* Fpg protein, acts as a DNA glycosylase towards duplex DNA containing 8-oxoG/C base pair, Fapy and has associated AP lyase activity.

Crystal structure data at 2.1 Å resolution of the core domain of hOGG1 complexed with a 8-oxoG/C containing oligonucleotide was reported by Verdine's laboratory. This structure reveals that hOGG1 recognizes in the DNA helix both 8-oxoG (using amino acids F319, Q315, G42, C253) and the cytosine opposite 8-oxoG (using amino acids N149, R154, R204, Y203). The modified base is fully extruded from the helix and inserted into an extra-helical active-site pocket on the enzyme.

Interestingly, mutations of residue, which recognises the cytosine, does not alter the activity on 8-oxoG/C but significantly increases the activity toward 8-oxoG opposite other bases than cytosine. Physiologically, such mutants become both pro-mutagenic by repair of 8-oxoG/A and anti-mutagenic by repair of 8-oxoG/C. Study of search intermediates of hOGG1 protein in the presence of high molecular weight DNA by atomic force microscope shows that enzyme scans

DNA at undamaged sites by inducing drastic kinks. The base excision repair pathway for 8-oxoG involves the resynthesis of a single nucleotide at the lesion site as shown by *in vitro* repair assays with mammalian cell extracts and more recently by reconstitution *in vitro* of this pathway by using human purified proteins. These data show that the BER short-patch pathway is predominant for the repair of major oxidative DNA damage, 8-oxoG. Activity of hOGG1 is greatly stimulated by Ape1 which binds to AP sites generated by the DNA glycosylase and enhances the enzymatic turnover. This observation suggests that Ape1 might preclude the AP lyase activity of OGG1. The finding that the dRP lyase activity of DNA polymerase is required in repair of oxidative lesions by mammalian cell extracts supports this mechanism.

The nuclear hOGG1 is associated with chromatin and the nuclear matrix during interphase and with condensed-chromatin during mitosis. The fraction of hOGG1 bound to chromatin is phosphorylated on a serine residue, and the kinase responsible for this post-translational modification of hOGG1 might be protein kinase C. Homozygous *ogg1*-/- null mice were generated by targeted disruption. These mice are viable and, despite an increase of potentially mutagenic DNA lesions in their genome and mitochondria, they do not display any particular phenotype.

The *ogg1*-/- null mice have an elevated spontaneous mutation rate in non- or slowly proliferating tissues with high oxygen metabolism, such as liver, but they do not develop malignancies. These data are at variance with several reports which associate mutations and polymorphism of the OGG1 gene, in particular the Ser326Cys polymorphism, with increased risk of cancer. Although the Ser326Cys polymorphism is slightly, or not associated with altered OGG1 activity, it might affect phosphorylation at this serine residue leading to adverse effects.

hNTH1: A human homolog of the *E. coli* Nth protein (endonuclease III) has been purified and cloned. This gene, called hNTH1 (human Nth homolog 1), encodes a 34.3 kDa protein that shares extensive homology with Nth including the conserved HhH-motif and the (4Fe-4S) cluster loop motif. hNTH1 protein acts as DNA glycosylase with an associated AP lyase activity, and has substrate specificity similar to the *E. coli* homolog towards damaged pyrimidine derivatives that result from ring saturation, ring fragmentation, ring contraction and unexpectedly ring-opened purine residues (Fapy). hNTH1 acts preferentially on 5-hydroxycytosines and AP sites when they are situated opposite to guanine and removes 8-oxoG opposite G. hNTH1 seems to be an exclusively nuclear protein and its expression is regulated during the cell cycle, showing a maximum in S-phase.

Homozygous *mNth1*-/- mutant mice do not have any detectable phenotype defect. Sensitivity of *mNth1*-/- mutant embryonic cells to H_2O_2 and menadione was not changed as compared to wild-type cells. Moreover, thymine glycol induced by ionising radiation is repaired, though more slowly than in wild type. The absence of a significant phenotype could be explained by the discovery of two novel thymine-glycol DNA glycosylases in *mNth1* mutant cells called TGG1 and TGG2. These ~40 kDa proteins localise in mitochondria and nucleus, respectively.

Thus, these data suggest the existence of several back-up DNA glycosylases in mammals to remove thymine glycol. The redundancy of such enzymatic activity highlights the importance of BER pathway for counteracting oxidative pyrimidine damage.

Although there is no definite evidence that BER enzymes act as multiprotein complexes, several reports suggest that addition of proteins can stimulate damage recognition and lesion processing. The NER enzyme XPG stimulates the hNTH1 activity *in vitro*. These data support the hypothesis that neurodegeneration in XP-patients could be due to defects in the repair of oxidative DNA damage. A yeast two-hybrid screen for other interacting partners of Nth1 led to the isolation of the DNA-binding protein B (DbpB)/Y box binding protein (YB-1). YB-1 greatly stimulates the hNTH1 activity. Altogether these data suggest that the BER pathways could be regulated by a number of non-BER proteins.

hNEH1: Human homologs of *E. coli* Nei/MutM were identified by database search of the genome and named hNEH1 and 2 (human Nei Homolog 1 and 2). hNEH1 was biochemically characterized, it encodes a 44 kDa protein which excises Fapy, oxidised purines and 8-oxoG when present in DNA. hNEH1 shows a tissue-specific expression with an S-phase specific increase at both RNA and protein level. Tissue-specific mRNA levels of OGG1 and NEH1 are distinct. These data suggest that the activities of OGG1 and NEH1 are not redundant at the tissue level and NEH1 might be involved in the replication-coupled repair of oxidative DNA damage.

Monofunctional DNA glycosylases ANPG, hTDG, MYH

ANPG: The ANPG protein is the human counterpart of the *E. coli* AlkA protein. This protein releases 3-meA and 7-meG from methylated DNA, but it is also able to excise hypoxantine and 1,N^6-ethenoadenine more efficiently than the AlkA protein. In contrast to the AlkA protein, the ANPG protein does not release 5-formyluracil and RT residues from DNA, however, there are activities in human cell-free extracts that can excise these lesions.

It should be noted that the human protein does not share significant amino acid sequence homology with the bacterial AlkA. Recently, it has been shown that ANPG efficiently excises 1,N^2-ethenoguanine (1,N^2-G) when present in DNA. Immunofluorescent staining for the ANPG protein in normal breast cells showed its nuclear localisation. Two alternatively spliced transcripts of human ANPG have been isolated from human cells. The gene encoding for ANPG maps to chromosome 16. The full-length cDNA first isolated by Samson and colleagues (ANPG70, 293 AA, 1991) differs from the splice variant only in the sequence of the first 8 and 13 N-terminal amino acids respectively. Two truncated versions of human ANPG have also been described: ANPG40 lacks the first 63 amino acids and ANPG80 the first 73 amino acids from the N-terminus, when compared to APNG70. It has been concluded that the non-conserved, N-terminal part apparently contributes little to their damage recognition and N-glycosylase activity.

Surprisingly, it has been shown that the ANPG80, which lacks 73 amino acid residues at the N-terminus, is unable to excise 1,N^2-G from the duplex oligonucleotide and that the ANPG40 has a reduced activity as compared to the ANPG70. These observations indicate that the non-conserved, N-terminal part of ANPG is essential for 1,N^2-G-glycosylase activity but dispensable for the release of A, hypoxanthine and N-methylpurines. The ANPG activity toward hypoxanthine-containing substrates is slightly activated in the presence of the hHR23 protein.

Although the core part of the ANPG70 and APDG proteins display 85% sequence identity, it seems that human enzyme repairs 1,N^2-G somewhat more efficiently than its rat counterpart. In fact, a similar difference between human and mouse ANPG in the recognition of 3-methylguanine and 7-methylguanine residues has been reported. Interestingly, the human full-length and splice variants of ANPG have a direct repeat in N-terminal amino acid sequence, which is absent in the rat and murine homologues. These observations further support the role of the non-conserved N-terminal part of mammalian ANPG in substrate specificity.

To assess the role of this repair enzyme *in vivo*, mice deficient in 3-meAde-DNA glycosylase (*APNG/Aag* null mice) have been generated. Unexpectedly, these APNG knockout mice exhibit neither any particular sensitivity to alkylating agents nor any significant increase in the spontaneous mutation rate except for splenic T lymphocytes. Furthermore, Roth and Samson have shown that *Aag*-/- myeloid progenitor bone marrow cells are more resistant than wild-type cells to alkylating agents. This result suggests that the initiation of base excision repair could be more lethal to the cell than leaving the damaged bases unrepaired.

The crystal structures of ANPG80 (a truncated form of the ANPG protein) complexed to a DNA duplex containing pyridine and A have been established. ANPG80 is a single domain protein of mixed/ structures, which forms a distinct structural group that does not resemble any other BER protein. The enzyme bends the DNA by about 20° and intercalates into the minor groove of DNA causing the abasic pyrrolidine and A to slip into the enzyme active site. The structure of the ANPG80 nucleotide-binding pocket changes little upon flipping in the A base. The structure of the base-binding pocket of ANPG reveals the lack of specific interactions for methylated bases. This fact probably contributes to its remarkable multifunctionality, providing glycosylase activity within the same enzyme for deamination, methylation and exocyclic base damage.

hTDG: The human thymine DNA glycosylase (hTDG) was first biochemically characterised for its ability to remove T mispaired with G. A more complete characterisation shows that hTDG exhibits a broad substrate specificity: it can excise T from DNA not only when mispaired with G but also from other T-containing mispairs except T/A. Moreover, the hTDG protein more efficiently removes U from a G/U but not from a U/A mispair and $3,N^4$-ethenocytosine opposite to all four natural bases. The hTDG gene is localised on the chromosome 12, it codes for several mRNAs which were cloned. These cDNA encode a single 46 kDa polypeptide (410 amino acids) which is homologous to the *E. coli* MUG protein. The hTDG protein has a high affinity for AP site and stays tightly bound to it after removal of T from a G/T mispair.

Ape1 activates hTDG by increasing the dissociation rate of hTDG from AP site probably through direct interaction. The stimulation of hTDG turnover by Ape1 is responsible for the efficient processing of $3,N^4$-ethenocytosine by hTDG. A second factor facilitating hTDG enzymatic turnover was recently identified. Hardeland have shown that SUMOylation of hTDG by SUMO-1 and SUMO-2/3 drastically reduces its affinity for AP sites, increases its enzymatic turn-over towards G/U and reduces its processing activity on G/T substrate. SUMOylation also enhances the stimulatory effect of Ape1 on hTDG.

Using the yeast one-hybrid screen, hTDG was found to interact with retinoic acid receptor (RAR) and retinoic X receptor (RXR) which are ligand-dependent transcription factors. Interaction between RXR, RAR/RXR and hTDG enhances the binding of the former on their responsive elements. These data suggest that hTDG has a dual function: repair enzyme and transcriptional activator. hTDG is also a transcriptional repressor since it interacts with the thyroid transcription factor 1 (TTF-1) and then strongly inhibits the expression of TTF-1 responsive genes.

Recently, hTDG has been found to interact with the transcriptional co-activator CBP/p300. The hTDG/CBP/p300 complex is competent for both BER and histone acetylation. hTDG enhances the CBP/p300 transcriptional activity and these proteins acetylate it. hTDG acetylation regulates the recruitment of Ape1 by hTDG. These data highlight the complexity of repair processes that are coupled to transcription, tightly regulated by post-translational modifications (SUMOylation and acetylation) and by protein-protein interactions.

SMUG1: The single-strand mismatch-specific uracil-DNA glycosylase (SMUG1) has been identified only in vertebrates by expression cloning. SMUG1 is a nuclear protein of 31 kDa which in fact efficiently removes uracil from double stranded DNA when paired with A and G, it also removes U from single-stranded substrates but less efficiently. Accumulation of 5-hydroxymethyluracil is associated with a cancer risk. The mammalian 5-hydroxymethyluracil DNA N-glycosylase activity is associated with SMUG1.

Like several DNA glycosylases, SMUG1 has a low turnover rate that could be due to its inhibition by AP sites. This could explain the Ape1-dependent stimulation of the hSMUG1 activity on both single and double-stranded DNA substrates. Targeted disruption of UNG gene in mice (*ung-/-*) does not result in relevant increase of spontaneous mutation rates, contrary to bacteria and yeast *ung*$^-$ mutants. It seems that hSMUG1 has an important role in the maintenance of genome stability since it represents the major uracil-DNA glycosylase activity in *ung-/-* cells.

hMYH: The human adenine-DNA glycosylase activity, called hMYH protein (human MutY homolog) was first purified from cellular extracts. Later using homology search, the genomic DNA and cDNA encoding hMYH were identified. Functional expression of hMYH in *E. coli* complements the mutator phenotype of *mutY* cells. The hMYH gene is localised on chromosome 1. It encodes a 59 kDa monofunctional DNA glycosylase which can excise A when opposite to 8-oxoG and to a lesser extent opposite to G. hMYH, similar to OGG1, is able to excise 2-OH-A, a mutagenic oxidative adduct. hMYH interacts with several proteins involved in the long-patch repair pathway such as PCNA, Ape1 and RPA.

In addition, hMYH interacts with the mismatch repair proteins hMSH2/hMSH6. Importantly, this interaction enhances the hMYH activity toward A/8-oxo-G. hMYH DNA glycosylase activity is equally enhanced by Ape1 on A/8-oxoG substrates. Unexpectedly this stimulation is independent of Ape1 AP endonuclease activity. Instead, Ape1 acts by stimulation of the formation of hMYH/DNA complexes. These data suggest that different repair pathways may cooperate through protein/protein interactions when counteracting oxidative DNA damage.

Ten different forms of mRNAs coding for hMYH were isolated. These mRNA encodes three proteins of 52, 53 and 57 kDa respectively. The 52 kDa and 53 kDa forms localise in the nucleus and the 57 kDa form in the mitochondria. Expression of hMYH is cell cycle-regulated with a maximum in S phase. Nuclear forms of hMYH co-localise with PCNA at DNA replication foci, suggesting a role of hMYH in replication-associated base excision repair.

Recently biochemical evidence has been provided that human cell extracts perform base excision repair of 8oxo-G/A mismatches on both strands. Similarly to what is reported in yeast, following adenine excision a cytosine is preferentially inserted opposite 8oxoG. This is followed by excision repair of 8oxoG in 8oxoG/C pair.

Interestingly, repair synthesis on either strand is completely inhibited by aphidicolin, suggesting that a replicative DNA polymerase is involved in the gap filling reaction. DNA polymerases/ are likely to be involved in this replication-associated BER. This was confirmed *in vivo* using murine MYH-deficient cells. *MYH-/-* cells repair A/8oxoG mispairs inefficiently but expression of wild-type MYH in these cells increases the repair. Interaction with PCNA is critical, since expression of a functional MYH lacking its PCNA-binding domain had no effect on the repair efficiency of A/8oxoG in *MYH-/-* cells. It has been speculated that hMYH might be considered as a cancer predisposing gene in humans since several mutations in this gene are linked with somatic mutations in *adenomatous polyposis coli* gene (APC).

AP endonucleases Ape1, Ape2: The major human AP endonuclease (HAP-1/Ape1/Apex), homologous to *E. coli* Xth protein (exonuclease III), was independently discovered as an AP-endonuclease and as redox-regulator of the DNA binding domain of Fos-Jun, Jun-Jun, AP-1 proteins and several other transcription factors including NF-kappa B, Myb and members of the ATF/CREB family. Ape1 also activates. Ape1 is a 35.5 kDa nuclear protein, which shows sequence homology to the *E. coli* Xth protein. Targeted homozygous disruption of Apex gene (*Apex-/-*) results in embryonic lethality in mice.

The heterozygous *Apex+/-* mice, so far, did not reveal significant differences in phenotypes associated with oxidative stress as compared to the wild-type. Due to the dual role of Ape1, the early embryonic lethality could be associated either with the BER defect or with the defective regulation of several transcription factors. Ape1 can substitute for exonuclease III in *E. coli* and for Apn1 in *S. cerevisiae*). Beside its AP endonuclease activity, it exhibits other enzymatic activities: 3'5' exonuclease, phosphodiesterase, 3' phosphatase and RNase H. It should be noted that these additional activities are 3-4-fold weaker than its AP endonuclease activity. In addition, Chou and Cheng have shown that Ape1 has a 3'-mismatch exonuclease activity and so it might be considered as a proof-reading enzyme.

Ape1 plays a central role in BER. It is implicated in both short-patch and long-patch repair pathway. It could be also an important regulator of BER. Ape1 interacts with DNA polymerase *in vivo* which is a key element for BER, this interaction permits the loading of pol on DNA at the AP site and the enhancement of its dRPase activity. Ape1 interacts with other BER proteins such as the scaffold protein XRCC1, which stimulates its activity, and with PCNA and the Flap endonuclease 1 (FEN1).

The $p21^{WAF1/CIP1}$ is involved in the regulation of cell cycle progression, DNA replication and DNA repair. It can bind to PCNA inhibiting DNA replication and PCNA stimulation of long-patch BER pathway.

Ape1 seems to be implicated in the compensation of the inhibitory effect of p21 on BER by stimulating and co-ordinating the long-patch BER. In addition, Ape1 is an activator of DNA glycosylases. Altogether these data suggest that BER pathways are co-ordinated and regulated by protein-protein interactions and that Ape1 plays an important role in these interactions. Moreover, it has been shown that Ape1 and hnRNP-L are the components of the nCARE-B2 element binding complexes in the *ape1* gene promoter. This result suggests that expression of the *ape1* gene is down regulated by its own product. Ape1/ref-1 could also be involved in cell response to chemotherapy since it was found that nuclear accumulation of Ape1 is inversely proportional to that of p53 in head-and-neck cancer.

A second AP endonuclease, Ape2, homologous to the *S. cerevisiae* Apn2, was identified in mammals. The *ape2* gene is localised on the X chromosome and Ape2 transcripts are ubiquitously expressed. Ape2, a 57.3 kDa protein, exhibits only a weak ability to complement *E. coli* and *S. cerevisiae* AP endonuclease deficient cells. Ape2 localises predominantly in the nucleus but also in the mitochondria. In the nucleus, Ape2 co-localises with PCNA and interacts with it *in vitro*. These data suggest that Ape2 is implicated in both mitochondrial and nuclear BER and also that the nuclear role of Ape2 is PCNA-dependent.

OTHER ORGANISMS

Evidences have accumulated that oxidative DNA damage are repaired also through BER pathway in plants and insects. Homologues of the *Escherichia coli* Nth and Fpg proteins were characterized in *Arabidopsis thaliana*. Previously it was suggested that *Drosophila* has no BER pathway. However, an homologue of the human OGG1 protein in *Drosophila melanogaster* has been identified and characterized.

A new DNA glycosylase from *A. thaliana* ROS1 is a 1393 aminoacids protein, with an endonuclease III domain at the C-terminus. The recombinant ROS1 is able to incise oxidatively damaged DNA and also DNA containing 5 meC, suggesting a role in controling DNA methylation levels. This agrees with the pleiotropic developmental defects of the ros1 mutants suggesting regulatory properties of this protein. These results suggest that this protein could be a new class of DNA glycosylases, perhaps only present in plants..

ALTERNATIVE REPAIR PATHWAYS

So far, several alternatives to the BER repair pathways for oxidative DNA damage have been described.

Nucleotide Incision Repair

The BER pathway requiring sequential action of two enzymes for incision of DNA raises theoretical problems for the efficient repair of oxidative DNA damage because it generates genotoxic intermediates such as abasic (AP) sites and/or blocking 3'-termini groups that must be eliminated by additional steps before initiating DNA repair synthesis. In addition, biological evidence hints at the existence of an alternative repair pathway. *E. coli* and mouse mutants deficient in the DNA glycosylases that remove oxidised bases are not sensitive to reactive oxygen species.

A recent finding that Nfo and Nfo-like endonucleases nick DNA on the 5' side of various oxidatively damaged bases, generating 3'-hydroxyl and 5'-phosphate termini, provides an alternative pathway to the classic BER. It was proposed to name it the nucleotide incision repair (NIR) pathway. The proposed NIR pathway eliminates the genotoxic intermediates, explains the genetic data and most likely identifies the physiological and original target as the modified base and not an artificial reduced abasic site, for the long patch repair pathway described in human cells.

This alternative repair pathway is evolutionarily conserved from *E. coli* to human. The NIR pathway directly generates proper ends for DNA replication on one side and on the other side the site for elimination of the lesion by specific nucleases. This mechanistic characteristic creates an advantage as compared to BER.

NUCLEOTIDE EXCISION REPAIR

Oxidative DNA damage is a major source of genetic mutation, and it is implicated in several human disorders including *Xeroderma pigmentosum* (XP). Human disorder XP is characterised by defect in nucleotide excision repair, the cells from XP patients fail to remove pyrimidine dimers caused by sunlight and, as a consequence, develop multiple cancers in areas exposed to light.

Despite the fact that neural tissue is shielded from sunlight-induced DNA damage, a fraction of XP patients display progressive neurologic degeneration secondary to a loss of neurones. In fact, it was shown that in *E. coli* the nucleotide excision repair pathway could be involved in repair of oxidative DNA damage. Furthermore, in human cells two major oxidative DNA lesions, 8-oxoguanine and thymine glycol, are excised from DNA *in vitro* by the same enzyme system responsible for removing pyrimidine dimers and other bulky DNA adducts. These results support the idea that the NER pathway could be directly involved in repair of oxidative DNA damage.

Transcription-Coupled Repair (TCR)

Transcription-coupled repair (TCR) is a pathway of DNA excision repair that preferentially removes the DNA lesions present in transcribed sequences of expressed genes. TCR was originally observed for DNA damage repaired by NER pathway. However, recent studies demonstrated that Cockayne syndrome (CS) cells which are deficient for TCR cannot remove 8oxo-G in transcribed sequence, despite its proficient repair in non-transcribed sequence. Furthermore, it has been demonstrated that the breast and ovarian cancer susceptibility genes, BRCA1 and BRCA2, may participate in the repair of the 8oxo-G residues specifically located on the transcribed DNA strand in human cells.

Translesional DNA Synthesis

In *E. coli*, in addition to BER pathway, the translesion DNA synthesis system can contribute to the repair of chromosomal AP sites *in vivo*. In the last few years, new human DNA polymerases Pol-, Pol-, and Pol-, which are able to promote replication through DNA lesions were discovered. These polymerases catalyse translesion DNA synthesis (TLS) in a distributive manner and tend to show lower replication fidelity than other polymerases when unmodified DNA is used as a template. The yeast and human Pol- replicate DNA containing 8-oxo-G efficiently and accurately by inserting a cytosine across from the lesion. In support of the biochemical data, genetic studies show that synergistic increase in the rate of spontaneous mutations occurs in the absence of Pol- in the yeast *ogg1* mutant.

Both error-free and error-prone synthesis were observed in DNA synthesis across A residue catalysed by Pol- and Pol-. It has been found that Pol- preferentially misinsert G opposite to uracil thus providing a mechanism whereby mammalian cells can decrease the mutagenic potential of lesions formed via the deamination of cytosine. Altogether these observations suggest an additional role for TLS-specific DNA polymerases in the prevention of the mutagenic replication of oxidative DNA damage.

DNA Mismatch Repair (MMR) Pathway

DNA mismatch-repair system is involved in the repair of mispaired bases formed during replication, genetic recombination and as a result of DNA damage. Studies in *S. cerevisiae* indicate the involvement of the mismatch repair pathway in prevention of genotoxic effect of oxidative DNA damage. It was shown that mutations in MSH2 and MSH6 caused a synergistic increase in mutation rate in combination with mutations in OGG1, resulting in a 140- to 218-fold increase in the G/C-to-T/A transversion rate. Consistent with this, the MSH2-MSH6 complex binds to 8oxoG:A mispairs and 8oxo-G/C base pairs with high affinity and specificity.

Another important source of 8oxo-G in DNA is the dNTP pool. Recently it has been shown that incorporated 8oxo-dGMP contributes significantly to the mutator phenotype of MMR-deficient cells. Increased expression of MTH1 in MSH2-/- cells produces a significant decrease of DNA 8oxo-G levels and is associated with a drastic reduction of the mutation rate. These findings indicate that MMR excises incorporated 8oxoGMP from newly synthesised DNA strand, thus providing cells with an additional protective strategy from oxidative stress.

Homologous Recombination Repair Pathway

It is well established that homologous recombination (HR) system is pivotal for the repair of an important form of oxidative DNA damage induced by ionising radiation - the double-strand break (DSB). Moreover, *in vivo* replication forks often encounter template DNA damage that can lead to replication fork collapse generating double-strand break. It is hypothesised that DNA recombination functions as a system for reactivation of the collapsed replication fork.

The importance of this function to the formation of lethal DNA damage is underscored by the uncontrollable fragmentation of chromosomes in Rad51-deficient vertebrate cell lines. Another important characteristic of the HR repair pathway is that it mainly provides a non-mutagenic replicative bypass of the blocking lesion using sister chromatid or homologous chromosome as an undamaged DNA template.

Recently, it has been found that defects in the homologous recombination pathway are associated with cancer-prone clinical syndromes, in particular *ataxia telangiectasia*, hereditary breast cancer, Bloom's syndrome and Werner's syndrome. These observations support the idea that the HR system could be an alternative to excision/incision repair pathways in counteraction of genotoxic effects of oxidative DNA damage.

Nucleotide Pool-sanitising Enzymes

Besides the modification of DNA by ROS, these species also oxidise the nucleotide pool. The modified nucleotides could be introduced in DNA and also in RNA during their synthesis leading to mutations. The *E. coli* mutT mutant shows a 100- to 10 000-fold increase in A/TG/C transversion rates as compared to wild-type. This effect is due to the accumulation of mutagenic 8oxo-GTP in the cellular dNTP pool which is incorporated opposite A and C during DNA synthesis, thus leading to A/TG/C transversions. The *E. coli* MuT protein is a 8oxo-GTPase that cleanses the cellular nucleotide pool from 8oxoGTP.

In addition, MutT hydrolyses ribonucleotide 8oxo-GTP, suggesting the importance of this enzyme in maintaining transcription fidelity. Mammalian cells contain MutT homolog, the MTH1 protein (MutT Homolog 1). The corresponding cDNAs were cloned in human and murine cells, and protein was characterised. MTH1 displays only 23% identity to the *E. coli* counterpart but the highly conserved region essential for the 8oxo-GTPase activity contains identical residues.

Therefore, expression of MTH1 human and rodent origins in *E. coli mutT* reverts to the mutator phenotype. These results indicate that mammalian proteins may have a similar antimutagenic activity *in vivo* as *E. coli* MutT protein. Like MutT, MTH1 hydrolyses 8oxo-GTP and oxidised ribonucleotides 8oxo-GTP. Interestingly, MTH1 possess an additional 2-hydroxyadenosine (2-OH-A) triphosphate pyrophosphorylase activity. Since hMYH (MutY homolog) is able to remove 2-OH-A from DNA, these data suggest that mammalian cells possess a pathway to sanitise oxidised forms of both guanine and adenine.

Seven types of mRNA transcripts, coded by the MTH1 gene, have been identified in human Jurkat cells, which lead to the synthesis of, at least, three polypeptides of 18, 21 and 22 kDa. MTH1 polypeptides localise mainly in the cytoplasm and in mitochondria.

The MTH1 knockout cell lines and mice were constructed by gene targeting. MTH1-/- cells exhibit a twofold increase of mutation rate as compared to MTH1+/+ cells which is in contrast to the increase of mutation rate in *E. coli mutM* cells. Therefore, it seems likely that mammalian cells have other enzyme(s) and/or pathway(s) capable of hydrolysing oxidised nucleotides. MTH1-/- mice exhibit slightly elevated tumour incidence than wild-type mice but their survival at 1.5 years is not affected.

Several studies on human cancers show that mutations in hMTH1 gene, so far are not involved in these pathologies. Unexpectedly, MTH1 and translesional DNA polymerase kappa (Pol-) expressions were found to significantly increase and decrease respectively in two mammary carcinoma cell lines characterised by frequent A/TG/C transversions. So far, no human diseases have been linked to defects in proteins involved in the BER pathway. DNA repair genes functionally expressed in mammalian cells and now in transgenic mice having a null mutation in the gene coding for BER proteins are very important tools to ascertain the biological role of these proteins *in vivo*.

It has been very astonishing to notice that, beside a few examples of targeted deletion of genes encoding the AP-endonuclease 1 and DNA polymerase in mice leading to embryonic lethality, the genotype of the other BER protein knockout mice (such as a number of DNA glycosylases) does not show any striking particularity in terms of predisposition to cancer and premature aging.

Furthermore, examination of 6200 *S. cerevisiae* genes transcript levels after exposure to various genotoxic agents reveals that the DNA repair genes are only modestly induced, which is in contrast to adaptive and SOS response in *E. coli*. These observations suggest the possibility of back-up repair pathway(s) for oxidative DNA damage that have to be characterized. One could expect important breakthroughs from crosses between different strains to produce double knockouts to identify the possible back-up systems, the processes involved in the regulation and the interactions of the different pathways. Detailed understanding of the mechanisms leading to the co-ordination of various proteins involved in the molecular reaction of BER is of paramount importance for gaining insights into the efficiency and fidelity of this key pathway for genome stability, prevention of cancer, resistance to chemotherapeutic agents, degenerative diseases and more recently in some aspects of teratogenicity.

4

RNA Structure

We now turn our attention to RNA, which differs from DNA in three respects. First, the backbone of RNA contains ribose rather than 2'-deoxyribose. That is, ribose has a hydroxyl group at the 2' position. Second, RNA contains uracil in place of thymine. Uracil has the same single-ringed structure as thymine, except that it lacks the 5' methyl group. Thymine is in effect 5'methyl-uracil. Third, RNA is usually found as a single polynucleotide chain.

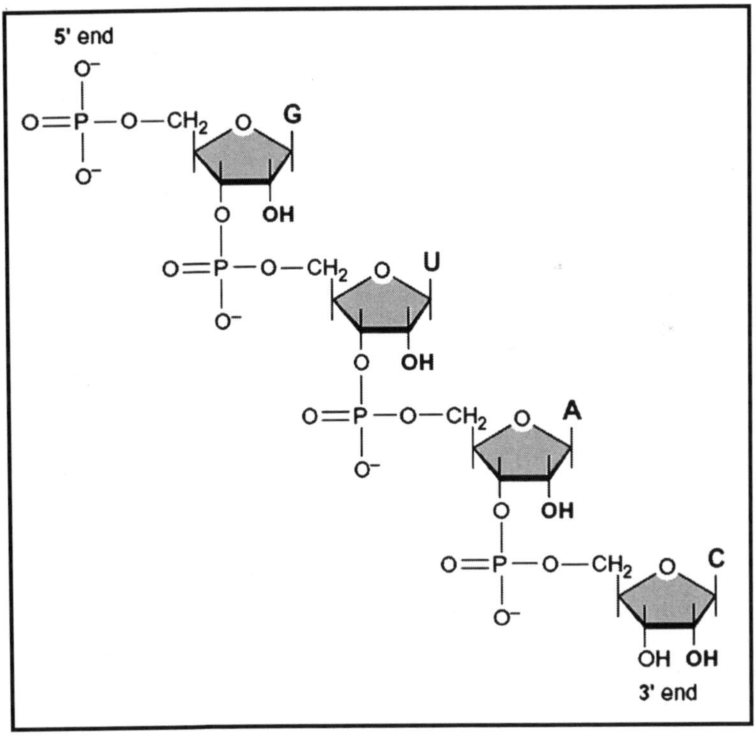

Fig. Structural Features of RNA

Except for the case of certain viruses, RNA is not the genetic material and does not need to be capable of serving as a template for its own replication.

Rather, RNA functions as the intermediate, the mRNA, between the gene and the protein-synthesizing machinery. Another function of RNA is as an adaptor, the tRNA, between the codons in the mRNA and amino acids. RNA can also play a structural role as in the case of the RNA components of the ribosome.

Yet another role for RNA is as a regulatory molecule, which through sequence complementarity binds to, and interferes with the translation of, certain mRNAs. Finally, some RNAs (including one of the structural RNAs of the ribosome) are enzymes that catalyze essential reactions in the cell. In all of these cases, the RNA is copied as a single strand off only one of the two strands of the DNA template, and its complementary strand does not exist. RNA is capable of forming long double helices, but these are unusual in nature.

RNA CHAINS FOLD BACK ON THEMSELVES TO FORM LOCAL REGIONS

Despite being single-stranded, RNA molecules often exhibit a great deal of double-helical character. This is because RNA chains frequently fold back on themselves to form base-paired segments between short stretches of complementary sequences. If the two stretches of complementary sequence are near each other, the RNA may adopt one of various stem-loop structures in which the intervening RNA is looped out from the end of the double-helical segment as in a hairpin, a bulge, or a simple loop. The stability of such stem-loop structures is in some instances enhanced by the special properties of the loop.

For example, a stem-loop with the "tetraloop" sequence UUCG is unexpectedly stable due to special base-stacking interactions in the loop. Base pairing can also take place between sequences that are not contiguous to form complex structures aptly named pseudoknots. The regions of base pairing in RNA can be a regular double helix or they can contain discontinuities, such as noncomplementary nucleotides that bulge out from the helix.

A feature of RNA that adds to its propensity to form double-helical structures is an additional, non-Watson-Crick base pair. This is the G:U base pair, which has hydrogen bonds between N3 of uracil and the carbonyl on C6 of guanine and between the carbonyl on C2 of uracil and N1 of guanine. Because G:U base pairs can occur as well as the four conventional, Watson-Crick base pairs, RNA chains have an enhanced capacity for self-complementarity. Thus, RNA frequently exhibits local regions of base pairing but not the long-range, regular helicity of DNA. The presence of 2'-hydroxyls in the RNA backbone prevents RNA from adopting a B-form helix. Rather, double-helical RNA resembles the A-form structure of DNA. As such, the minor groove is wide and shallow, and hence accessible, but recall that the minor groove offers little sequence-specific information. Meanwhile, the major groove is so narrow and deep that it is not very accessible to amino acid side chains from interacting proteins. Thus, the RNA double helix is quite distinct from the DNA double helix in its detailed atomic structure and less well suited for sequence-specific interactions with proteins (although some proteins do bind to RNA in a sequence-specific manner).

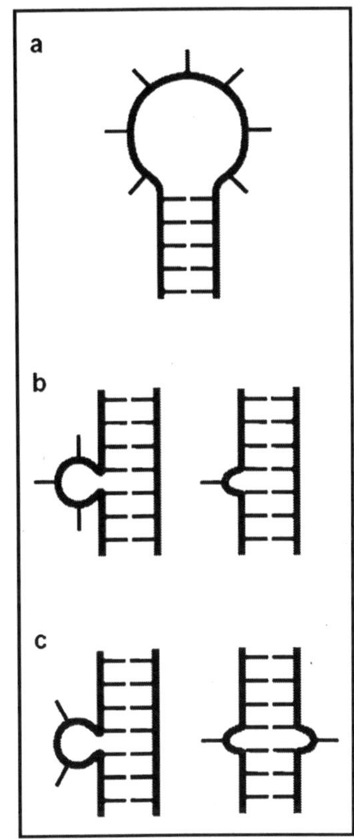

Fig. Double Helical Characteristics of RNA.

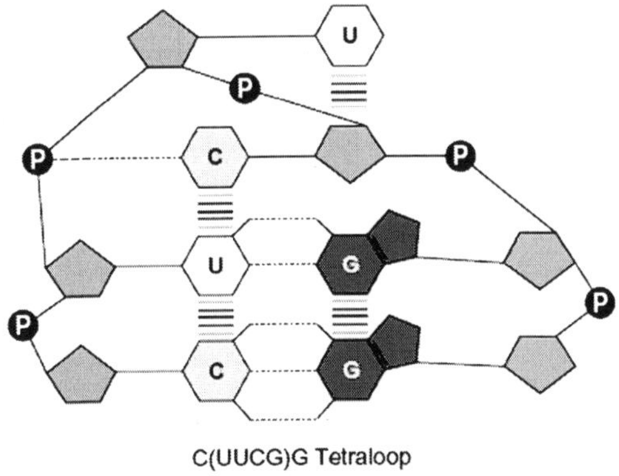

Fig. Tetraloop.

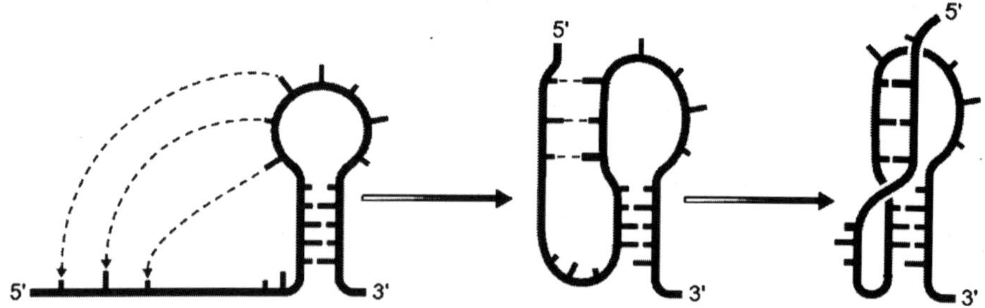

Fig. Pseudoknot

RNA CAN FOLD UP INTO COMPLEX TERTIARY STRUCTURES

Freed of the constraint of forming long-range regular helices, RNA can adopt a wealth of tertiary structures. This is because RNA has enormous rotational freedom in the backbone of its non-base-paired regions. Thus, RNA can fold up into complex tertiary structures frequently involving unconventional base pairing, such as the base triples and base-backbone interactions seen in tRNAs. Proteins can assist the formation of tertiary structures by large RNA molecules, such as those found in the ribosome.

Proteins shield the negative charges of backbone phosphates, whose electrostatic repulsive forces would otherwise destabilize the structure. Researchers have taken advantage of the potential structural complexity of RNA to generate novel RNA species (not found in nature) that have specific desirable properties. By synthesizing RNA molecules with randomized sequences, it is possible to generate mixtures of oligonucleotides representing enormous sequence diversity.

For example, a mixture of oligoribonucleotides of length 20 and having four possible nucleotides at each position would have a potential complexity of 420 sequences or 1012 sequences! From mixtures of diverse oligoribonucleotides, RNA molecules can be selected biochemically that have particular properties, such as an affinity for a specific small molecule.

Some RNAs Are Enzymes

It was widely believed for many years that only proteins could be enzymes. An enzyme must be able to bind a substrate, carry out a chemical reaction, release the product and repeat this sequence of events many times. Proteins are well suited to this task because they are composed of many different kinds of amino acids and they can fold into complex tertiary structures with binding pockets for the substrate and small molecule cofactors and an active site for catalysis.

Now we know that RNAs, which as we have seen can similarly adopt complex tertiary structures, can also be biological catalysts. Such RNA enzymes are known as ribozymes, and they exhibit many of the features of a classical enzyme, such as an active site, a binding site for a substrate and a binding site for a cofactor, such as a metal ion. One of the first ribozymes to be discovered was RNase P, a ribonuclease that is involved in generating tRNA molecules from larger, precursor RNAs. RNase P is composed of both RNA and protein; however, the RNA moiety alone is the catalyst. The protein moiety of RNase P facilitates the reaction by shielding the negative charges on the RNA so that it can bind effectively to its negatively charged substrate. The RNA moiety is able to catalyze cleavage of the tRNA precursor in the absence of the protein if a small, positively charged counter ion, such as the peptide spermidine, is used to shield the repulsive, negative charges. Other ribozymes carry out trans-esterification reactions involved in the removal of intervening sequences known as introns from precursors to certain mRNAs, tRNAs, and ribosomal RNAs in a process known as RNA splicing.

THE HAMMERHEAD RIBOZYME

Before concluding our discussion of RNA, let us look in more detail at the structure and function of one particular ribozyme, the hammerhead. The hammerhead is a sequence-specific ribonuclease that is found in certain infectious RNA agents of plants known as *viroids*, which depend on self-cleavage to propagate. When the viroid replicates, it produces multiple copies of itself in one continuous RNA chain. Single viroids arise by cleavage, and this cleavage reaction is carried out by the RNA sequence around the junction.

One such self-cleaving sequence is called the *hammerhead* because of the shape of its secondary structure, which consists of three base-paired stems surrounding a core of non-complementary nucleotides required for catalysis. The tertiary structure of the ribozyme, however, looks more like a wishbone. To understand how the hammerhead works, let us first look at how RNA undergoes hydrolysis under alkaline conditions.

At high pH, the 2' hydroxyl of the ribose in the RNA backbone can become deprotonated, and the resulting negatively charged oxygen can attack the scissile phosphate at the 3' position of the same ribose. This reaction breaks the RNA chain, producing a 2', 3' cyclic phosphate and a free 5' hydroxyl. Each ribose in an RNA chain can undergo this reaction, completely cleaving the parent molecule into nucleotides. Many protein ribonucleases also cleave their RNA substrates via the formation of a 2', 3' cyclic phosphate. Working at normal cellular pH, these protein enzymes use a metal ion, bound at their active site, to activate the 2' hydroxyl of the RNA. The hammerhead is a sequencespecific ribonuclease, but it too cleaves RNA via the formation of a 2', 3' cyclic phosphate.

Hammerhead-mediated cleavage involves a ribozyme-bound Mg'' ion that deprotonates the 2' hydroxyl at neutral pH, resulting in nucleophilic attack on the scissile phosphate. Because the normal reaction of the hammerhead is self-cleavage, it is not really a catalyst; each molecule normally promotes a reaction one time only, thus having a turnover number of one. But the hammerhead can be engineered to function as a true ribozyme by dividing the molecule into two portions—one, the ribozyme, that contains the catalytic core and the other, the substrate, that contains the cleavage site. The substrate binds to the ribozyme at stems I and III. After cleavage, the substrate is released and replaced by a fresh uncut substrate, thereby allowing repeated rounds of cleavage.

LIFE EVOLVE FROM AN RNA

The discovery of ribozymes has profoundly altered our view of how life might have evolved. We can now imagine that there was a primitive form of life based entirely on RNA. In this world, RNA would have functioned as the genetic material and as the enzymatic machines. This RNA world would have preceded life as we know it today, in which information transfer is based on DNA, RNA, and protein.

A hint that the protein world might have arisen from an RNA world is the discovery that the component in the ribosome that is responsible for the formation of the peptide bond, the peptidyl transferase, is an RNA molecule. Unlike RNase P, the hammerhead, and other previously known ribozymes which act on phosphorous centers, the peptidyl transferase acts on a carbon centre to create the peptide bond.

It thus links RNA chemistry to the most fundamental reaction in the protein world, peptide bond formation. Perhaps then the ribosome ribozyme is a relic of an earlier form of life in which all enzymes were RNAs. DNA is usually in the form of a right-handed double helix. The helix consists of two polydeoxynucleotide chains. Each chain is an alternating polymer of deoxyribose sugars and phosphates that are joined together via phosphodiester linkages. One of four bases protrudes from each sugar: adenine and guanine, which are purines, and thymine and cytosine, which are pyrimidines. While the sugar phosphate backbone is regular, the order of bases is irregular and this is responsible for the information content of DNA.

Each chain has a 5' to 3' polarity, and the two chains of the double helix are oriented in an antiparallel manner—that is, they run in opposite directions. Pairing between the bases holds the chains together. Pairing is mediated by hydrogen bonds and is specific: Adenine on one chain is always paired with thymine on the other chain, whereas guanine is always paired with cytosine. This strict base-pairing reflects the fixed locations of hydrogen atoms in the purine and pyrimidine bases in the forms of those bases found in DNA. Adenine and cytosine almost always exist in the amino as opposed to the imino tautomeric forms, whereas guanine and thymine almost always exist in the keto as opposed to enol forms. The complementarity between the bases on the two strands gives DNA its self-coding character. The two strands of the double helix fall apart (denature) upon exposure to high temperature, extremes of pH, or any agent that causes the breakage of hydrogen bonds. Upon slow return to normal cellular conditions, the denatured single strands can specifically reassociate to biologically active double helices (renature or anneal). DNA in solution has a helical periodicity of about 10.5 base pairs per turn of the helix. The stacking of base pairs upon each other creates a helix with two grooves. Because the sugars protrude from the bases at an angle of about 120°, the grooves are unequal in size. The edges of each base pair are exposed in the grooves, creating a pattern of hydrogen bond donors and acceptors and of van der Waals surfaces that identifies the base pair.

The wider—or *major*—groove is richer in chemical information than the narrow (*minor*) groove and is more important for recognition by nucleotide sequence-specific binding proteins. Almost all cellular DNAs are extremely long molecules, with only one DNA molecule within a given chromosome. Eukaryotic cells accommodate this extreme length in part by wrapping the DNA around protein particles known as nucleosomes. Most DNA molecules are linear but some DNAs are circles, as is often the case for the chromosomes of prokaryotes and for certain viruses. DNA is flexible. Unless the molecule is topologically constrained, it can freely rotate to accommodate changes in the number of times the two strands twist about each other.

DNA is topologically constrained when it is in the form of a covalently closed circle, or when it is entrained in chromatin. The linking number is an invariant topological property of covalently closed circular DNA. It is the number of times one strand would have to be passed through the other strand in order to separate the two circular strands. The linking number is the sum of two interconvertible geometric properties: twist, which is the number of times the two strands are wrapped around each other; and the writhing number, which is the number of times the long axis of the DNA crosses over itself in space. DNA is relaxed under physiological conditions when it has about 10.5 base pairs per turn and is free of writhe. If the linking number is decreased, then the DNA becomes torsionally stressed, and it is said to be negatively supercoiled.

DNA in cells is usually negatively supercoiled by about 6%. The left-handed wrapping of DNA around nucleosomes introduces negative supercoiling in eukaryotes. In prokaryotes, which lack histones, the enzyme DNA gyrase is responsible for generating negative supercoils. DNA gyrase is a member of the type II family of topoisomerases. These enzymes change the linking number of DNA in steps of two by making a transient break in the double helix and passing a region of duplex DNA through the break. Some type II topoisomerases relax supercoiled DNA, whereas DNA gyrase generates negative supercoils. Type I topoisomerases also relax supercoiled DNAs but do so in steps of one in which one DNA strand is passed through a transient nick in the other strand. RNA differs from DNA in the following ways: its backbone contains ribose rather than 2'-deoxyribose; it contains the pyrimidine uracil in place of thymine; and it usually exists as a single polynucleotide chain, without a complementary chain.

As a consequence of being a single strand, RNA can fold back on itself to form short stretches of double helix between regions that are complementary to each other. RNA allows a greater range of base pairing than does DNA. Thus, as well as A:U and C:G pairing, U can also pair with G. This capacity to form a non-Watson-Crick base pair adds to the propensity of RNA to form doublehelical segments. Freed of the constraint of forming longrange regular helices, RNA can form complex tertiary structures, which are often based on unconventional interactions between bases and between bases and the sugarphosphate backbone.

Some RNAs act as enzymes—they catalyze chemical reactions in the cell and in vitro. These RNA enzymes are known as ribozymes. Most ribozymes act on phosphorous centers, as in the case of the ribonuclease RNase P. RNase P is composed of protein and RNA, but it is the RNA moiety that is the catalyst. The hammerhead is a self-cleaving RNA, which cuts the RNA backbone via the formation of a 2', 3' cyclic phosphate in a reaction that involves an RNA-bound Mg'' ion. Peptidyl transferase is an example of a ribozyme that acts on a carbon centre.

This ribozyme, which is responsible for the formation of the peptide bond, is one of the RNA components of the ribosome. The discovery of RNA enzymes that can act on phosphorous or carbon centers suggests that life might have evolved from a primitive form in which RNA functioned both as the genetic material and as the enzymatic machinery.

5

Cell Cycle

Microorganisms typically face the world as single cells rather than as the multi-cellular assemblies of higher organisms. Each single cell must therefore contain all the structures necessary for managing its internal state and dealing with the outside environment. Not surprisingly, this evolutionary process results in the use of rather similar structures and processes to solve similar need in different microorganisms.

However, prokaryotes have been on this Earth for a long period of time and this has allowed them to differentiate into a dizzying number of different species. Eukaryotic microbes are not quite so diverse, but they still display a remarkable range of properties. The diversity of the microbial organisms also means that this survey of structures is not exhaustive. No one cell contains all the structures that we describe here, but we will explore the more common structures that have been observed by scientists in the past 150 years.

A distinction in this discussion must be made between the two types of prokaryotes: the Archaea and their cousins, the Bacteria. We will initially focus on the Bacteria, since that is what we know the most about. Many of the structures we will examine are found in both the Bacteria and the Archaea, but there are some significant differences and these will be covered at the end of the chapter. Finally, we will talk about the features that are distinctive among the microbial eukaryotes. So how did scientists find out so much about such very small organisms? As you might guess, many techniques come into play when tackling a subject as complex as bacterial structure. Electron microscopes have been important, of course, but so have genetics, molecular biology and biochemistry. Microscopes help scientists to visualize where these structures are located and how they are arranged spatially in the microbe. Bacterial genetics and molecular biology identify and analyse the genes necessary for the synthesis and regulation of these structures. Biochemistry permits the detailed examination of each part separately, with implications for its role in the living bacterium. The powerful combination of these disciplines has provided a deep understanding of how a bacterium is put together, but there is still much to learn.

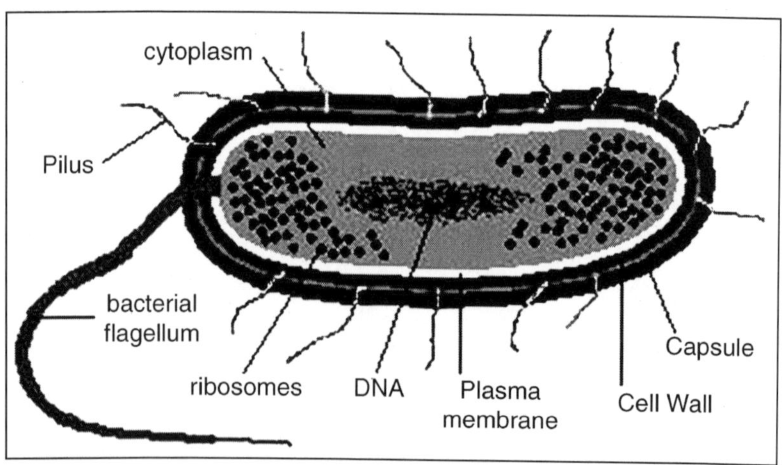

Fig. The Generalized Bacterium

This chapter on microbial structure is separated into two sections. In the first section, we describe the chemical nature of the types of molecules and polymers that are important in carrying out the business of biology.

In the second section, we examine the functional units in the cell, describing how the various chemical structures in the cell interact to carry out important cellular functions. In this discussion we assume that the student has had an introductory chemistry course (at least in high school) and is somewhat familiar with chemical notation.

First and foremost, it is important to point out that there are some universal structures that all living cells contain. They are the basic building blocks of life: DNA, RNA, protein and cellular membranes. Most, but not all, bacteria also possess cell walls. Beyond these essentials, the frequency of the rest of the structures we mention here ranges in the bacterial world from quite common to very rare.

CELL

Cells are structural units that make up plants and animals, also there many single cell organisms. What cells all have in common is they are small 'sacks' composed mostly of water. The 'sacks' are made from a phospholipid bilayer. The membrane is semi-permeable (allowing some things to pass in or out of the cell and blocking others), there are also other methods of transport that we will get into later.

So what is in a cell? The cell as we mentioned is a fluid like membrane that surrounds the contents of the cell. Each component will be discussed in more detail later.

Cells are 90 per cent fluid (cytoplasm) which consists of free amino acids, proteins, glucose, and numerous other molecules. The cell environment (ie. the contents of the cytoplasm, and the nucleus, as well as, they way the DNA is packed) affect the gene expression/regulations, and thus are VERY important parts of inheritance, below are approximations of other components:

ELEMENTS

- 59 per cent Hydrogen (H)
- 24 per cent Oxygen (O)
- 11 per cent Carbon (C)
- 4 per cent Nitrogen (N)
- 2 per cent Others - Phosphorus (P), Sulphur (S), etc.

As far as molecules that make up the cell:

- 50 per cent protein
- 15 per cent nucleic acid
- 15 per cent carbohydrates
- 10 per cent lipids
- 10 per cent Other

What is inside the cell is the cytoplasm which is:

- Cytosol—a lot of water all except the organelles.
- Organelles (which also have membranes) in 'higher' eukaryote organisms:
- *Nucleus (in eukaryotes)*: Where genetic material (DNA) is located, RNA is transcribed.
- *Endoplasmic Reticulum (ER)*: Important for protein synthesis. It is a transport network for molecules destined for specific modifications and locations. There are two types:
- *Rough ER*: Has ribosomes, and tends to be more in 'sheets'.
- *Smooth ER*: Does not have ribosomes and tends to be more of a tubular network.
- *Ribosomes*: Half are on the Endoplasmic Reticulum, the other half are 'free' in the cytosol, this is where the RNA goes for translation into proteins.
- *Golgi Apparatus*: Important for glycosylation, secretion.
- *Lysosomes*: Digestive sacks - the main point of digestion, these are only found in animal cells.
- *Peroxisomes*: Use oxygen to carry out catabolic reactions, in both plant and animals.
- *Microtubules*: made from tubulin, and make up centrioles,cilia,etc.
- *Cytoskeleton*: Microtubules, actin and intermediate filaments.
- *Mitochondria*: convert foods into usable energy. (ATP production) A mitochondrion does this through aerobic respiration. They have 2 membranes, the inner membranes shapes differ between different types of cells, but they form projections called cristae. The mitochondrion is about the size of a bacteria, and it carries its own genetic material and ribosomes.

- *Vacuoles*: More commonly associated with plants. Plants commonly have large vacuoles.
- Found in Plants and not in animals:
- *Chloroplasts*: Convert light/food into usable energy. (ATP production)
- Plastids
- *Cell Wall*: Found in prokaryotic plants and it provides structural support and protection.

SIZE OF CELLS

Eukaryotes are typically 10 times the size of prokaryotic cells. Plant cells are on average some of the largest cells, which may be because of the large water filled vacuoles in some plant cells. So, you ask, what are the relative sizes of biological molecules and cells?

These are All Approximations:

- 0.1 nm (nanometer) diameter of a hydrogen atom
- 0.8 nm Amino Acid
- 2 nm Diameter of a DNA Alpha helix
- 4 nm Globular Protein
- 6 nm microfilaments
- 10 nm thickness cell membranes
- 11 nm Ribosome
- 25 nm Microtubule
- 50 nm Nuclear pore
- 100 nm Large Virus
- 200 nm Centriole
- 200 nm (200 to 500 nm) Lysosomes
- 200 nm (200 to 500 nm) Peroxisomes
- 1 um (micrometer)
- (1 - 10 um) the general sizes for Prokaryotes
- 1 um Diameter of human nerve cell process
- 2 um E.coli - a bacterium
- 3 um Mitochondrion
- 5 um length of chloroplast
- 6 um (3 - 10 micrometers) the Nucleus
- 9 um Human red blood cell
- 10 um
- (10 - 30 um) Most Eukaryotic animal cells
- 90 um Amoeba
- 100 um Human Egg
- 1 mm (1 millimeter, 1/10th of a centimeter)
- 1 mm Diameter of the squid giant nerve cell
- 2 mm Diameter of a frog egg

DIFFERENCE BETWEEN ELEMENTS/COMPOUNDS

The various elements that make up the cell are:

- 59 per cent Hydrogen (H)
- 24 per cent Oxygen (O)
- 11 per cent Carbon (C)
- 4 per cent Nitrogen (N)
- 2 per cent Others - Phosphorus (P), Sulphur (S), etc.

The difference between these elements is their respective weights, electrons and in general their properties. A given element can only have so many other atoms attached. For instance carbon (C) had 4 electrons in its outer shell and thus can only bind 4 atoms, Hydrogen only has 1 electron and thus can only bind to one other atom. An example would be Methane which is CH_4. Oxygen only has 2 free electrons, but will some times form a double bond, which is a 'ester' (which typically smell good or bad).

Methane	**Water**	**Methanol (Methyl Alcohol)**
H | H-C-H | H	H H \ / O	H | H-C-O-H | H

As far as molecules that make up the cell:

- 50 per cent Protein
- 15 per cent Nucleic acid
- 15 per cent Carbohydrates
- 10 per cent Lipids
- 10 per cent Other

Here is a list of Elements, symbols, weights and biological roles.

Element	(Symbol)	Atomic Weight	Biological Role
Calcium	(Ca)	40.1	Bone; muscle contraction
Carbon	(C)	12.0	Constituent(backbone) of organic molecules
Chlorine	(Cl)	35.5	Digestion and photosynthesis
Copper	(Cu)	63.5	Part of Oxygen-carrying pigment of mollusk blood.
Fluorine	(F)	19.0	For normal tooth enamel development

Hydrogen	(H)	1.0	Part of water and all organic molecules
Iodine	(I)	126.9	Part of thyroxine (a hormone)
Iron	(Fe)	55.8	Hemoglobin, oxygen caring pigment of many animals
Magnesium	(Mg)	24.3	Part of chlorophyll, the photo-synthetic pigment; essential to some enzymes.
Manganese	(Mn)	54.9	Essential to some enzyme actions.
Nitrogen	(N)	14.0	Constituent of all proteins and nucleic acids.
Oxygen	(O)	16.0	Respiration; part of water; and in nearly all organic molecules.
Phosphorus	(P)	31.0	High energy bond in ATP.
Potassium	(K)	39.1	Generation of nerve impulses.
Selenium	(Se)	79.0	For the working of many enzymes.
Silicon	(Si)	28.1	Diatom shells; grass leaves.
Sodium	(Na)	23.0	Part of Salt; nerve conduction
Sulfur	(S)	32.1	Constituent of most proteins. Important in protein structure: Sulfide bonds are strong.
Zinc	(Zn)	65.4	Essential to alcohol oxidizing enzyme.

WHAT IS LIVING

This is a topic that is been of many long discussions and it depends on your initial definitions.

Some definitions are:

- The quality that distinguishes a vital and functional being from a dead body or purely chemical matter.
- The state of a material complex or individual characterized by the capacity to perform certain functional activities including metabolism, growth, and reproduction.
- The sequence of physical and mental experiences that make up the existence of an individual.

Under these varying definitions life may or may not include a virus that is only 'alive' if it can insert its genetic material into a living cell. To me live is the substance that can react to its environment, grow, improve and reproduce. To have less of a definition would include to much to have more would not include some cells.

INTERESTING ABOUT CELL BIOLOGY

What makes cell biology particularly interesting is that there is so much that is not understood. Cells are a complex system in and of themselves. And when you add to a individual cell its environment, whether that is the single celled organism or multicellular, there is a complex web reactions. One organism, like the human, can have the same genetic material in every cell, yet, there are over 200 types of cells in the human, that are different shapes, sizes and and carry out very different functions. And ALL of these cells were developed from 1 (one) cell.

- Complexity
- Inter-relations of cells
- Intra-relations of cells
- The cell and its environment.
- Its ability to Live and reproduce.
- Its ability to grow and change.
- It is what makes up you and the food you eat.

TYPES OF CELLS

The Major differences between Prokaryotic and Eukarotic cells are that prokaryotes don't have a nucleus and rarely have membrane bound organelles, (the only exception I have heard of is bacteria with vacuoles). The both do have DNA for genetic material, have a exterior membrane, have ribosomes, accomplish similar functions, and are very diverse. For instance, there are over 200 types of cells in the human body, that very greatly in size, shape, and function.

PROKARYOTES

- Prokaryotes are cells without a nucleus. They have genetic materials but are not enclosed within a membrane.These include bacteria and cyanophytes. The genetic material is a single circular DNA and is contained in the cytoplasm, since there is no nucleus. Recombination happens through transfers of plasmids (short circles of DNA that pass from one bacterium to another). They do not engulf solids nor do they have centrioles or asters. There are pictures of 2 prokaryotes below. Prokaryotes have a cell wall made up of peptidoglycan.

EUKARYOTES

- These are cells with a nucleus, this is where the genetic material is surrounded by a membrane much like the cells membrane. Eucaryotic cells are found in humans and other multicellular organisms (plants and animals) also algae, protazoa. They have both a cellular membrane and a nuclear membrane, also the genetic material forms multiple chromosomes, that is linear and complexed with proteins that help it 'pack' and is involved in regulation.

Eukaryotes are composed of both plant and animal cells. Plants vary from animal cells in that they have large vacuoles, cell wall, chloroplasts, and a lack of lysosomes, centrioles, pseudopods, and flagella or cilia. Animal cells do not have the chloroplasts, and may or may not have cilia, pseudopods or flagella, depending on the type of cell.

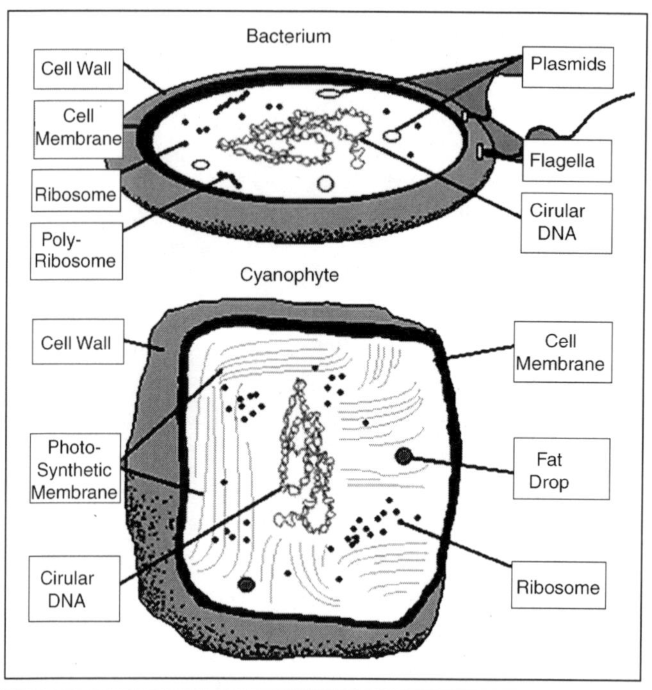

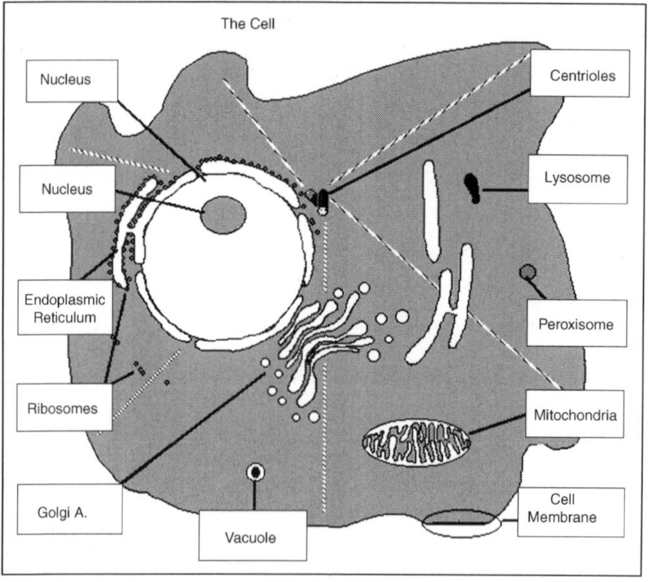

PARTS OF THE CELL

Some interesting things are, what are the organelles with DOUBLE lipid bilayer membranes?

- Nucleus
- Mitochondria
- Chloroplasts

CELL MEMBRANES

Phospholipids

Phospholipids are made up of a hydrophilic head and a hydrophobic tail. The head group has a 'special' region that changes between various phospholipids. This head group will differ between cell membranes [types of cells] or different concentrations of specific 'head groups'. The fatty acid tails call also differ, but there is always one saturated and one unsaturated 'leg' of the tail.

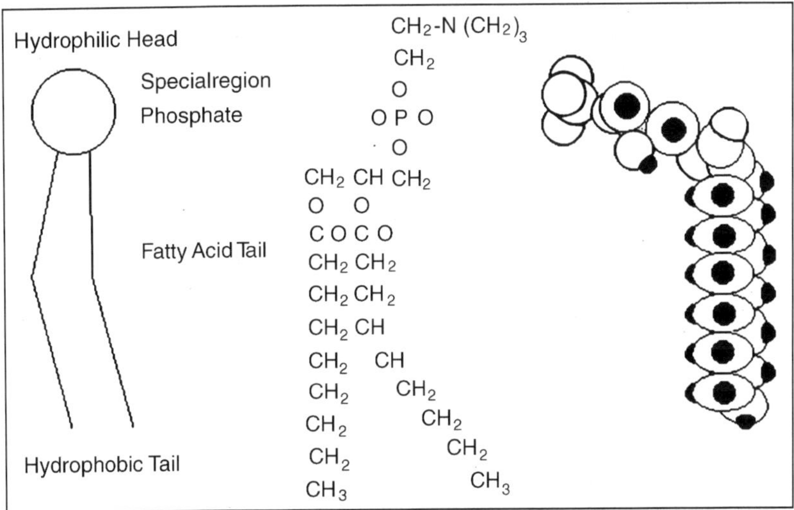

Phospholipids are 2 fatty acids one saturated and one unsaturated (shown by the double bond) that are linked to a glycerol.

Cholesterol

I have symbolized cholesterol as:

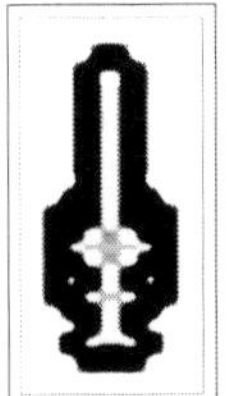

Cholesterol is a major component of cell membranes and serves many other functions as well. Cholesterol helps to 'pack' phospholipids in the membranes, thus giving more rigidity to the membranes. Also cholesterol serves diverse functions such as: it is converted to vitamin D (if irradiated with Ultra Violet light, modified to form steroid hormones, and is modified to bile acids to digest fats.

Semi-Permiable/Osmosis

The membranes of cells are a fluid, they are semi-permeable, which means some things can pass through the membrane through osmosis or diffusion. The rate of diffusion will vary depending on the its: size, polarity, charge and concentration on the inside of the membrane versus the concentration on the outside of the membrane. When something is permeable it means that something can spread throughout, like (*The perfume is permeating the room.*). Here is a list of some molecules and how they relate to passing through the membrane without assistance, in other words, through osmosis.

Hydrophobic Molecules

- O_2 - Oxygen
- N_2 - Nitrogen
- Benzene

Small Uncharged Polar Molecules

- H_2O - Water
- Urea
- Glycerol
- CO_2 - Carbon Dioxide

Large Uncharged Polar Molecules

- Glucose
- Sucrose

Ions:

- H^+ - Hydrogen ion
- Na^+ - Sodium ion
- K^+ - Potassium ion
- Ca^{2+} - Calcium ion
- Cl^- - Chloride ion

Various substances will pass through the membranes at varying rates through osmosis.

Membrane Proteins

One role of proteins in cells is for transport of molecules/ions into or out of cells. Three methods of doing this are through active, facilitated or passive transport. Other roles are in cell recognition, receptors, cell to cell communication. There is more information on membrane proteins and other proteins in later sections.

Hydrophobic/Hydrophilic

A very simplistic idea of what these characteristics are is: Hydrophilic and hydrophobic are, respectively, the like and dislike. Hydrophilic areas of a phospholipid, or a protein are 'attracted' to water, and hydrophobic regions are repelled by water.

Self-Assembly

Self-assembly occurs due to the thermodynamics, if the phospholipids are in a water (or other polar solution) the tails will want to be 'away' from the solution. The could all go to the top (like oil on water), or they could have the tails point towards each other. With the tails pointing towards each other, this could form 2 different formations. First would be a micelle which would like like a ball with the phospholipid heads on the outside and the tails pointing together like this or in the form of a lipid bilayer:

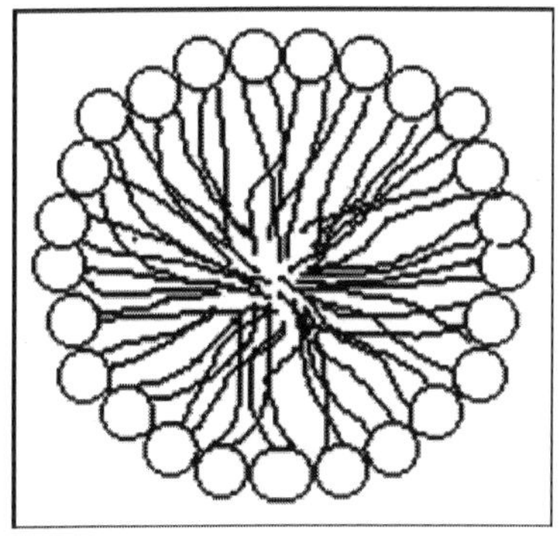

UNITS OF CELL

Now that you have had an introduction to the chemicals that make up the typical cell, we will now look at how this chemistry combines to form major functional units. These can be thought of as the organizations that carry out the major business of the cell: growth, replication, feeding and movement. We will first start with membranes because so many things interact with them. Next internal structures in the cytoplasm will be described and finally structures outside of the membrane. First, however, we should describe an important evolutionary hypothesis that will make sense of much of the following details. As you will see, there are a number of curious similarities and differences in the details of cellular structure among bacteria, archaea and eukaryotes.

In general much of the machinery in a eukaryotic nucleus and in the cytoplasm looks rather a lot like what is present in the archaea. However, the organelles of eukaryotes, such as the mitochondria and chloroplasts, have properties that are much more similar to those of bacteria. How is this possible? One clue comes from observing organisms in nature. It is very common to find cooperative relationships between different species and this is also true in the microbial world. In some instances these relationships involved close physical contact between their participants, sometimes with one participant engulfing the other. In 1968 Dr. Lynn Margulis extended this observation and proposed that some of the organelles found in eukaryotes, specifically mitochondria and chloroplasts, were originally endosymbionts of their host. Originally these two microbes probably were able to live independently, but over time, the endosymbiont lost functionality that its host was already providing and then became dependent.

Over the years ample evidence has accumulated to support this exceptional insight:

- Both mitochondria and chloroplasts contain DNA that resembles the chromosomes of bacteria.
- Both organelles are surrounded by two membranes reminiscent of gram-negative cell wall structure observed in a class of microbes.
- Mitochondria and chloroplasts divide by a method that resembles binary fission.
- Much of the internal structure and biochemistry of the photosynthetic organelle inside chloroplasts is very similar to that observed in cyanobacteria (a photosynthetic microbe).
- The ribosomes of mitochondria and chloroplasts resemble those found in microbes and analysis of the sequence of the 16S rRNA of these ribosomes showed that the organelles are in fact closely related to proteobacteria (mitochondria) and cyanobacteria (chloroplasts) This last bit of proof is very strongest evidence substantiating Dr. Margulis's hypothesis.

In summary, it seems clear that the eukaryotic cell was born by the merger of an archaeal cell with a gram-negative proteobacteria. Photosynthetic eukaryotes arose from a second endosymbiosis, where the eukaryotic cell engulfed a cyanobacterium. Clearly, eukaryotes, including us, were the result of the cooperation of several bacterial species in the long distant past.

Membranes Surround the Cell and Hold it in

- Membranes are thin, highly conserved structures found in all cells.
- Membranes are stabilized by hydrophobic interactions, hydrogen bonds and ionic interactions.
- Membranes are composed of lipids and proteins. The lipids form a bilayer, with the hydrophilic portion of the lipids facing the aqueous environment, and the hydrophobic portions clustering together inside

the membrane. The majority of membrane proteins are involved in moving small molecules across the membrane or in energy generation.

General Properties

The cytoplasmic membrane immediately surrounds the inside of the cell and is perhaps the most conserved structure in living cells. Membranes are thin structures, measuring about 8 nm thick and every living thing on this planet has some type of membrane They are the major barrier separating the inside of the cell from the outside and allow cells to selectively interact with their environment.

Membranes are highly organized and asymmetric. This asymmetry comes from the fact that the membrane that faces the environment performs very different functions than does the side that faces the cytoplasm. Membranes are also dynamic, constantly adapting to changing environmental conditions.

Physical Structure

Membranes are composed of lipids and proteins. The majority of lipids are phospholipids as described earlier, but about 50 per cent of all know bacterial species also contain hopanoids as shown in Figure. These molecules have a similar structure to sterols found in eukaryotic membranes and serve to help stabilize the membrane. Proteins are more numerous in bacterial membranes than in eukaryotic membranes. This is because bacteria in general only contain a single membrane in contact with the cytoplasm and this has to carry out all the functions of the cell. In eukaryotes these functions are divided amongst the cytoplasmic membrane and the other organelles. The chemical structure and space-filling model of a hopanoid, which is found in many different bacterial membranes.

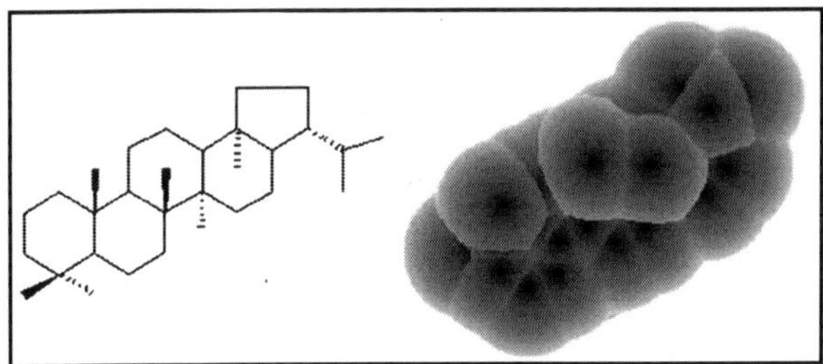

Fig. A hopanoid

Much of the general behaviour of membranes is dictated by the behaviour of lipids in water. Because phospholipids are amphipathic, they tend to congregate when placed in an aqueous environment. This is done in a very specific fashion such that the hydrophilic portions face the water and the hydrophobic portions are buried inside. Under the cell's direction lipids are organized into a bilayer, where there are two sheets of lipids oriented so that the hydrophobic faces of each sheet face each other as shown in Figure.

Lipid bilayers can be almost any size and can form vesicles spontaneously, if lipids are placed in an aqueous environment. In the cell, however, the synthesis of membranes is performed by specific enzymes and is tightly controlled.

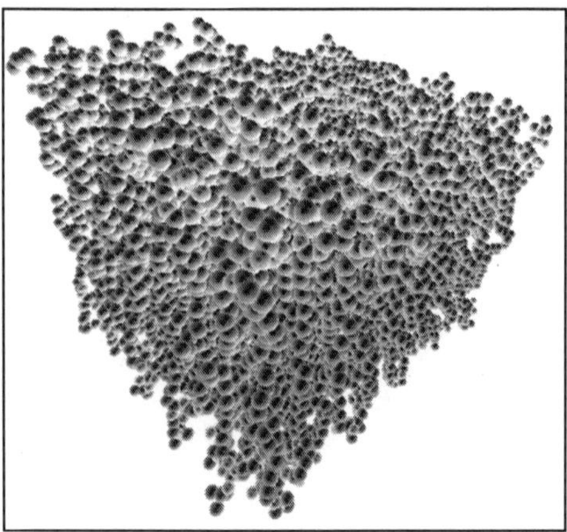

Fig. A Model of a Lipid Bilayer

The cytoplasmic membrane is held together by a number of forces. Hydrophobic interactions between the alkyl chains of neighbouring lipids are a major component of membrane stability. Hydrogen bonds between lipids and between membrane proteins and lipids also hold a membrane together. Further stability comes from negative charges on proteins that form ionic interactions with divalent cations such as $Mg+^2$ and $Ca+^2$ and the hydrophilic head of lipids. The hydrophobic region of the membrane provides a critical function: it prevents polar compounds, such as ions and most biological molecules, from passing through it. This allows the cell to create and maintain gradients of ions and small molecules across the membrane by mechanisms described below. So how do polar molecules ever cross this membrane, since this represents an important interaction between the cell and its environment? This transfer across the membrane comes about through the specific functioning of proteins that are imbedded in the membrane. Some proteins span the membrane while others are exposed on the outside or the inside. These proteins may move within the plane of the membrane or they may be anchored to structures in or near the membrane.

Many of the membrane-spanning proteins are involved in transport of the polar molecules that must pass through the membranes. A subset of these proteins are also involved in energy generation as discussed below. The membrane is fluid and has the consistency of a light grade oil. It has been termed a fluid mosaic: "fluid" because the lipids are free to move about on each side of the membrane and "mosaic" because there is a definite pattern to it. Lipids do not generally switch from outside to inside or vice versa, because of the problem of trying to move the hydrophilic group through the hydrophobic core of the bilayer.

Membranes are a Selective Barrier

- Membranes are semi-permeable structures that prevent most common biological compounds from moving across them.
- Biological compounds can be moved across the membrane by facilitated diffusion, group translo-cation, and active transport. All of these functions are mediated by proteins in the membrane.

The concentration of solutes, sugars, and most ions is generally much higher within the cell than outside. A fundamental principle of nature is that different concentrations of a given solute tend to equilibrate across the boundary due to diffusion. However, the cell boundary is the membrane and its hydrophobic core prevents this diffusion for polar molecules. Compounds such as amino acids, organic acids and inorganic salts must therefore be specifically transported across the membrane by proteins and once inside these molecules cannot escape. The cell can therefore control the nature and amount of these compounds that enter or leave the cell.

Though hydrophilic, water is not very polar and can flow freely across the membrane, as can some small non-polar molecules. This creates a serious problem. The inside of the cell is full of many types of solutes: proteins, nucleic acids, other small molecules and ions. In comparison the outside environment, in most cases, is very dilute. Because of this there is a higher concentration of water outside the cell than inside the cell. Nature hates imbalances such as this and in an effort to correct the problem; water tends to flow into the cell, by a process called osmosis. Osmosis causes a high pressure against the cell membrane.

This pressure would rapidly cause lysis of most cells and one of the major purposes of the peptidoglycan of the cell wall is to prevent the cell membrane from bursting. For molecules that are soluble in both the lipid membrane and the surrounding aqueous environment, the law of simple diffusion directs transport. The membrane is not a barrier for such molecules. These types of molecules are uncommon since solubility in both a hydrophobic and a hydrophilic environment is unusual. There is no transport protein for such compounds, so there is no specificity of control or energy cost. The cell cannot create a concentration gradient of these molecules. One important example is water. Water can pass freely into and out of cells.

There are three basic types of transport systems:

1. Facilitated Diffusion
2. Group Translocation
3. Active Transport

Many of the proteins in the membrane function to help carry out selective transport, particularly of polar compounds. These proteins typically span the entire membrane, making contact with the outside environment and the cytoplasm. They often require the expenditure of energy to help compounds move across the membrane, though cells can also use concentration gradients of these compounds to generate energy, as described below

Facilitated Diffusion

This process involves a protein that binds the molecule to be transported and physically moves that compound through the membrane. Binding of the molecule to the protein causes a conformational change in the protein so that the molecule now faces the opposite side from where it was.

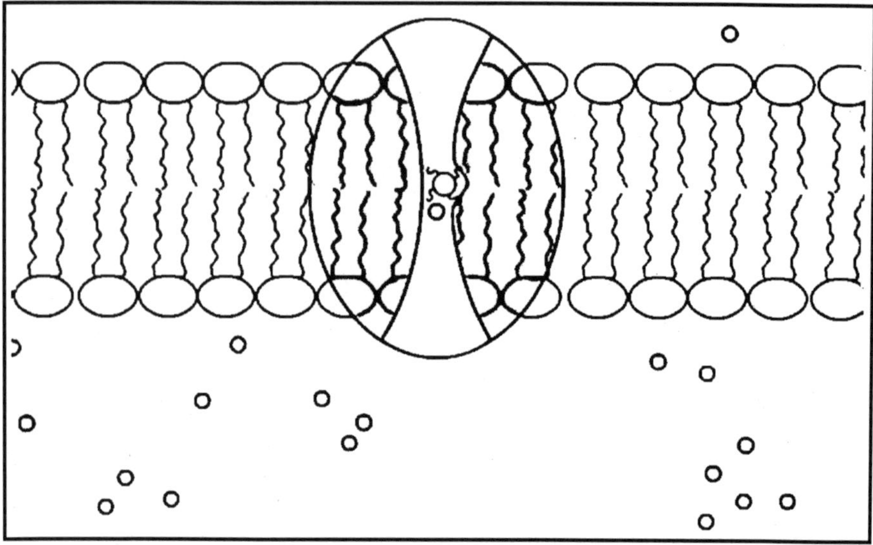

Fig. Facilitated Difusion

Facilitated diffusion, as shown in Figure, is therefore specific because a protein must bind the molecule. However, these small molecules are readily moved in and out of the cell, so a gradient cannot be formed nor is energy required. One example of a protein involved in facilitated diffusion is the glycerol facilitator protein. In E. coli this enzyme binds glycerol and a few other polyalcohols and allows their diffusion into the cell. Once inside, the glycerol is immediately phosphorylated, preventing its diffusion back outside the cell. Figure of the migration of solutes in and out of the cell as facilitated by a protein. Notice that this mechanism does not lead to a solute concentration inside the cell that is higher than outside. Rather, it leads to an equilibrium of that solute across the gradient.

Group Translocation

In this process, a protein specifically binds the target molecule and transports it inside the cell while simultaneously modifying it chemically. Most group translocations require energy and tend to be unidirectional, unlike facilitated diffusion.

The substrates of catabolic pathways, such as sugars, are sometimes transported by group translocation. This is an efficient way to both bring substrate into the cell and begin the breakdown process. Figure shows an of group translocation.

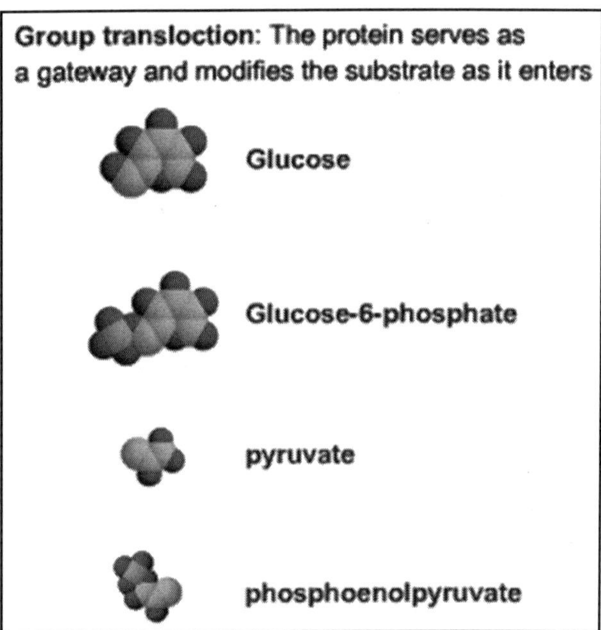

The glucose molecule that is being transported into the cell is modified by the addition of a phosphate from phosphoenolpyruvate to form glucose-6-phosphate.

Active Transport

In active transport, energy is expended to transport the small molecules, but they are not chemically altered. The process is efficient enough to cause the internal concentration in the cell to reach many times its external concentration. Active transport proteins are molecular pumps that expend energy to pump their substrates against a concentration gradient.

This energy comes in two forms: ATP and ion gradients (both ATP and ion gradients are made by central metabolism and we will cover their formation in the chapters on metabolism). In ATP-based active transport, ATP hydrolysis is coupled to the movement of the small molecule across the membrane.

One large group of proteins involved in this type of transport is the ATP binding cassette (ABC) transporters. ABC transporters have been found in all living species with 80 identified in the E. coli genome and 48 in the human genome. The mechanism of ABC transporters is exemplified by the maltose binding protein of E. coli.

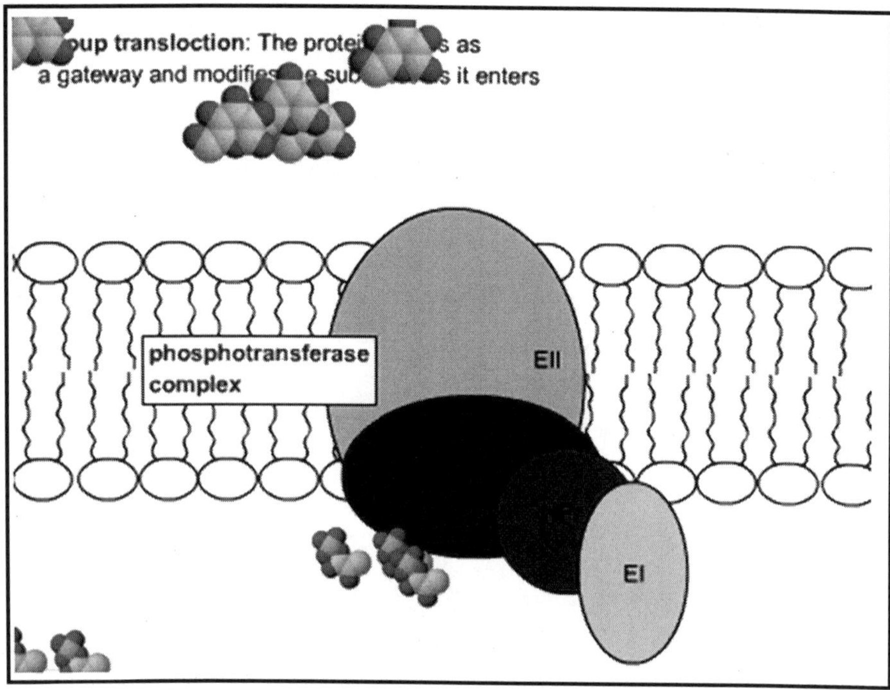

Fig. Group Translocation

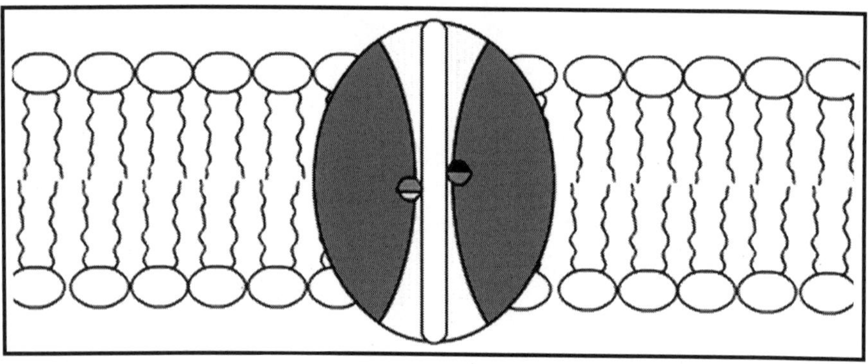

Fig. Active Transport

Three separate types of transport are shown. In the first, an antiporter moves two different small molecules across the cell membrane, but in the opposite directions. In the second, a symporter moves two or more different molecules to move into the cell simultaneously. Typically, the desired molecule is being concentrated against a gradient and that transport is driven by the transport of the other molecule, which is moving with a gradient. Finally, a uniporter binds and transports the target molecule only. Energy is required for these processes and the cell can accumulate molecules inside the cell using this mechanism.

Ion gradient active transport uses the energy of one chemical gradient, that of the specific ion, to drive the creation of a different gradient, the uptake of the small molecule. The ion gradient that supports the work is at a higher concentration on one side of the membrane than the other. The transport protein both binds its small molecule to transport and provides a gateway for this ion to fall down its concentration gradient. When the ion moves through its gateway, it causes a conformational change in the protein and this change is used to transport the target small molecule into the cell. Active transport proteins may be highly specific for only one molecule or may be able to carry a class of chemically related molecules. The in Figure shows several different types of transport molecules.

An example of a more general transport protein is the branch chain amino acid transporter of Pseudomonas aeruginosa, which transports leucine, valine, and isoleucine. Figure summarizes the various properties of transport mechanisms.

Table. Properties of Various Transport Systems

Property	Passive Diffusion	Facilitated Diffusion	Active Transport	Group Translocation
Carrier Mediated	-	+	+	+
Concentration Against Gradient	-	-	+	Not Applicable
Specificity	-	+	+	+
Energy Expended	-	-	+	+
Solute Modified During Transport	-	-	-	+

Membranes can help Generate Energy

- In many microorganisms a collection of proteins are involved in generating energy. These proteins move high energy electrons down an electron transport chain and in the process move protons from the inside to the outside of the membrane.
- This proton gradient can do work by driving other proteins that form ATP, that transport molecules across the membrane, or that move the cell.

Many cells use respiratory processes to obtain their energy. During respiration, organic or inorganic compounds that contain energy are oxidized, releasing electrons to do work. In many microorganisms these electrons find their way to the membrane where they are passed down a series of electron carriers as shown in Figure. During this operation, protons are transported outside the cell. This creates a gradient of protons across the cell membrane, energizing it, in a fashion similar to charging a battery. The energy of this gradient can then be used to do work directly, a process known as the proton motive force, or can be channeled into a special protein known as ATP synthase. ATP synthase can convert: ADP to ATP, and the ATP can itself do work. Membranes are critical in many cells for the generation of usable energy. The cartoon shows the various membrane proteins involved in converting high-energy electrons from photoreceptors into useful energy. They do so by forming a proton gradient across the cell membrane, termed a proton motive force, which is in turn used by other proteins to synthesize ATP. This will be discussed in greater detail in the chapter on metabolism.

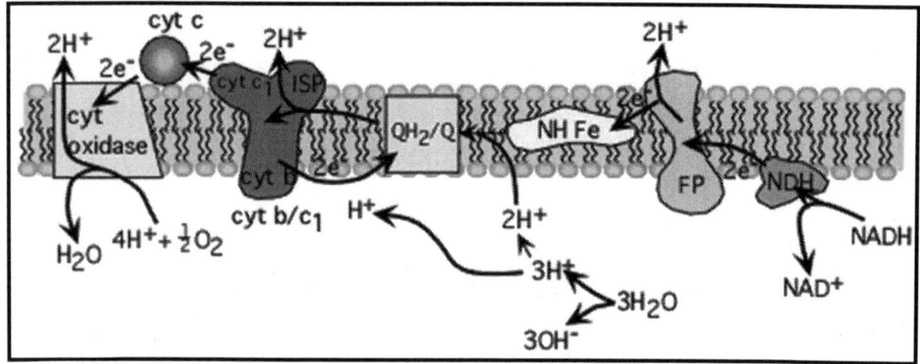

Fig. Generating the Proton Motive Force

The prokaryotic cells performing photosynthesis have membrane systems specific to that process. Light excites electrons found in pigmented proteins in the membrane and the electrons are again passed down through a series of electron carriers. As above, a proton motive force is generated and used to synthesize ATP. The specifics of these systems are discussed in the chapter on metabolism.

CHEMICAL COMPONENTS OF CELL

Biochemical analysis of whole tissues was the early approach to the study of chemical composition of the cell like as the liver, brain, skin, or plant meristem. This method had limited cytologic value, because the material analysed was generally a mixture of different cell types and, in addition, contained extra cellular material. In recent years the development of cell fractionation methods and various micro methods has led to the isolation of different subcellular particles and thus to more important and precise information about the molecular architecture of the cell.

The chemical components of the cell may be classified as *inorganic* (water and mineral ions) and *organic* (proteins carbohydrates, nucleic acids, lipids, and so forth). The protoplasm of a plant or animal cell contains 75 to 85 per cent water, 10 to 20 per cent protein 2 to 3 per cent lipid, 1 per cent carbohydrates, and 1 per cent inorganic material.

Salts and Ions

Salts dissociated into anions (*e.g.*, Cl^-) and cations (*e.g.*, Na^+ and K^+) are important in maintaining *osmotic pressure* and the *acid-base equilibrium* of the cell. Retention of ions produces an increase in osmotic pressure and thus the entrance of water. Some of the inorganic ions, such as magnesium, are indispensable as co-factors in enzymatic activities; others, such as inorganic phosphate, form adenosine triphosphate (ATP), the chief supplier of chemical energy for the living processes of the cell, through oxidative phosphorylation. The concentration of various ions in the intracellular fluid differs from that in the interstitial fluid.

For example, the cell has a high concentration of K^+ and Mg^{2+}, while Na^+ and Cl^- are localized mainly in the interstitial fluid. The dominant anion in cells is phosphate; some bicarbonate is also present. *Calcium ions* are exist in the circulating blood and in cells. In bone they combine with phosphate and carbonate ions to form a crystalline arrangement.

Phosphate occurs in the blood and tissue fluids as a free ion, but much of the phosphate of the body is bound in the form of phospholipids, nucleotides, phosphoproteins, and phosphorylated sugars. As primary phosphate (H_2PO_4) and secondary phosphate (HPO_4^{2-}), phosphate contributes to the acid-base equilibrium, thereby buffering the pH of the blood and tissue fluids. Other ions found in tissues are sulfate, carbonate, bicarbonate, magnesium, and amino acids. Certain *mineral components* are also found as part of larger macromolecules. For example, *iron,* bound by metal-carbon linkages, is found in hemoglobin, ferritin, the cytochromes, and some enzymes.

Water

Expect bone and enamel, water is the most abundant component of cells. It serves as a natural solvent for mineral ions and other substances and also as a dispersion medium of the colloid system of protoplasm. For example, from micro-injection experiments it is known that water is readily miscible with protoplasm. Furthermore, water is must require for metabolic activity, since physiologic processes occur exclusively in aqueous media. Water molecules also participate in many enzymatic reactions in the cell and can be formed as a result of metabolic processes.

In cell, water found in two forms: *free* and *bound. Free water* represents 95 per cent of the total cellular water and is the principal part used as a solvent for solutes and as a colloid dispersion medium. *Bound water,* which represents only 4 to 5 per cent of the total cellular water, is loosely held to the proteins by hydrogen bonds and other forces. Because of the asymmetrical distribution of charges a water molecule acts as a *dipole.* Because of this polarity, water can bind electro statically to both positively and negatively charged groups in the protein.

Thus, each amino group in a protein molecule is capable of binding 2.6 molecules of water. Water can also be used to instance eliminate substances from the cell and to absorb heat—by virtue of its high specific heat coefficient—thus preventing drastic temperature changes in the cell. The water content of an organism is related to the organism's age and metabolic activity. For instance, it is highest in the embryo (90 to 95 per cent) and decreases progressively in the adult and in the aged.

Macromolecules

Structural and other properties of the cell are intimately related to large molecules made of repeating units linked by covalent bonds. Such units are known as monomers and outcome macromolecule is called a *polymer.* Molecules having an increasing number of monomers possess widely different characteristics.

For example, among the hydrocarbons, methane and ethane are gases, while butane and octane are liquids; further polymerization (20 or more monomers) produces oils, and finally solids, such as paraffins.

The three basic examples of polymers in living organisms are as follows:

1. *Nucleic acids* result from the repetition of four different units called *nucleotides.* The repetition of the four nucleotides in the DNA molecule is the primary source of genetic information.
2. *Polysaccharides* can be polymers of monosaccharides, forming starch, cellulose, or glycogen, or may also involve the repetition of other molecules, forming more complex polysaccharides.
3. *Proteins* and *polypeptides* consist of the association in various proportions of some 20 different amino acids linked by peptide bonds. The order in which these 20 monomers can be linked gives rise to an astounding number of combinations in various protein molecules. This can determine not only their specificity, but in certain cases, their biological activity.

Proteins

The Chains of Amino Acids Linked by Peptide are called Proteins

The building blocks of proteins are the amino acids. Essentially, an amino acid is an organic acid in which the carbon next to the—COOH group (called an alpha carbon) is bound to an amino group (—NH_2). In addition, the alpha carbon is bound to a side-chain (R) which is different in the various amino acids.

The amino acids differ from one another only in the side-chain; for example, the R in alanine has one carbon, while in leucine it has four carbons. The properties of the various amino acids depend on the chemical composition, of their side-chains; for example, lysine and arginine are basic because their sidechains contain an extra amino group, and the acidic amino acids (glutamic and aspartic acids) contain an extra carboxyl group. Because of the simultaneous presence of acidic (carboxyl) and basic (amino) groups, amino acids can have both positive and negative charges and are therefore *amphoteric molecules* or zwitterions.

The ionized form of an amino acid is:

$$\begin{array}{c} \text{H} \\ | \\ {}^{+}\text{H}_3\text{N}-\text{C}-\text{COO}^{-} \\ | \\ \text{R} \end{array}$$

The condensation of amino acids to form a protein molecule occurs in such a manner that the acidic group of one amino acid combines with the basic group of the adjoining one, with the simultaneous loss of one molecule of water. The linkage—NH—CO—is known as the *peptide linkage* or *peptide bond.*

The formed molecule preserves its amphoreric character, since an acidic group is always at one end and a basic group is at the other, in addition to the lateral residues (radicals) that can be either basic or acidic. A combination of two amino acids is *dipeptide;* of three, a *tripeptide.* When a few amino acids are linked together, the structure is an *oligopeptide.* A *polypeptide* contains a large number of amino acids. The distance between two peptide links is about 0.35 nm. A protein with a molecular weight of 30,000 consisting of 300 amino acid residues, if fully extended, should have a length of 100 nm, a width of 1.0 nm, and a thickness of 0.46 nm.

The term *protein* means that all basic functions in living organisms depend on specific proteins. They constitute the enzymes and the contractile machinery of the cell, and are present in the blood, and other intercellular fluids. Some long-chain proteins, such as *collagen* and *elastin* play an important role in the organization of the extracellular framework of tissues.

Student may refer the biochemistry book for detailed study of the classification of Proteins however, it is important to stress that the properties of proteins vary considerably. For instance, *keratin* and *collagen,* are insoluble and fibrous; the *globular proteins e.g.*, egg albumin and serum proteins, are soluble in water or salt solutions and are spherical rather than threadlike molecules.

The *conjugated proteins* are attached to a nonprotein moiety, the so-called *prosthetic* group. To such a group belong the *nucleoproteins* associated with nucleic acids, the *glycoproteins,* the *lipoproteins* (*e.g.*, blood lipoproteins), and the *chromoproteins,* which have a pigment as the prosthetic group, such as hemoglobin, haemocyanin, and the cytochromes. Hemoglobin and myoglobin (present in muscle) contain the prosthetic group *heme,* an iron-containing organic compound that combines with oxygen.

Primary Structure

The amino acid sequence of the polypeptide chain is known as the *primary structure* of the protein molecule. It is the most important and specific structure and determines the so-called *secondary* and *tertiary* structures. Groups of protein subunits containing secondary and tertiary structures constitute the *quaternary* structure; many proteins are formed by several polypeptide subunits. In the protein molecule, amino acids are arranged like beads on a string, and their sequence is of great biological importance. In the hemoglobin molecule a change in a single amino acid produces profound biological changes.

Secondary Structure

In a protein made by several hundred amino acids, the chain may sometimes be linear, but more frequently it assumes different shapes that constitute the secondary *structure.* Fibrous proteins are often arranged in an orderly manner that can be analysed by x-ray diffraction methods. This technique has facilitated classification of proteins into three structural types or groups.

The β-*keratin type* has an identity period of about 0.72 nm, and the adjacent chains are disposed in a *pleated sheet structure.* In which the side-chains of the amino acid residues stick out perpendicular to the plane of the chain. By hydrogen bond the individual chains are chains are held together which formed peptide grid

The α-*helix structure* found in the *a-keratin type* is produced when the polypeptide chain forms a helical structure, like a spiral winding around an imaginary cylinder, in such a way that hydrogen bonds are established within the molecule and not with an adjacent molecule. For the *collagen group,* a model made of three helical chains has been proposed.

Tertiary Structure

In the d globular *proteins* the polypeptide chain is held together in a definite way to form a compact structure in three dimensions. The arrangement in space of such chains is very complex but may be resolved by X-ray diffraction. In globular proteins the chains are folded in a compact way with the polar groups towards the surface thus leaving little space in the interior for water molecules (the hydrophobic amino acids are in the centre of the molecule).

These proteins have regions of a -helix or b configurations, and other parts exist as a *random coil, i.e.*, in a flexible structure that, may change at random. The tertiary structure is given by the way in which these helical or random-coil segments arrange with respect to each other. The spatial arrangement is predetermined by the sequence of amino acids in the primary structure and by the bonds that can be established among some of the residues. Enzyme activity and antigenicity *i.e.*, series of biological properties are related to the tertiary structure. The *denaturation* of a protein is brought about by high temperatures or other nonphysiologic conditions, and consists of a disruption of the tertiary structure. This is usually accompanied by loss of biological activity. In the case of the enzyme ribonuclease, denaturation can be achieved by treatment with b-mercaptoethanol and high concentrations of urea. Mercaptoethanol is a reducing agent, which can disrupt—S—S—bridges (disulfide bonds), reducing them to —SH groups; while urea disrupts weak interactions (see later discussion). The tertiary structure of ribonuclease is held together by four disulfide bonds, which are established between pairs of cysteines (an amino acid that contains an —SH group).

After denaturation, the enzyme can be correctly refolded into its natural conformation *(renaturation) by* removing the urea and mercaptoethanol gradually by dialysis; after this, the enzymatic activity is recovered. There are 105 possible combinations in which eight cysteines can pair to produce four disulfide bridges, but only the biologically active natural conformation is produced after careful renaturation, because it is thermodynamically the most stable structure. This is a clear demonstration that all the information needed to produce the complex folding of a protein molecule is contained in its amino acid sequence.

Quaternary Structure

Protein Subunits: The primary, secondary, or tertiary structures, are concern a single polypeptide chain but quaternary structure contain two or more chains These chains may or may not be identical, but in both cases they are linked by weak bonds. For example, the hemoglobin molecule is composed of four polypeptides or subunits, two designated as a and two as b. Separation and association of the subunits may occur spontaneously. Hemoglobin may be broken into two half molecules (two a and two b) by urea.

When urea is removed, they reassemble, forming complete, functional molecules. This binding is highly specific and takes place only between the half molecules. This is called the *principle of selfassembly.* This principle also applies to the building up of more complex cellular structures, such as the cell membrane, microtubules, and so forth.

Weak Interactions-Essential for Protein Structure

In the structure of protein many types of bonds are involved The primary structure is fully determined by *covalent bonds* (peptide bonds). *Disulfide bonds* (—S—S—bridges) are also covalent and are established between the —SH groups of two cysteine residues. Disulfide bonds can be reversibly dissociated by the action of *reducing* agents (such as mercaptoethanol and dithiothreitol-DTT). In addition, several *weak interactions* are very important in the establishment of the secondary and tertiary structures. All these weak bonds are non-covalent; the main types are: *Ionic or electrostatic bonds,* which result from the attractive force between ionized groups of opposite charge.

Hydrogen bonds result from a H^+ (proton) that is shared between two neighbouring electronegative atoms. The H^+ can be shared between nitrogen or oxygen atoms which are close to each other.

Hydrogen bonds' are very important in biology and are the main force that holds the two strands of DNA together. Hydrogen bonds are essential for the specific pairing between nucleic acid bases, which is the basis for the coding of genetic information.

In *hydrophobic interactions* water tends to be excluded by non-polar groups, which associate with each other so that they are not in contact with water. In globular proteins the sidechains of the most hydrophobic amino acids (those with long non-polar side-chains, such as leucine, isoleucine, valine, and phenylalanine) tend to aggregate inside the molecule, and the polar (charged) groups protrude from the surface of the tertiary structure. The hydrophobic residues tend to repel the water molecules that surround the protein, thereby causing the globular structure to be more compact.

Van der Waals interactions occur only when two atoms come Very close together. The closeness of two molecules can induce charge fluctuations which may produce dipoles and mutual attraction at very short range. The essential difference between a covalent and a non-covalent bond is in the amount of energy needed to break the bond d.

Although each individual bond is weak, large numbers of them can produce very stable structures, as in the case of doublestranded DNA. A more detailed discussion will show that covalent bonds are generally broken by the intervention of enzymes, whereas non-covalent bonds are easily dissociated by physicochemical forces.

Electric Charges of Proteins

In addition to the terminal —NH_3^+ and—COO charged groups, proteins contain dicarboxylic- and diamino-amino acids which dissociate as follows:

1. The acidic groups lose protons and become negatively charged. For example, in aspartic and glutamic acids, the free carboxyl group dissociates into—COO + H^+
2. The basic groups, by gaining protons, become positively charged—NH_2 + H^+ —NH_3^+. This is found in amino acids with two basic groups, such as lysine or arginine. AH these so-called *ion-producing groups* contribute to the acid-base reactions of proteins and to the electrical properties of protein molecules.

The actual charge of a protein molecule is the result of the sum of all single charges. Because dissociation of the different acidic and basic groups takes place at different hydrogen ion concentrations of the medium, pH greatly influences the total charge of the molecule. In an acid medium, amino groups capture hydrogen ions and react as bases,

$$(—NH_2 + H^+ \rightarrow —NH_3^+);$$

in an alkaline medium the reverse takes place and carboxylic groups dissociate

$$(— COOH \rightarrow COO^- + H^-).$$

For every protein there is a definite pH at which the sum of positive and negative charges is zero. This pH is called the *isoelectric point* (pI) At the isoelectric point, proteins placed in an electric field do not migrate to either of the poles, whereas at a lower pH they migrate to the negative pole *(cathode)* and at a higher pH, to the positive pole *(anode).* This migration is called *electrophoresis,* and it provides a very useful technique for the separation of cellular proteins.

Separation of Cell Proteins

Each protein has a characteristic isoelectric point, and this property can be used in the separation of proteins. In the technique called *isoelectric focusing,* proteins are subjected to electrophoresis on a pH gradient. Each protein moves until it reaches a pH equal to its individual isoelectric point. At that moment, migration in the electric field stops because the net charge of the protein is zero. Recently the techniques of isoelectric focusing and polyacrylamide gel electrophoresis have been combined to produce *two dimensional* separation of proteins. Several hundred cellular proteins can be resolved from one another.

This technique is increasingly used in Cell Biology, and its great resolving power is due to the use of two *independent properties* of proteins. The proteins are first separated by isoelectric focusing (this is the first dimension), which separates proteins according to their *charge* (isoelectric point). The proteins are subsequently separated by electrophoresis (this is the second dimension) in polyacrylamide gels containing SDS, which separates proteins according to their *size* (molecular weight). This technique results in a series of spots distributed throughout the polyacrylamide gel (if the same property of proteins had been used in both dimensions, the spots would be distributed along a diagonal).

When the detergent sodium dodecyl sulfate (SDS) is used with electrophoresis, the proteins are separated mainly according to their molecular weight. This is because SDS binds to the proteins, giving them large numbers of negative charges due to the sulfate. Thus, most of the protein charges will come from the SDS, minimizing the role of charge differences between individual proteins (differences which would otherwise affect electrophoretic mobility), and all the proteins migrate according to their size. The larger proteins move more slowly than the smaller ones because they encounter more resistance when traversing the molecular pores within the polyacrylamide gel used for electrophoresis. SDS electrophoresis is widely used as a method for determining molecular weights of proteins.

THE CELL WALL SURROUNDS AND HOLDS IN THE MICROBE

- The cell wall in bacteria contains peptidoglycan, a polymer of N-acetyl glucosamine, N-acetyl muramic acid and amino acids.
- Gram-positive cell walls contain a thick layer of peptidoglycan that encircles the cell.
- Gram-negative cell walls contain a thin layer of peptidoglycan between the cytoplasmic membrane and the outer membrane.

This section will restrict itself to the bacterial cell wall, but at the end of the chapter we will compare this to archaeal cell walls. The cell wall is essential to the survival of most microorganisms. Many microbes live in environments in relatively dilute environments and the cell wall's most important function is to prevent the cell from bursting due to the osmotic stress placed upon it as discussed previously. The cell wall also determines the shape of the cell. Any cell that has lost its cell wall, either artificially or naturally, becomes roughly spherical and lyses due to osmotic pressure, unless placed in certain concentrated solutions. Finally, the cell wall helps to support any structure that penetrates from the cell out into the environment.

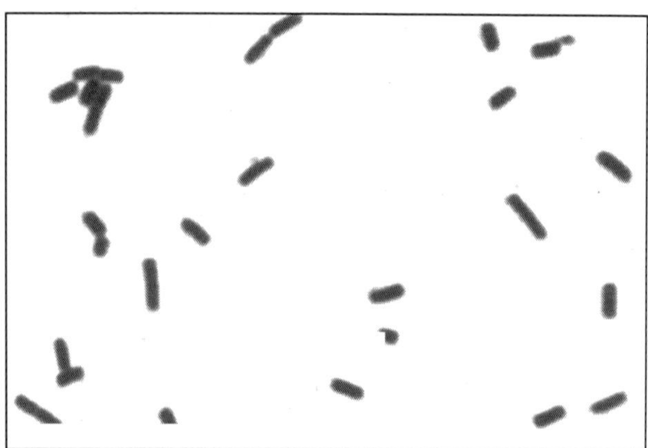

Fig. A Gram-Positive Bacterium

A Gram stain of the gram-positive bacterium Bacillus cereus The structure and synthesis of prokaryotic cell walls is unique and many compounds found in the bacterial cell wall are found nowhere else in nature. It is true that plants also make cell walls, but they are chemically and structurally different. There are two basic types of bacterial cell wall structures that have been studied in detail: gram-positive and gram-negative. These two classes of bacterial cells look very different following staining with the Gram stain and this has been a standard basis for starting to identify different bacterial species. Figures show

Gram stains of gram-positive and gram-negative bacteria, respectively. A Gram stain of the gram-negative bacterium Serratia marcescens. When the Gram stain was developed by Hans Christian Gram in 1884 the molecular basis of the stain was unknown. In fact very little was understood about bacteria in general.

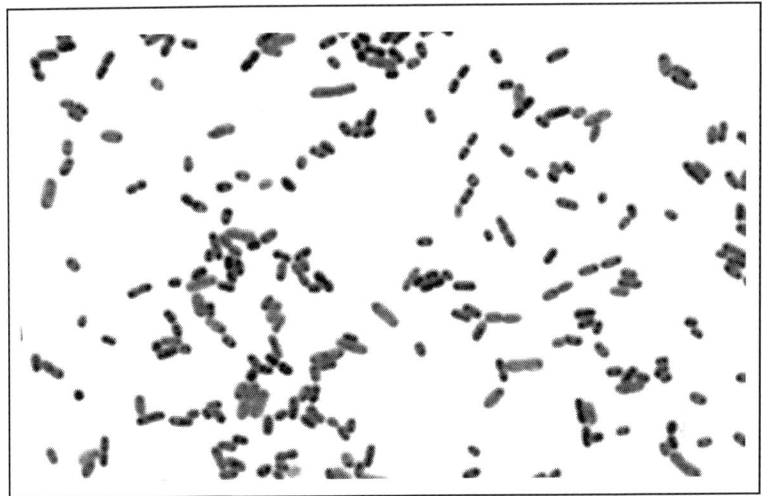

Fig. A Gram-Negative Bacterium

He just determined empirically that when bacterial smears were run through a four-step staining procedure using two different dyes, some cells retained the first dye and stained purple, while other only retained the second dye and stained pink. Years later it was discovered that the basis for this differential reaction relates to the cell wall as shown in Figure.

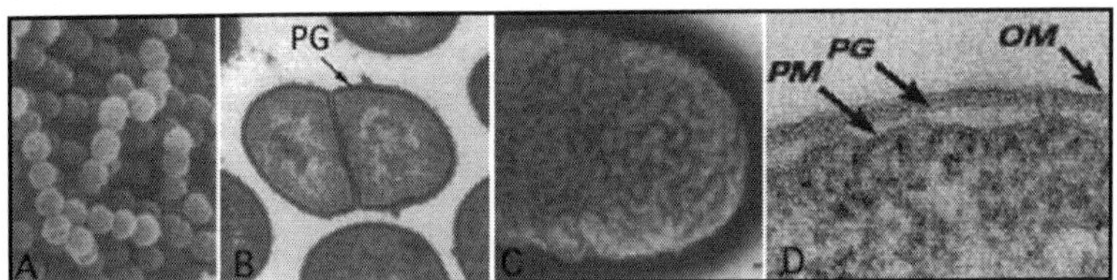

Fig. A comparison of the Ultrastructure of Gram-Positive and Gram-Negative Cells

The different Gram reactions occur because of structural differences between the bacterial cell walls. Gram-positive cells (Group B streptococci) appear smooth in a scanning electron micrograph (A) and are composed of a single layer of peptidoglycan (B). Gram-negative cells (E. coli) have an undulating surface and have three layers (C and D). As shown in Figure, the gram-negative cell has an additional layer and the outside of the cell appears convoluted when compared to the gram-positive cell. The gram-positive wall is much thicker than is the gram-negative wall and its external appearance is smoother. gram-positive and gram-negative cells do share one thing in common that is unique to bacteria - peptidoglycan.

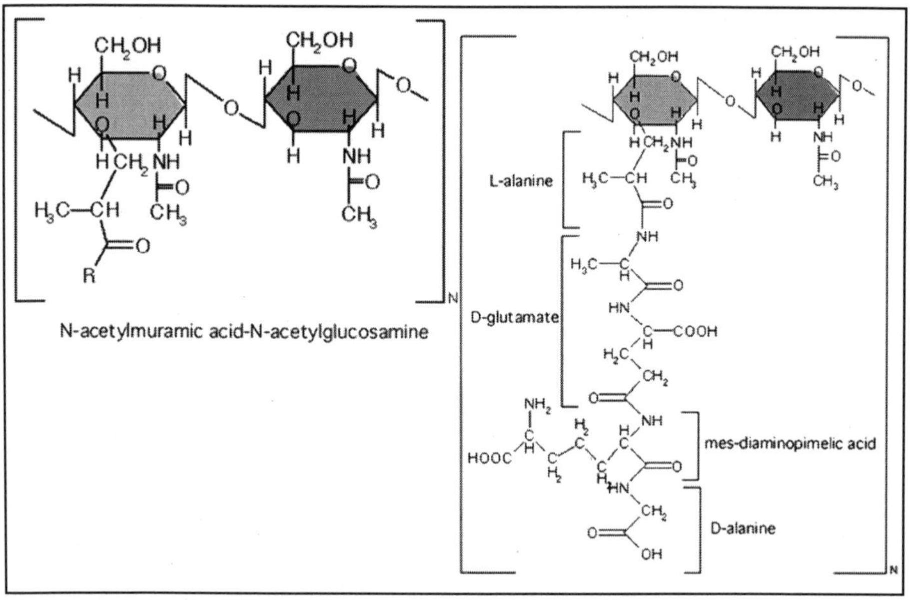

Fig. The Chemical Structure of Peptidoglycan

We will talk about the structure of this and then move on to examine the various structures found in each cell wall type. Peptidoglycan is a thick rigid layer composed of an overlapping lattice of two sugars, N-acetyl glucosamine (NAG) and N-acetyl muramic acid (NAM), that are cross-linked by amino acid bridges as shown in Figure. The exact molecular makeup of these cross-bridges is species-specific. NAM is only found in the cell walls of bacteria and nowhere else. Attached to NAM is a side chain generally composed of four amino acids. In the best-studied bacterial cell walls (E. coli) the cross-bridge is most commonly composed of L-alanine, D-alanine, D-glutamic acid and diaminopimelic acid (DPA). The generalized peptidoglycan monomer showing the two sugars that make up the backbone. The R group consists of four amino acids, with the best-studied cell walls containing L-alanine, D-alanine, D-glutamic acid and diaminopimelic acid. Note that peptidoglycan contains D-amino acids, which are different than the L-amino acids found in proteins. D-amino acids have the identical composition as L-amino acids, but are their mirror images. The use of D-amino acids is unusual in biology and bacteria have enzymes called racemases to convert between D and L forms specifically for this use. The NAM, NAG and amino acid side chain form a single peptidoglycan unit that can link with other units via covalent bonds to form a repeating polymer. The polymer is further strengthened by covalent bonds between cross-bridges and the degree of cross-linking determines the degree of rigidity. In the E. coli, the penultimate D-alanine of one unit is linked to DPA of the next cross-bridge.

In some gram-positive microbes there is a peptide composed of various amino acids that serves as a link between the cross-bridges. For example, in Staphylococcus aureus strains, five glycines make up the linker between peptidoglycan monomers. The sequence of these linkers varies considerably between species. The completed peptidoglycan layer forms a strong mesh that can be thought of as a chain link fence.

The complete cell wall contains one or more layers of peptidoglycan one atop the other, providing much of the strength of the cell wall. While both gram-negative and gram-positive bacteria have peptidoglycan, its physical arrangement in the cell wall is different. In gram-positive cells the peptidoglycan is a heavily cross-linked woven structure that encircles the cell in many layers. It is very thick with peptidoglycan accounting for 50 per cent of weight of cell and 90 per cent of the weight of the cell wall. Electron micrographs show the peptidoglycan to be 20-80 nm thick. In gram-negative bacteria the peptidoglycan is much thinner with only 15-20 per cent of the cell wall being peptidoglycan and it is only intermittently cross-linked. In both cases peptidoglycan is not a barrier to solutes, as the openings in the mesh are large enough for most molecules including proteins to pass through. Figure shows an artist's rendering of what the structure might look like.

The peptidoglycan polymers then crosslink with other peptidoglycan chains to form a complex mesh that wraps the cell in a structure a kin to chicken wire. There are numerous antibacterial agents that target the bacterial cell wall because mammals do not synthesize walls and therefore are not susceptible to the toxic effects of these agents. Penicillin inhibits the linking of the amino acid side chains of peptidoglycan units, which therefore weakens the stability of the wall eventually, causing the cells to rupture.

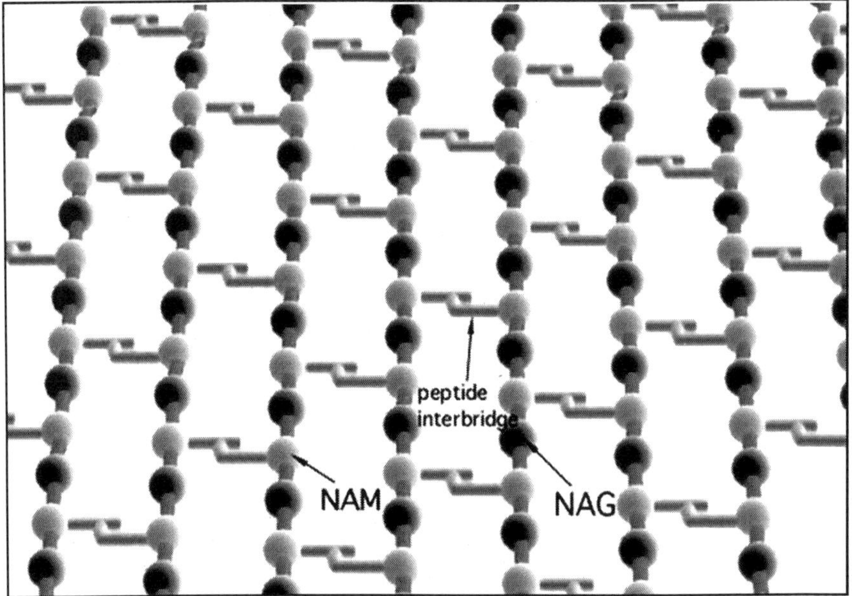

Fig. A Cartoon of the Peptiodglycan Mesh

Humans and other animals even synthesize an enzyme that specifically attacks bacterial cell walls. The enzyme lysozyme is found in many body fluids and hydrolyzes the NAM-NAG bond in the cell wall. It serves as a critical part of the mammalian defence against bacterial invasion. Figure shows a depiction of the gram-positive cell wall.

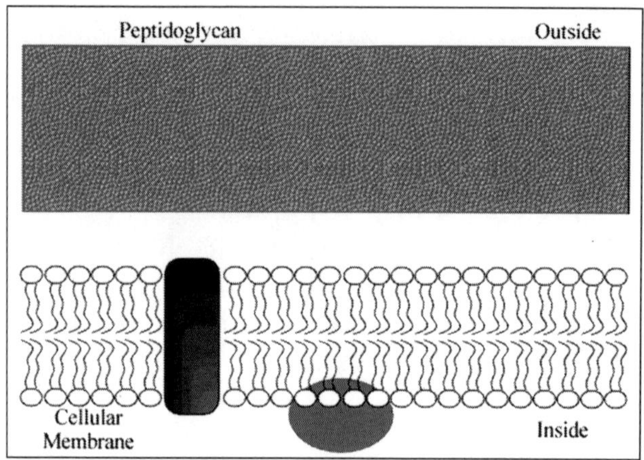

Fig. The Gram-Positive Cell Wall

The cell wall is made mostly of peptidoglycan, interspersed with teichoic acid which knits the different layers together. The amount of crosslinking is higher and the wall is thicker than in gram-negative cell walls.

CELL CYCLE REGULATION

Many of the most important discoveries about the mechanisms that control events of the cell cycle were elucidated using yeasts which are single cell eukaryotes. By analysis of various mutants that inactivated genes encoding essential components of cell cycle control systems in yeast many important control genes were identified. These genes were identified as cell division cycle genes or cdc genes. Thus, many cell cycle control genes in mammalian cells are also called cdc genes. Much of the control of the progression through the phases of a cell cycle are exerted at checkpoints. There are many such checkpoints but the two most critical are those that occur near the end of G1 prior to S-phase entry and those near the end of G2 prior to mitosis.

As indicated above, there is the need for cell cycle control mechanisms to exert their influences at specific times during each transit through a cell cycle. The heart of this timing control is the responsibility of a family of protein kinases that are called cyclin-dependent kinases, CDKs. The kinase activity of these enzymes rises and falls as the cell progresses through a cell cycle. Different CDKs operate at different points in the cell cycle. As would be expected, the oscillating changes in the activity of CDKs leads to oscillating changes in the phosphorylation of various intracellular proteins.

These phosphorylations alter the activity of the modified proteins which then effect changes in events of the cell cycle. The cyclical activity of each CDK is controlled by a complex series of proteins, the most important of which are the cyclins, hence the name of the enzymes as cyclin-dependent kinases. The CDKs are absolutely dependent upon their interaction with the cyclins for activity. Unless they are tightly bound CDKs have no kinase activity. The cyclins were originally idenitified because they undergo a cycle of synthesis and degradation at specific points in each cell cycle. Thus, whereas the levels of the various CDKs remain fairly constant throughout the cell cycle, their activities changes in concert with the fluctuations of the cyclins.

Four different classes of cyclins have been defined on the basis of the stage of the cell cycle in which they bind and activate CDKs. These four classes are G1-cyclins, G1/S-cyclins, S-cyclins, and M-cyclins. The cyclin nomenclature and associated CDK in mammalian cells are listed in the following Table.

Cyclin-CDK Complex	Cyclin	CDK Partner
G1-CDK	cyclin D*	CDK4, CDK6
G1/S-CDK	cyclin E	CDK2
S-CDK	cyclin A	CDK2
M-CDK	cyclin B	CDK1**

- There are three D cyclins in mammals: D1, D2, and D3
- CDK1 is the same as CDC2 in fission yeast and CDC28 in budding yeast

The G1-cyclins are not found in all eukaryotic cells but in those where they are synthesized they promote passage through a restriction point in late G1 called Start. The G1/S-cyclins bind to their cognate CDKs at the end of G1 and it is this interaction that is required to commit the cell to the process of DNA replication in S-phase. The S-cyclins bind to their cognate CDKs during S-phase and it is this interaction that is required for the initiation of DNA synthesis. The M-cyclins bind to their cognate CDKs and in so doing promote the events of mitosis.

Although CDKs are inactive unless bound to a cyclin, there is more to the activation process than just the interaction of the two parts of the complex. When cyclins bind to CDKs they alter the conformation of the CDK resulting in exposure of a domain that is the site of phosphorylation by another kinase called CDK-activating kinase (CAK). Following phosphorylation the cyclin-CDK complex is fully active. In addition to control of CDK kinase activity by cyclin binding and CAK phosphorylation, control is exerted to inhibit CDK activity through interaction with inhibitory proteins as well as by inhibitory phosphorylation events.

Thus, there is extremely tight control on the overall activity of each CDK. One of the inhibitory kinases that phosphorylates CDKs is called Wee1. The inhibitory phosphorylations are removed through the action of a phosphatase called CDC25. The action of these two regulatory enzymes on CDK activity is most important at the level of the M-CDK activity at the onset of mitosis. Proteins that bind to and inhibit cyclin-CDK complexes are called CDK inhibitory proteins (CKI, for cyclin-kinase inhibitor).

Mammalian cells express two classes of CKI. These are called CIPs for CDK inhibitory proteins and INK4 for inhibitors of kinase 4. The CIPs bind and inhibit CDK1, CDK2, CDK4, and CDK6 complexes, whereas the INK4s bind and inhibit only the CDK4 and CDK6 complexes. There are at least three CIP proteins in mammalian cells and these are identified as $p^{21Cip1/WAF1}$ (gene symbol=CDKN1A), p^{27KIP1} (gene symbol=CDKN1B), and p^{57KIP2} (gene symbol=CDKN1C). The expression of each of these CIPs is controlled by specific events that may have occurred during cell cycle transit. For example p^{21Cip1} expression is induced in response to DNA damage.

This induction is under the control of the action of the tumor suppressor protein p53. There are at least four INK4 proteins that are each identified by their molecular weights: $p^{15INK4B}$, $p^{16INK4A}$, and $p^{18INK4C}$ (these were the first 3 characterized) as well as p^{19INK4}. The $p^{16INK4A}$ protein is also a tumor suppressor since loss of its function leads to cancer. All the INK4 proteins contain 4 tandem repeats of a sequence of amino acids that were first identified in ankyrin and are thus referred to as ankyrin repeats.

As indicated above, many cells reside in a resting or quiescent state but can be stimulated by external signals to re-enter the cell cycle. These external growth promoting signals are the result of growth factors binding to their receptors. Most growth factors induce the expression of genes that are referred to as early and delayed-response genes. The activation of early response genes occurs in response to growth factor receptor-mediated signal transduction resulting in phosphorylation and activation of transcription factor proteins that are already present in the cell.

Many of the induced early response genes are themselves transcription factors that in turn activate the expression of delayed-response genes. In the context of the cell cycle, these delayed-response genes encode proteins of the G1-CDK complexes. One such early response gene is the proto-oncogene MYC. With respect to the cell cycle some of the genes turned on by activation of MYC are cyclin D, proteins of the ubiquitin ligase complex called SCF (Skp1/cullin/F-box protein) and the members of the E2F transcription factor family.

There are six members of the E2F family: E2F1 through E2F6). The synthesis of cyclin D will result in the activation of G1-CDK complexes. The synthesis of components of SCF leads to the degradation of p27KIP1 which normally inhibits G1-CDK complexes. The synthesis of E2F family members results in increased synthesis of proteins involved in DNA synthesis as well as the synthesis of the S-phase cyclins A and E and CDK2. Regulation of E2F activity by the tumor suppressor pRB will be discussed below. The cyclical degradation of the cyclins is effected through the action of several different ubiquitin ligase complexes. The action of ubiquitin ligases in protein turn-over is discussed in more detail in the Protein Modifications page.

There are two important ubiquitin ligase complexes that control the turn-over of cyclins and other cell cycle regulating proteins. One is the SCF complex which functions to control the transit from G1 to S-phase and the other is called anaphase promoting complex (APC) which controls the levels of the M-phase cyclins as well as other regulators of mitosis.

One important function of APC is to control the initiation of sister chromatid separation which begins at the metaphase-anaphase transition. The attachment of the sister chromatids to the opposite poles of the mitotic spindles occurs early during mitosis. The ability of the sister chromatids to be pulled apart is initially inhibited because they are bound together by a protein complex termed cohesin complex. The cohesin complex is deposited along the chromosomes as they are duplicated during S-phase. Anaphase can only begin with the disruption of the cohesin complex.

The breakdown of the cohesin complex is brought about as a consequence of the activation of the ubiquitin ligase activity of the APC. APC targets a protein called securin. Securin functions to inhibit the protease called separase and the action of separase is to degrade the proteins of the cohesin complex, thus allowing sister chromatid separation.

THE MECHANICS OF CELL DIVISION

The process of cell division must occur in a highly ordered and accurate manner so as to ensure that each daughter cell receives an identical copy of the parental cells genome. Obviously the first step in this process is the accurate replication of the genome as described in the DNA Metabolism page. The processes described above relate to the regulation of the progressive steps taken as a cell progresses to cytokinesis. This section will discuss the biochemical processes undertaken to effect accurate separation of the duplicated DNA (the sister chromatids) and cytokinesis.

Following duplication of the chromosomes the sister chromatids are held together through multisubunit protein complexes referred to as cohesins. These complexes are found all along the length of each chromatid as the DNA is replicated. Following DNA replication the chromosomes condense and this is the role of proteins called condensins. Chromosome condensation is the first easily identifiable sign that a cell is about to enter M-phase. Cohesins and condensins are structurally related and act in concert to prepare the chromosomes for mitosis.

The task of separating the sister chromatids such that each daughter cell receives one copy of each chromosome is carried out by the mitotic spindle which is composed of microtubules and several proteins that interact with them.

An additional cytoskeletal structure is required for the actual separation of the cell into two new cells. This structure is referred to as the contractile ring. The process of mitosis occurs through a highly ordered series of five steps as outlined above. The actual separation of the parental cell into two daughter cells can be considered the sixth step in mitosis. Whereas, prophase, prometaphase, metaphase, anaphase, and telophase occur in a strictly controlled sequential fashion, cytokinesis begins in anaphase and continues until the cell divides.

During interphase the microtubule machinery is in a constant state of dynamic instability. Individual microtubules are growing or shrinking at any given moment. During prophase the activation of M-CDK complexes initiates a change in the microtubule structures to one where there are a large number of shorter microtubules surrounding each centrosome. The centrosomes are cytoplasmic nucleation sites for the mitotic spindles. M-CDK complexes initiate these changes via the phosphorylation of microtubule motor proteins and microtubule-associated proteins (MAPs). During prometaphase the nuclear envelope abruptly breaks down as a consequence of M-CDK complexes phosphorylating the nuclear lamina. The dissolution of the nuclear membrane allows the microtubules to access the mitotic spindles. When microtubules attach to the mitotic spindle they become stabilized. The microtubules eventually become attached at the kinetochore which is a complex protein structure that assembles onto the highly condensed DNA at the centromere.

The chromosomes are pulled back and forth by the microtubules eventually becoming aligned equidistant from the two spindle poles. The alignment of the chromosomes forms the metaphase plate. The chromosomes oscillate about the metaphase plate awaiting the signal that will induce the sister chromatids to separate. This phase of mitosis is referred to as the spindle-attachment checkpoint and it ensures that cells do not enter anaphase until all of the chromosomes are attached to both poles of the mitotic spindles.

As described above, sister chromatids begin to separate with the activation of the APC. As each sister chromatid is pulled towards one of the poles of the mitotic spindle the kinetochore microtubules depolymerize. By the end of anaphase, the daughter chromosomes have separated to opposite ends of the cells and have begun to decondense which signals the onset of telophase.

Telophase is denoted by the reassembly of the nuclear envelope around each group of daughter chromosomes. During this process the lamins that make of the nuclear lamina are dephosphorylated allowing them to re-associate with the nuclear envelope. Following the formation of the new nuclear envelopes, the chromosomes decondense into their interphase state and transcriptional activity begins anew. The cell is now ready for the final process, complete separation into two daughter cells.

TUMOR SUPPRESSORS AND CELL CYCLE REGULATION

Tumor suppressors are so called because cancer ensues as a result of a loss of their normal function, *i.e.* these proteins suppress the ability of cancer to develop. It would seem obvious, therefore, that one import function of tumor suppressors would be control of the progression of a cell through a round of the cell cycle. If cells are able to synthesize damaged DNA before it is repaired or to divide when the DNA is damaged then the resulting daughter cells can pass on the resultant DNA damage to their progeny.

The result can be catastrophic resulting in cancer. For this reason, the two most important check points in the eukaryotic cell cycle are the G1-S transition and the entry into mitosis. The former prevents DNA replication prior to repair of damaged DNA and the latter prevents damage that may have occurred to the DNA during replication to propagated into daughter cells during mitosis. Following the isolation and characterization of two tumor suppressor genes in particular it was found that they function to control the ability of cells to progress through these two important checkpoints. The protein encoded by the retinoblastoma susceptibility gene (pRB) and the p53 protein are both tumor suppressors. The function of pRB is to act as a brake preventing cells from exiting G1 and that of p53 is to inhibit progression from S-phase to M-phase.

The best understood effect of G1-CDK activity is that exerted on transcription factors of the E2F family, hereafter referred to simply as E2F. In the context of the cell cycle regulation, E2F activates the expression of cyclin A, cyclin E and CDK2. These proteins are components of the S-CDK complexes necessary for progression through S-phase. The activity of E2F is itself controlled via interaction with pRB. When pRB binds E2F it can no longer function as a transcription factor as it is sequestered in the cytosol.

Interaction of pRB and E2Fcorrelates to the state of phosphorylation of pRB and the affinity between the two proteins is highest when pRB is hypophosphorylated. Phosphorylation of pRB is maximal at the start of S phase and lowest after mitosis and entry into G1. Stimulation of quiescent cells with mitogen induces phosphorylation of pRB, while in contrast, differentiation induces hypophosphorylation of pRB.

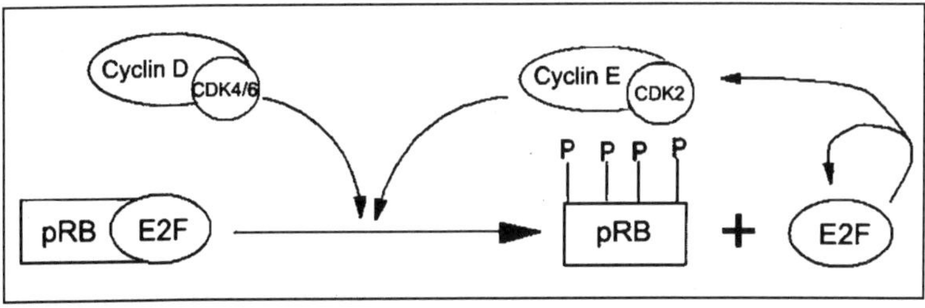

Fig. Regulation of E2F by pRB

One of the most significant substrates for phosphorylation by the G1 cyclin-CDK complexes is pRB. When pRB is phosphorylated by G1 cyclin-CDK complexes it releases E2F allowing E2F to transcriptionally activate its target genes. When E2F activates the expression of S-CDK complex proteins these complexes also target pRB for phosphorylation, thus maintaining the cell in a pro cell cycle progression state.

One major function of the p53 protein, which is active as a homotetrameric transcription factor, is to serve as a component of the checkpoint that controls whether cells enter as well as progress through S-phase. The action of p53 is induced in response to DNA damage. Under normal circumstances p53 levels remain very low due to its interaction with a member of the ubiquitin ligase family called MDM2.

MDM2 is so named since it was isolated as an amplified gene in the tumorigenic mouse cell line 3T3DM. In response to DNA damage, *e.g.* as a result of uv-irradiation or irradiation, cells activate several kinases including checkpoint kinase 2 (CHK2) and ataxia telangiectasia mutated (ATM). One target of these kinases is p53.

ATM also phosphorylates MDM2. When p53 is phosphorylated it is released from MDM2 and can carry out its transcriptional activation functions. One target of p53 is the cyclin inhibitor p21Cip1 gene. Activation of p21Cip1 leads to increased inhibition of the cyclin D1-CDK4 and cyclin E-CDK2 complexes thereby halting progression through the cell cycle either prior to S-phase entry or during S-phase.

As a consequence of p53-induced synthesis of p21 expression, there is a convergence between the roles of p53 and pRB in regulation of cyclin-CDK complexes.

In either case the aim is to allow the cell to repair its damaged DNA prior to replication or mitosis. Even given the limited discussion of the functions of pRB and p53 it is still easy to understand how loss of either protein function can lead to aberrant cell cycle progression and the potential for the development of cancer.

REGULATORY CASCADE OF CYCLIN GENE EXPRESSION

When cells traverse the G0 to G1 phase to the S-phase transition, a series of cyclin-dependent kinases is activated. The addition of serum growth factors to quiescent cells promotes transcription of the cyclin D1 gene. Cyclin D1 then associates with pre-existing cdk4 to form an active complex. The kinase activity associated with this complex can phosphorylate specific sites on the retinoblastoma protein (pRb), leading to inactivation of pRb and the activation cyclin E transcription by E2F. Activation of the cyclin E gene can be blocked by the cdk inhibitor p16.

Cyclin E associates with existing cdk2 and this active complex regulates the function of several sets of target proteins. First, cyclin E/cdk2 complexes associate with E2F/p107 complexes to activate expression of the cyclin A gene. Also, cyclin E/cdk2 complexes cooperate with cyclin D1 to amplify the phosphorylation of pRb. Cyclin A associates with cdk2 to form a kinase complex that phosphorylates downstream targets involved in the initiation of DNA replication.

The p53 Signaling Pathway

The tumor-suppressor protein p53 exhibits sequence-specific DNA-binding, directly interacts with various cellular and viral proteins, and induces cell cycle arrest in response to DNA damage. In response to signals generated by a variety of genotoxic stresses, *e.g*, UV irradiation or DNA damage, p53 is expressed and undergoes post-translational modification that results in its accumulation in the nucleus.

The p53-dependent pathways help to maintain genomic stability by eliminating damaged cells either by arresting them permanently or through apoptosis.

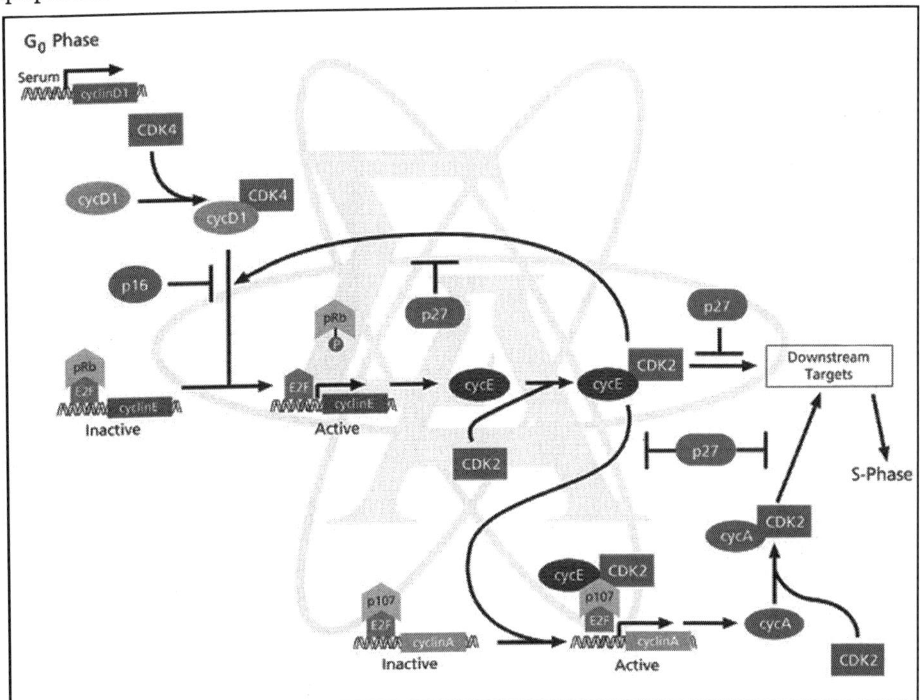

For example, ?-irradiation activates p53 to turn on the transcription of p21CIP1, which, in turn, binds to and inhibits cyclin-dependent kinases, causing hypophosphorylation of retinoblastoma (Rb), thus preventing the release of E2F and blocking the G1-S transition.

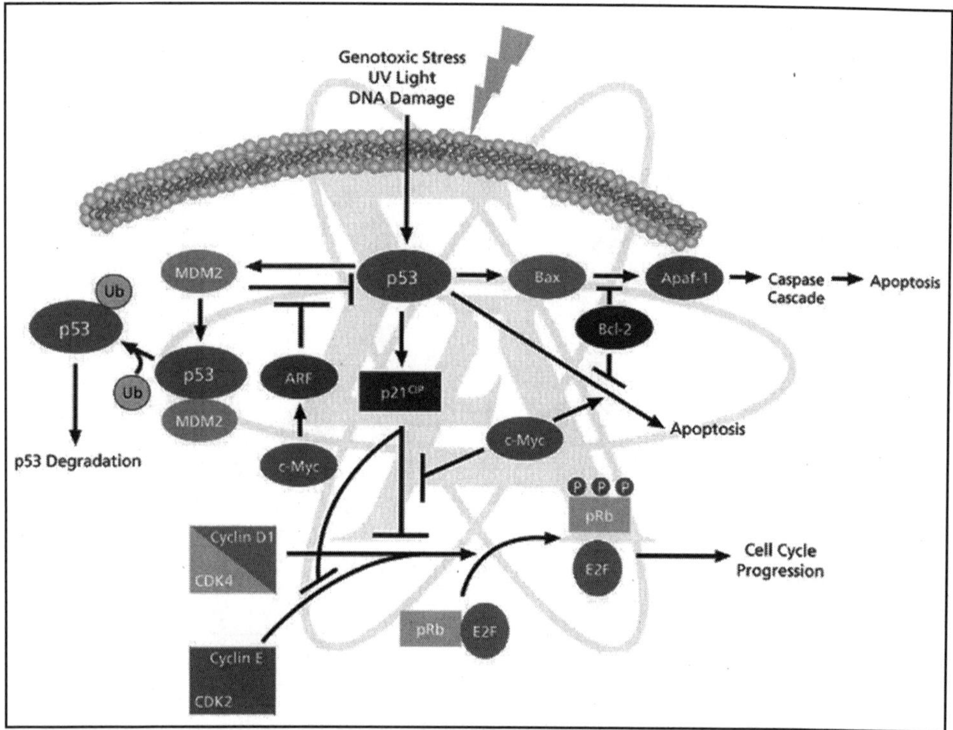

Some of the cellular effects of p53 can be blocked by the deregulated expression of c-Myc, Bcl-2, or E2F. p53 activity is controlled through an autoregulatory loop involving Mdm2.

APOPTOSIS AND CELL CYCLE

Apoptosis is a genetically controlled response by which eukaryotic cells undergo programmed cell death. This phenomenon plays a major role in developmental pathways (1), provides a homeostatic balance of cell populations, and is deregulated in many diseases including cancer. Control of cell number is determined by an intricate balance of cell death and cell proliferation.

Accumulation of cells through suppression of death can contribute to cancer and to persistent viral infections, while excessive death can result in impaired development and in degenerative diseases. Identification of genes that control cell death, and understanding of the impact of apoptosis in both development and disease has advanced our knowledge of apoptosis in the past few years. There appears to be a linkage between apoptosis and cell cycle control mechanisms.

Elucidating the mechanisms that link cell cycle control with apoptosis will be of key importance in understanding tumour progression and designing new models of effective tumour therapy.

PROGRAMMED CELL DEATH

Programmed cell death (PCD), or apoptosis, can be triggered by a wide range of stimuli, including cell surface receptors like Fas and FasL. It constitutes a system for the removal of unnecessary, aged, or damaged cells that is regulated by the interplay of proapoptotic and antiapoptotic proteins of the Bcl-2 family. The proapoptotic proteins Bax, Bad, Bid, Bik, and Bim contain an ?-helical BH3 death domain that fits the hydrophobic BH3 binding pocket on the antiapoptotic proteins Bcl-2 and Bcl-XL, forming heterodimers that block the survival-promoting activity of Bcl-2 and Bcl-XL. Thus, the relative abundance of proapoptotic and antiapoptotic proteins determines the susceptibility of the cell to programmed death.

The proapoptotic proteins act at the surface of the mitochondrial membrane to decrease the mitochondrial trans-membrane potential and promote leakage of cytochrome c. In the presence of dATP cytochrome c complexes with and activates Apaf-1. Activated Apaf-1 binds to downstream caspases, such as procaspase-9, and processes them into proteolytically active forms. This begins a caspase cascade resulting in apoptosis.

ACTIVATION AND INHIBITION OF APOPTOSIS

Several mechanisms have been identified in mammalian cells for the induction of apoptosis. These mechanisms include factors that lead to perturbation of the mitochondria leading to leakage of cytochrome c or factors that directly activate members of the death receptor family. Fas is a member of the tumor necrosis factor (TNF) receptor superfamily, a family of transmembrane receptors that include neurotrophin receptor (p75NTR), TNF-R1, and a variety of other cell surface receptors.

Fas Ligand (Fas L) transmits signals to Fas on a target cell by inducing trimerization of Fas. Activation of Fas causes the recruitment of Fas-associated protein with death domain (FADD) via interactions between the death domain of Fas and FADD and is followed by pro-caspase-8 binding to FADD via interactions between the death effector domains (DED) of FADD and pro-caspase-8 leading to the activation of caspase-8.

Activation of caspase-8 leads to the activation of other caspases, in effect beginning a caspase cascade that ultimately leads to apoptosis. Caspase-8 activation can also activate Bid, leading to activation of the apoptotic programme. Fas-induced apoptosis can be effectively blocked at several stages by either FLICE-inhibitory protein (FLIP), by Bcl-2, or by the cytokine response modifier A (CrmA).

In addition, activation of caspase-3 by caspase-9 can be blocked by inhibitor of apoptosis proteins (IAPs). Moreover, the protein kinase, Akt, can be activated by various growth factors and its activity can be blocked by PTEN. Akt functions to promote cell survival through two distinct pathways. Akt inhibits apoptosis by phosphorylating the Bcl-2 family member Bad, which then interacts with 14-3-3 and dissociates from Bcl-xL allowing for cell survival.

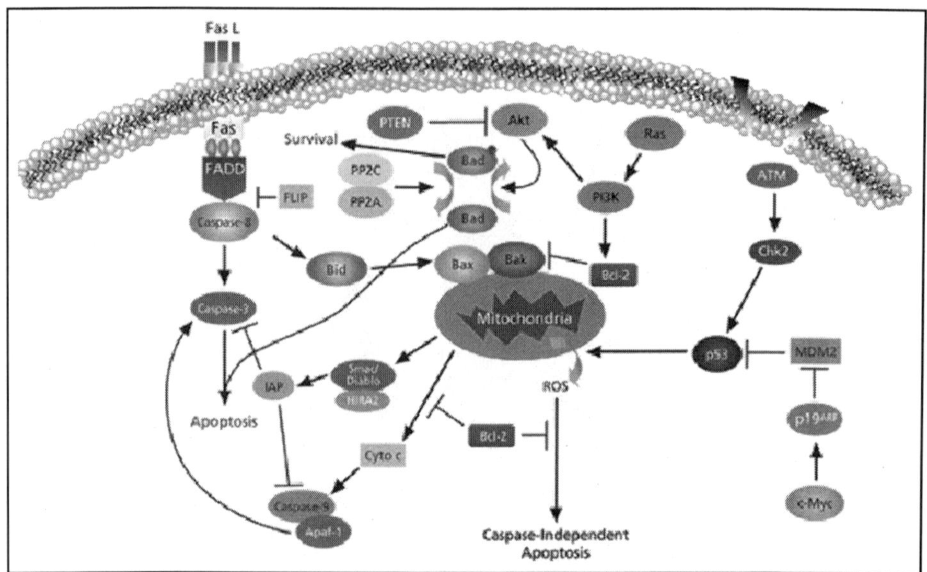

Alternatively, Akt activates IKK? that ultimately leads to NF?B activation and cell survival. Proapoptotic Bcl-2 family members, such as Bax and Bak can promote mitochondrial permeability, while Bcl-2 can inhibit their effects. Upon mitochondrial permeability, apoptogenic factors are released from the mitochondrial inter-membrane space and leak into the cytosol. One factor is cytochrome c, which induces the liberation of protease activators (caspases) that ultimately lead to apoptosis through nuclear damage (DNA fragmentation, DNA mutations). In addition, Smac/Diablo is released and can block IAP inhibition of capsase activity. Mitochondrial permeability is also related to the increased generation of reactive oxygen species (ROS), which plays a role in the degradation phase of apoptosis (*i.e.* plasma membrane alterations).

MITOCHONDRIA IN APOPTOSIS

Increases in cytosolic Ca2+ levels due to activation of ion channel-linked receptors, such as that for the excitatory amino acid neurotransmitter glutamic acid, can induce permeability transition (PT) of the mitochondrial membrane. PT constitutes the first rate-limiting event of the common pathway of apoptosis. Upon PT, apoptogenic factors leak into the cytoplasm from the mitochondrial intermembrane space. Two such factors, cytochrome c and apoptosis inducing factor (AIF), begin a cascade of proteolytic activity that ultimately leads to nuclear damage (DNA fragmentation, DNA mutations) and cell death.

Cytochrome C, a key protein in electron transport, appears to act by forming a multimeric complex with Apaf-1, a protease, which in turn activates procaspase

9, and begins a cascade of activation of downstream caspases. Smac/Diablo is released from the mitochondria and inhibits IAP (inhibitor of apoptosis) from interacting with caspase 9 leading to apoptosis. Bcl-2 and Bcl-X can prevent pore formation and block the release of cytochrome c from the mitochondria and prevent activation of the caspase cascade and apoptosis. PT is also related to the mitochondrial generation of reactive oxygen species which plays a role in the degradation phase of apoptosis (*i.e.* plasma membrane alterations).

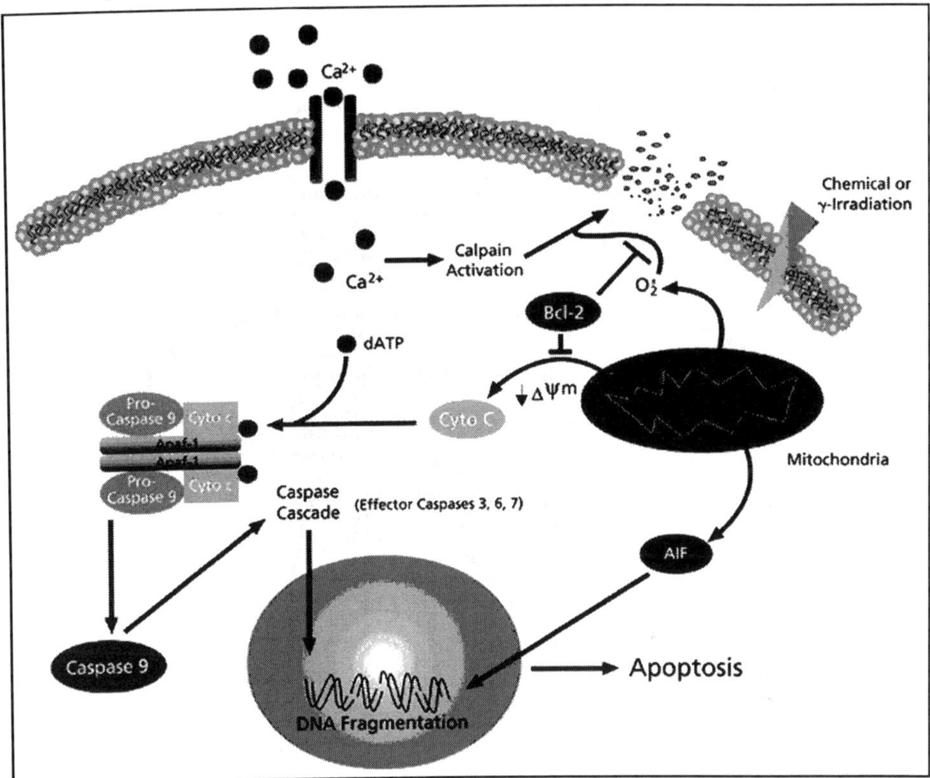

ATM/P53 SIGNALING PATHWAY

The p53 tumor suppressor protein is phosphorylated and activated by several DNA damage-inducible kinases, such as ATM, and is a key effector of the DNA damage response by promoting cell cycle arrest or apoptosis. Deregulation of the Rb-E2F1 pathway also results in the activation of p53 and the promotion of apoptosis, and this contributes to the suppression of tumor development.

Here, we describe a novel connection between E2F1 and the ATM DNA damage response pathway. In primary human fibroblasts lacking functional ATM, the ability of E2F1 to induce the phosphorylation of p53 and apoptosis is impaired. In contrast, ATM status has no effect on transcriptional activation of target genes or the stimulation of DNA synthesis by E2F1. Cells containing mutant Nijmegen breakage syndrome protein (NBS1), a component of the Mre11-Rad50 DNA repair complex, also have attenuated p53 phosphorylation and apoptosis in response to E2F1 expression.

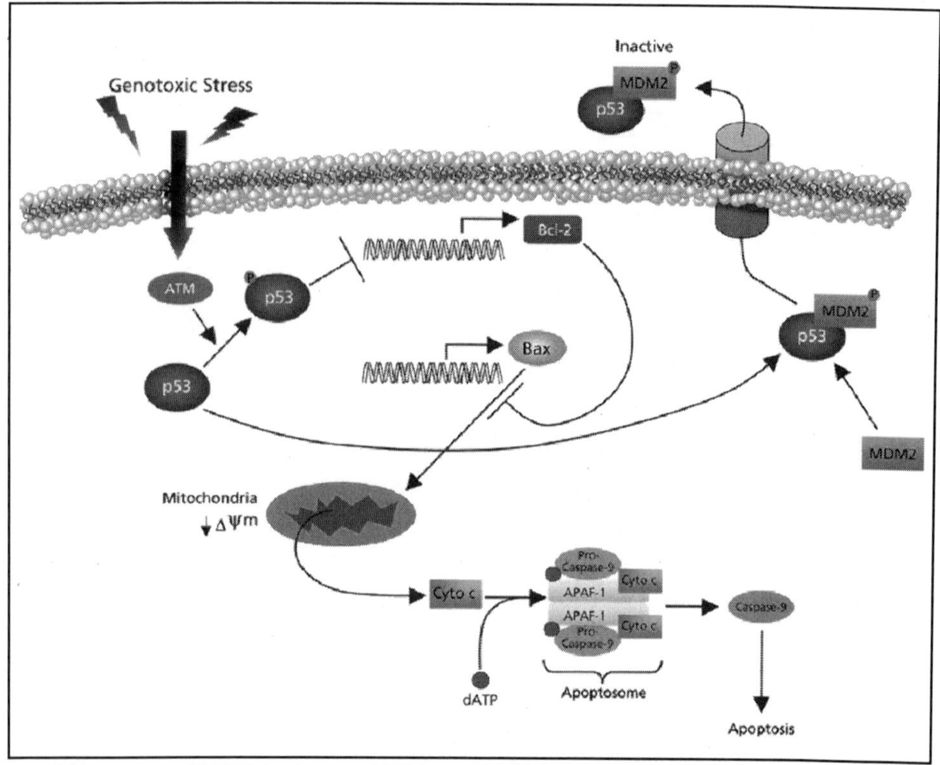

Moreover, E2F1 induces ATM- and NBS1-dependent phosphorylation of the checkpoint kinase Chk2 at Thr68, a phosphorylation site that stimulates Chk2 activity. Delayed γH2AX phosphorylation and absence of ATM autophosphorylation at Ser1981 suggest that E2F1 stimulates ATM through a unique mechanism that is distinct from agents that cause DNA double-strand breaks.

These findings identify new roles for several DNA damage response factors by demonstrating that they also participate in the oncogenic stress signaling pathway between E2F1 and p53. The ataxia telangiectasia-mutated gene (ATM) encodes a protein kinase that acts as a tumor suppressor. ATM activation, via IR damage to DNA, stimulates DNA repair and blocks cell cycle progression. One mechanism through which this occurs is ATM dependent phosphorylation of p53. p53 can cause growth arrest of the cell at a checkpoint to allow for DNA damage repair or can cause the cell to undergo apoptosis if the damage cannot be repaired. The critical role of p53 is evident by the fact that it is mutated in over 50 per cent of all human cancers.

CASPASE CASCADE

Apoptosis or programmed cell death is triggered by a variety of stimuli, including cell surface receptors like FAS, the mitochondrial response to stress, and factors released from cytotoxic T cells. The caspases comprise a class of cysteine proteases many members of which are involved in apoptosis. The caspases convey the apoptotic signal in a proteolytic cascade, with caspases cleaving and activating other caspases that subsequently degrade cellular targets that lead to cell death. The activating caspases include caspase-8 and caspase-9. Caspase-8 is the initial caspase activated in response to receptors with a death domain that interacts with FADD. The mitochondrial stress pathway begins with the release of cytochrome c from mitochondria, which then interacts with Apaf-1, causing self-cleavage and activation of caspase-9. The effector caspases, caspase-3, -6 and-7 are downstream of the activator caspases and act to cleave various cellular targets. Granzyme B and perforin, proteins released by cytotoxic T cells, induce apoptosis in target cells by forming transmembrane pores and triggering apoptosis, perhaps through cleavage of caspases. Caspase independent mechanisms of granzyme B-mediated apoptosis have been suggested.

CASPASE ACTIVATION INTRINSIC PATHWAY

Cytochrome c is released from the mitochondria of pre-apoptotic cells and binds to Apaf-1 in the presence of dATP/ATP. The interaction results in a conformational change in Apaf-1 allowing the molecules of Apaf-1 to associate with each other.

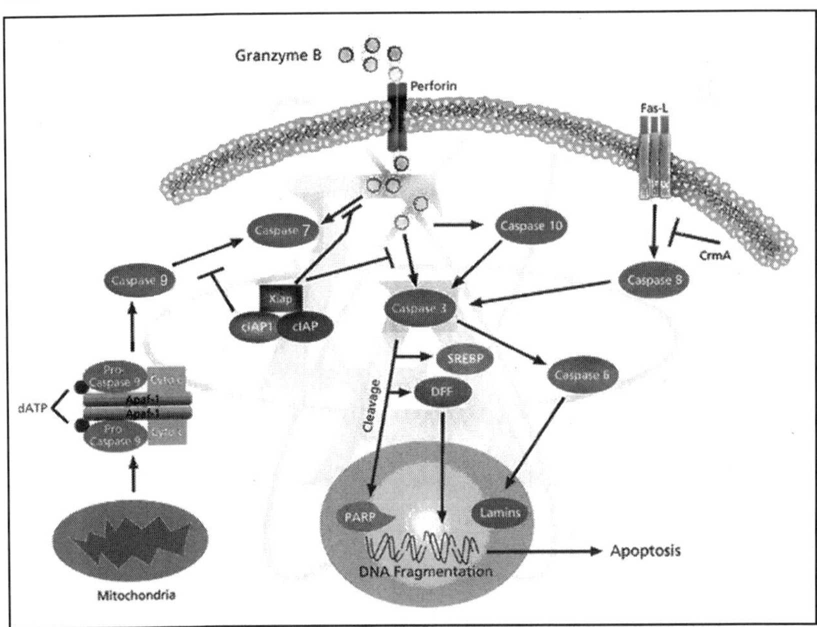

This formation results in a wheel-like structure that contains 7 molecules each of Apaf-1, cytochrome c and ATP.

Pro-Caspase-9 contains a caspase recruitment domain (CARD) that is used to mediate specific interactions with Apaf-1 CARD, which becomes exposed on the apoptosome during assembly. Pro-Caspase-9 is autoactivated and after cleavage, mature caspase-9 remains bound to the apoptosome where it is able to activate executioner caspases such as caspase-3 and caspase-7.

GRANZYME B

Granzymes are serine proteases that are released by cytoplasmic granules within cytotoxic T cells and natural killer cells. Their purpose is to induce apoptosis within virus-infected cells, thus destroying them. Cytotoxic T cells and natural killer cells release a protein called perforin, which attacks the target cells.

Researchers used to think that perforin creates pores within the cells, through which the granzymes can enter, inducing apoptosis. However, new evidence indicates that GrB complex (GrB, perforin, and another protein (glycopectin)) can enter a cell through the mannose 6-phosphate receptor (or another receptor found in tumor cells) and is enclosed in a vesicle (a sac).[Perforin then allows GrB to pass through the vesicle surface and into the cell, causing apoptosis by various pathways.

They do so by cleaving caspases (especially caspase-3), which in turn activates caspase-activated DNase. This enzyme degrades DNA, thus inducing apoptotic cascades.

Also, GrB cleaves the protein Bid, which recruits the protein Bax and Bak to change the membrane permeability of the mitochondria, causing the release of cytochrome c (which is one of the parts needed to form caspase-9), Smac/ Diablo and Omi/HtrA2 (which suppress the inhibitor of apoptosis proteins(IAPs)), among other proteins.

As well, GrB is shown to cleave many of the chemicals responsible for apoptosis without the aid of caspase, as proven by experiments on caspase knockout mice CTL cells incubated with other cells. Granzyme B and perforin, proteins released by an effector cell (cytotoxic T cell), can induce apoptosis in target cells by forming transmembrane pores and through cleavage of effector caspases such as caspase-3.

In addition, caspase-independent mechanisms of granzyme B-mediated apoptosis have been suggested. Caspase-activated DNAse (CAD) is activated through the cleavage of its associated inhibitor ICAD by caspase-3. CAD is then able to interact with components such as topoisomerase II (Topo II) to condense chromatin, leading to DNA fragmentation and ultimately apoptosis.

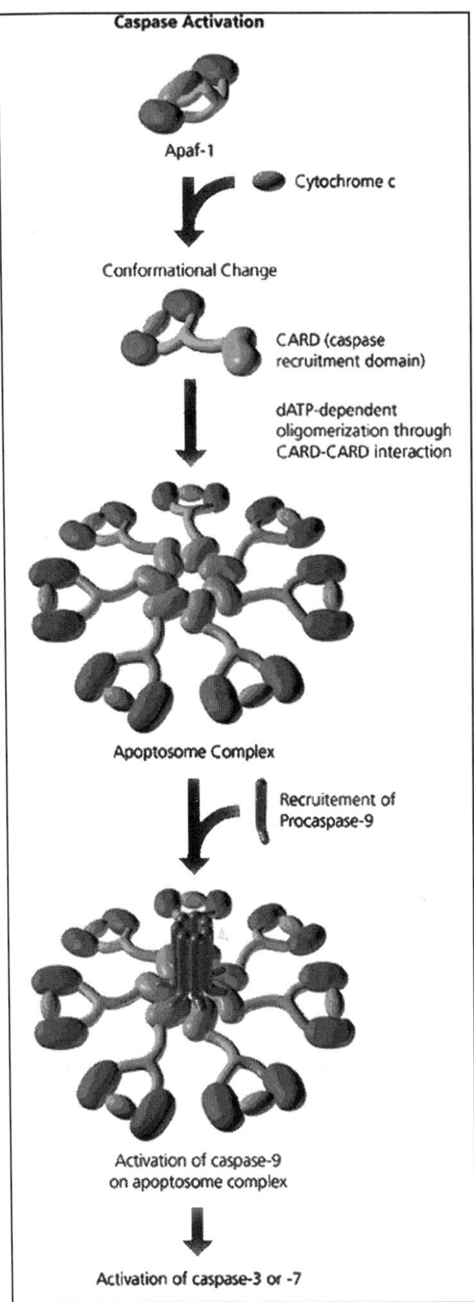

FAS AND RELATED PROTEINS WITH DEATH DOMAINS

Fas, a death domain-containing member of the tumor necrosis factor receptor family and its ligand FasL have been predominantly studied with respect to their capability to induce cell death. However, a few studies indicate a proliferation-inducing signaling activity of these molecules too. We describe here a novel signaling pathway of FasL and the tumor necrosis factor-related apoptosis-inducing ligand (TRAIL) that triggers transcriptional activation of the proto-oncogene c-fos, a typical target gene of mitogenic pathways. FasL- and TRAIL-mediated up-regulation of c-Fos was completely dependent on the presence of Fas-associated death domain protein (FADD) and caspase-8, but caspase activity seemed to be dispensable as a pan inhibitor of caspases had no inhibitory effect. Upon overexpression of the long splice form of cellular FADD-like interleukin-1-converting enzyme (FLICE) inhibitory protein (cFLIP) in Jurkat cells, FasL- and TRAIL-induced up-regulation of c-Fos was almost completely blocked.

The short splice form of FLIP, however, showed a rather stimulatory effect on c-Fos induction. Together these data demonstrate the existence of a death receptor-induced, FADD- and caspase-8-dependent pathway leading to c-Fos induction that is inhibited by the long splice form FLIP-L. The death domain (DD) of Fas (FADD) shows high-sequence homology to the Drosophila melanogaster protein reaper.

The death domain is also found in tumor necrosis factor receptor 1 (TNF-R1), TNF-R1-associated death domain (TRADD), Fas-associated death domain (FADD), and Fas receptor-interaction protein (RIP). Fas and FADD associate through their homologous DD while TNF-R1 and TRADD associate through their homologous DD.

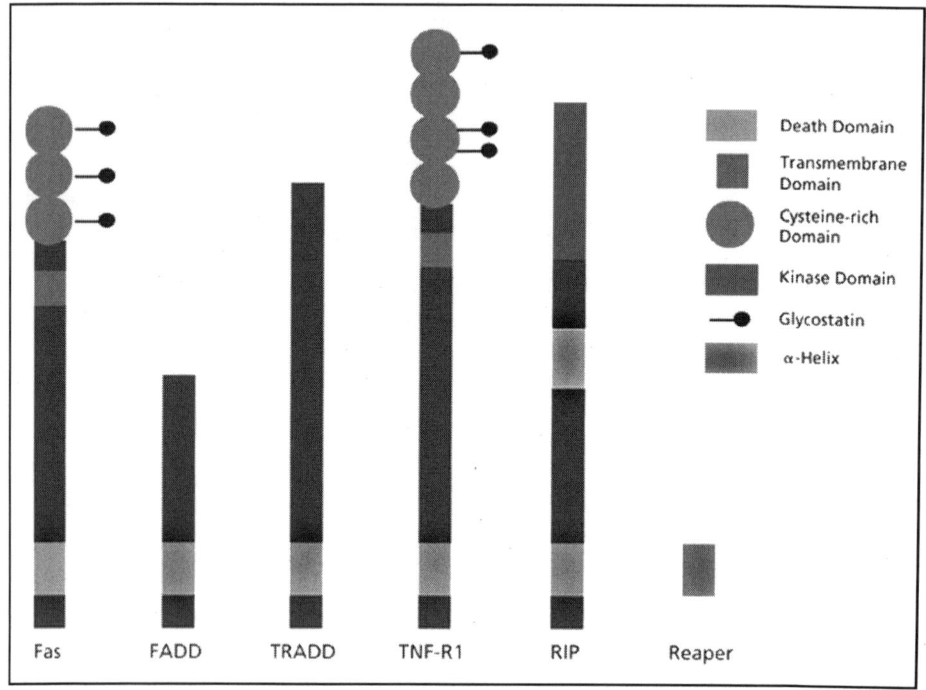

FAS SIGNALING PATHWAY

Fas/APO-1/CD95 (36 kDa) is a member of the tumor necrosis factor (TNF) receptor superfamily, a family of transmembrane receptors that also includes the p75 neurotrophin receptor, TNF-R1, and a variety of other cell surface receptors. Fas has been shown to be an important mediator of apoptotic cell death, as well as being involved in inflammation. Binding of the Fas ligand (Fas-L) induces trimerization of Fas in the target cell membrane. Activation of Fas causes the recruitment of Fas-associated protein with death domain (FADD) via interactions between the death domains of Fas and FADD. Procaspase 8 binds to Fas-bound FADD via interactions between the death effector domains (DED) of FADD and pro-caspase 8 leading to the activation of caspase 8. Activated caspase 8 cleaves (activates) other procaspases, in effect beginning a caspase cascade that ultimately leads to apoptosis. Caspases cleave nuclear lamins, causing the nucleus to break down and lose its normal structure. Fas-induced apoptosis can be effectively blocked at several stages by either FLICE-inhibitory protein (FLIP), by Bcl-2, or by the cytokine response modifier A (CrmA).

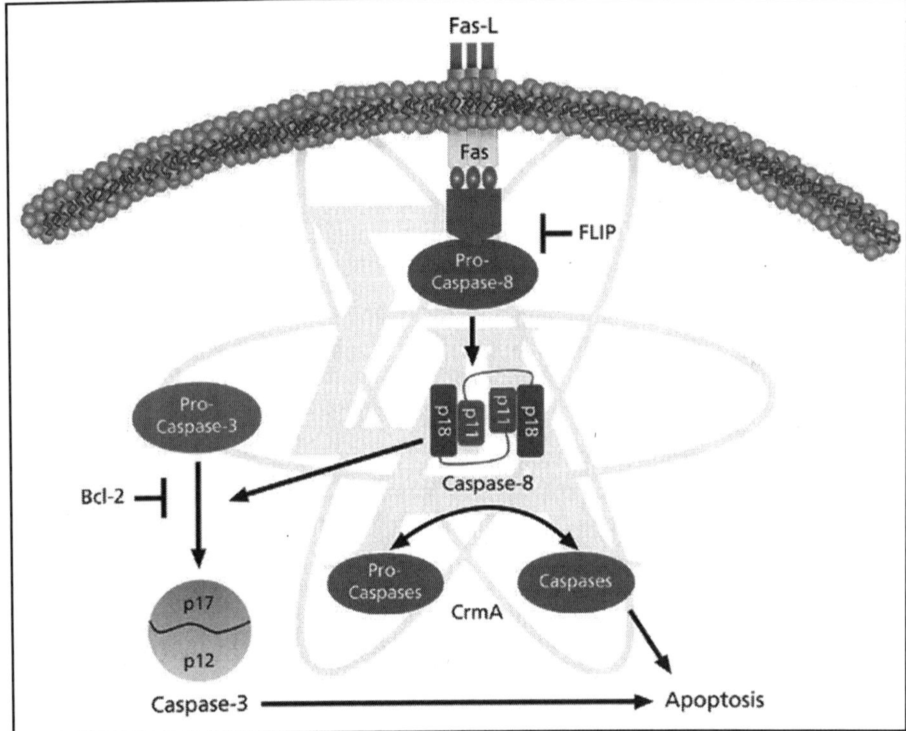

G 1 AND S PHASES OF THE CELL CYCLE

In proliferating cells, the cell cycle consists of four phases. Gap 1 (G1) is the interval between mitosis and DNA replication that is characterized by cell growth. The transition that occurs at the restriction point (R) in G1 commits the cell to the proliferative cycle. If the conditions that signal this transition are not present, the cell exits the cell cycle and enters G0, a nonproliferative phase during which growth, differentiation and apoptosis occur. Replication of DNA occurs during the synthesis (S) phase, which is followed by a second gap phase (G2) during which growth and preparation for cell division occurs. Mitosis and the production of two daughter cells occur in M phase. Passage through the four phases of the cell cycle is regulated by a family of cyclins that act as regulatory subunits for cyclin-dependent kinases (cdks). The activity of the various cyclin/cdk complexes that regulate the progression through G1-S-G2 phases of the cell cycle is controlled by the synthesis of the appropriate cyclins during a specific phase of the cell cycle. The cyclin/cdk complex is then activated by the sequential phosphorylation and dephosphorylation of the key residues of the complex, located principally on the cdk subunits. The cyclin cdk complex of early G1 is either cdk2, cdk4, or cdk6 bound to a cyclin D isoform. There are several proteins that can inhibit the cell cycle in G1. If DNA damage has occurred, p53 accumulates in the cell and induces the p21-mediated inhibition of cyclin D/cdk. Mdm2, by facilitating the nuclear export/inactivation of p53, becomes part of an inhibitory feedback loop that inactivates p21-mediated G1 arrest.

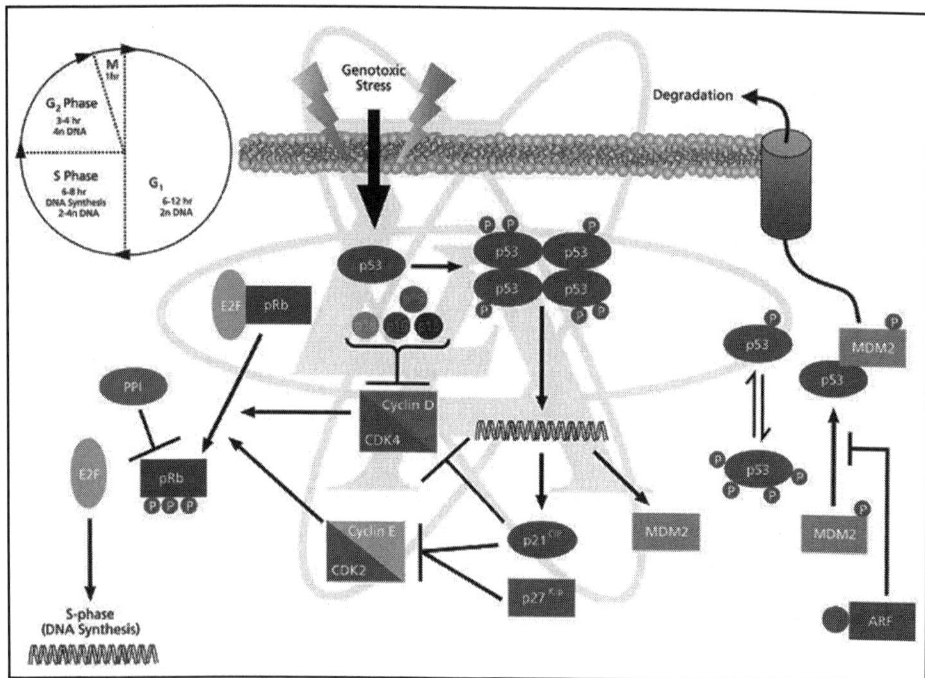

Similarly, activation of TGF-b receptors induces the inhibition of cyclin D/ cdk by p15, while cyclic AMP inhibits the cyclin D/cdk complex via p27. If the cyclin D/cdk complex is inhibited, retinoblastoma protein (Rb) is in a state of low phosphorylation and is tightly bound to the transcription factor E2F, inhibiting its activity. Passage through the restriction point and transition to S phase is triggered by the activation of the cyclin D/cdk complex, which phosphorylates Rb. Phosphorylated Rb dissociates from E2F, which is then free to initiate DNA replication. Cyclin E/cdk2 accumulates during late G phase and triggers the passage into S phase. The entire genome is replicated during S phase. The synthesis and accumulation of cyclin B/cdc2 also begins during S phase, but the complex is phosphorylated at Thr14 -Tyr15 and is inactive. Cyclin A/cdk2 accumulates during S phase and its activation triggers the transition to G2, a phase characterized by the accumulation of cyclin B/cdc2, the inhibition of DNA replication, cell growth and new protein synthesis.

G2 AND M PHASES OF THE CELL CYCLE

The transition from G2 phase to mitosis is triggered by the Cdc25-mediated activation (dephosphorylation) of the cyclin B/cdc2 complex (MPF). The activation of cyclin B/cdc2 that is necessary for G/M progression is currently the most well characterized step in the cell cycle. CyclinB/cdc2 is activated by phosphorylation of Thr160 and the dephosphorylation of Thr14 and Tyr15.

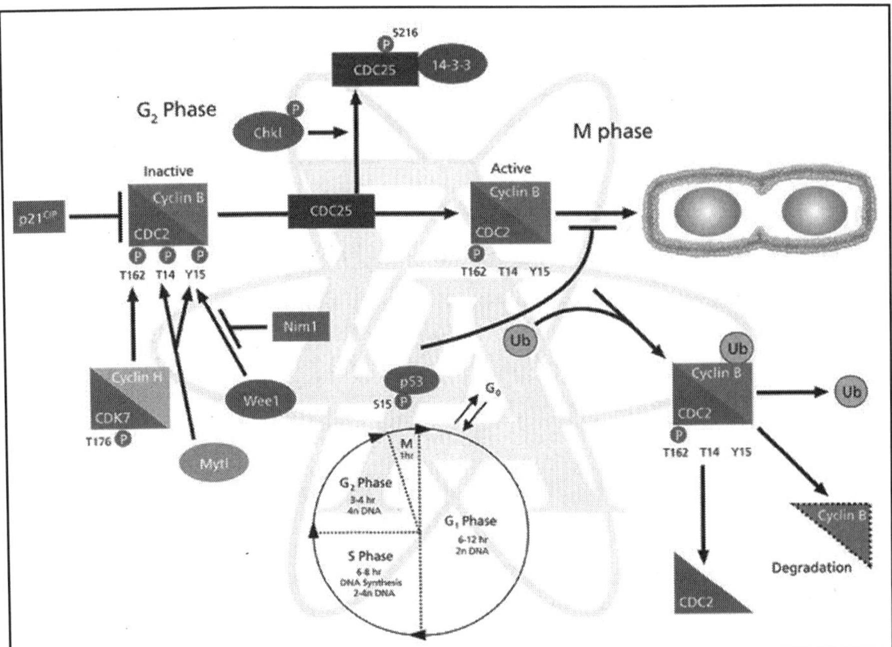

Thr160 is phosphorylated by cyclin activating kinase (CAK), following the activation of CAK by a cyclin activating kinase (CAKAK). However, the complex is kept in an inactive state due to the phosphorylation of Thr15, which is catalyzed by the Weel kinase. Cyclin B/cdc2 activation is triggered when Cdc25, a phosphatase, dephosphorylates Thr15. In turn, the activity of Cdc25 is regulated by both activating and inhibitory phosphorylations. Phosphorylation of Ser216 by Chk1 (a check point activated kinase that participates in the G2-arrest of cells with damaged DNA) leads to the inactivation of Cdc25, while phosphorylation by an M-phase activated kinase creates a positive feedback loop leading to the rapid activation of the cyclin B/cdc2 complex.

MPF catalyzes the phosphorylation of lamins and histone 1, and is involved in the regulation of events receding cell division, such as spindle formation, chromatin condensation, and fragmentation of the nuclear envelope and of organelles such as the Golgi and endoplasmic reticulum. The metaphase to anaphase transition is triggered by inactivation of MPF and the degradation of cyclin B. This induces the separation of chromatids and their movement to the poles of the mitotic spindle, after which the mitotic apparatus disappears, the nuclear membranes reform and the nucleoli reappear. During cytokinesis, the cytoplasm divides and the resulting daughter cells enter G1.

THE LIFE CYCLE OF CELLS

A cell in an adult organism can be viewed as a steady-state system. The DNA is constantly read out into a particular set of mRNAs, which specify a particular set of proteins. As these proteins function, they are also being degraded and replaced by new ones, and the system is so balanced that the cell neither grows, shrinks, nor changes its function. This static view of the cell, however, misses the all-important dynamic aspects of cellular life. The dynamics of a cell can best be understood by examining the course of its life. A new cell arises when one cell divides or when two cells, like a sperm and an egg cell, fuse. Either event sets off a cell-replication programme that is encoded in the DNA and executed by proteins. This programme usually involves a period of cell growth, during which proteins are made and DNA is replicated, followed by cell division, when a cell divides into two daughter cells. Whether a given cell will grow and divide is a highly regulated decision of the body, assuring that an adult organism replaces worn out cells or makes more cells in response to a new need. Examples of the latter are the growth of muscle in response to exercise or damage, and the proliferation of red blood cells when a person ascends to a higher altitude and needs more capacity to capture oxygen. However, in one major and devastating disease — cancer — cells multiply even though they are not needed by the body. To understand how cells become cancerous, biologists have intensely studied the mechanisms that control the growth and division of cells.

THE CELL CYCLE FOLLOWS A REGULAR TIMING MECHANISM

Most eukaryotic cells live according to an internal clock; that is, they proceed through a sequence of phases, called the cell cycle, during which DNA is duplicated during the synthesis (S) phase and the copies are distributed to opposite ends of the cell during mitotic (M) phase. Progress along the cycle is controlled at key checkpoints, which monitor the status of a cell, for instance, the internal amount of DNA or the presence of extracellular nutrients. When certain conditions are met, the cell proceeds to the next checkpoint. The cycle begins after the cell divides into two daughter cells, each containing an identical copy of the parental cell's genetic material. The cell cycle of prokaryotes is simple and fast. Replication of the single chromosome begins at a particular DNA sequence, the replication origin, which is anchored to the cell membrane. Once DNA replication is complete, assembly of new membrane and cell wall forms a septum, which eventually divides the cell in two. Because the origins of the two newly formed chromosomes are anchored to different membrane sites, each daughter cell receives one chromosome. In ideal growth conditions, the bacterial cell cycle is repeated every 30 minutes. Only a few types of eukaryotic cells can grow and divide as quickly as bacteria. Most growing plant and animal cells take 10 – 20 hours to double in number, and some duplicate at a much slower rate.

Many cells in adult animals, such as nerve cells and striated muscle cells, do not divide at all. They have temporarily exited from the cell cycle after mitosis and entered a "paused or quiescent" state called G_0. Because eukaryotic cells are larger and more complex than prokaryotic cells, a specialized mechanism coordinates their replication of genomic DNA, distribution of chromosomes, and cell division.

MITOSIS APPORTIONS

Mitosis is the mechanism in eukaryotes for partitioning the genome equally at cell division. To accomplish this complex task, plant and animal cells build a specialized machine, called the mitotic apparatus, which captures the chromosomes and then pushes and pulls them to opposite sides of the dividing cell.

Remarkably, the mitotic apparatus is a temporary structure that exists only during mitosis to distribute the genetic material. Although the events of mitosis unfold continuously, they are conventionally divided into four substages representing phases of chromosome movement. During the first substage, prophase, the replicated chromosomes, each comprising two identical chromatids, are condensed into compact packets and then released to the cytoplasm when the nuclear membrane breaks down.

During metaphase and anaphase, the chromosomes are sorted, and each chromatid of a pair moves to opposite sides of the cell. The end of mitosis is marked by re-formation of a membrane around each set of chromosomes (telophase). Division of the cytoplasm, called cytokinesis, then yields two daughter cells, each with a $2n$ complement of genetic material. Cell division in plant and animal cells differs mainly at cytokinesis. Animal cells divide in two by pinching of the cytoplasm. However, because a plant cell is surrounded by a rigid cell wall, daughter cells are formed by building a new cell membrane and cell wall between the two daughter nuclei, thereby cutting the cytoplasm into two portions.

CELL DIFFERENTIATION

The most complicated example of cellular dynamics occurs when a cell changes, or *differentiates,* to carry out a specialized function. This process often is marked by a change in the microscopic appearance, or *morphology,* of the cell. For example, the different structures of a nerve cell and a muscle cell reflect their respective functions in long-distance communication and contraction, highlighting the biological principle that "form follows function." Cell differentiation creates the diversity of cell types that arise during the development of an organism from a fertilized egg.

This is a process of extensive cell multiplication and differentiation. A mammal that starts as one cell becomes an organism with hundreds of diverse cell types such as muscle, nerve, and skin. Here we see at its most dramatic the power of DNA to control cellular behaviour: development is a DNA-orchestrated set of cellular changes (easily tens of thousands of them) that occur virtually without fail. The almost perfect resemblance of "identical" twins is a testament to the programme encoded by DNA to reproducibly direct the development of a human being.

Nowhere is the variety of cellular activities and responses better illustrated than in the body's immune system. It is there that many cell types come together in organized tissues specifically designed to allow the body to distinguish its own cells from those of foreign invaders. Within the immune system, we see both development of specialized cells that can recognize invading cells and formation of tissues from cells that originate in various parts of the body. The immune-system cells not only actively survey their environment with surface receptor proteins like antibodies, but also change their properties when they encounter a foreign substance, allowing the body to rid itself of invaders.

CELLS DIE BY SUICIDE

Unchecked cell growth and multiplication produce a mass of cells, a tumor. *Programmemed cell death* plays the very important role of population control by balancing cell growth and multiplication. In addition, cell death also eliminates unnecessary cells. For example, during embryogenesis, the digits of our fingers and toes are sculpted by the death of cells in the intervening spaces. If these cells remained alive, our hands and feet would become webbed. Thus the timing and location of cell death, as well as cell growth and division, must be precisely controlled.

Cell death follows an internal programme of events called apoptosis, in which all traces of a cell vanish. The first visible sign of apoptosis is condensation of the nucleus and fragmentation of the DNA. The cell soon shrivels and is consumed by macrophages. A cell is directed to commit suicide when an essential factor is removed from the extracellular environment or when an internal signal is activated. Thus, the default state of the cell is to remain alive. The discovery of genes that suppress the growth of tumors by activating cell death stimulated an exciting new line of cancer research that may lead to more effective treatment strategies

INTEGRATING CELLS INTO TISSUES

The evolution of multicellular organisms permitted specialized cells and tissues to form; a flowering plant has at least 15 cell types, and a vertebrate hundreds. In both plants and animals, cells that are specialized to carry out a particular task are found together in the tissues in which the task is performed: a xylem or meristem; a liver, a muscle, or a nerve ganglion. Different types of cells in a tissue are often arranged in precise patterns of staggering complexity. For instance, the hundreds of different types of neurons in the human brain are interconnected to one another through a network of some 10^{15} synaptic connections! The coordinated functioning of many types of cells within tissues, and of multiple specialized tissues, permits the organism as a whole to move, metabolize, reproduce, and carry out other essential activities.

A key step in the evolution of multicellularity must have been the ability of cells to contact tightly and interact specifically with other cells. Various integral membrane proteins, collectively termed cell-adhesion molecules (CAMs), enable many animal cells to adhere tightly and specifically with cells of the same, or similar, type; these interactions allow populations of cells to segregate into distinct tissues. Following aggregation, cells elaborate specialized cell junctions that stabilize these interactions and promote local communication between adjacent cells.

Animal cells also secrete a complex network of proteins and carbohydrates, the extracellular matrix (ECM), that creates a special environment in the spaces between cells. The matrix helps bind the cells in tissues together and is a reservoir for many hormones controlling cell growth and differentiation. The matrix also provides a lattice through which cells can move, particularly during the early stages of differentiation.

Defects in these connections lead to cancer and developmental malformations. The extracellular matrix has three major protein components: highly viscous proteoglycans, which cushion cells; insoluble collagen fibers, which provide strength and resilience; and soluble multiadhesive matrix proteins, which bind these components to receptors on the cell surface. Different combinations of these components tailor the strength of the extracellular matrix for different purposes. For example, animals contain many types of extracellular matrices, each specialized for a particular function such as strength (in a tendon), cushioning (in cartilage), or adhesion. In the case of smooth muscle cells that surround an artery, the extracellular matrix must provide strong but flexible connections.

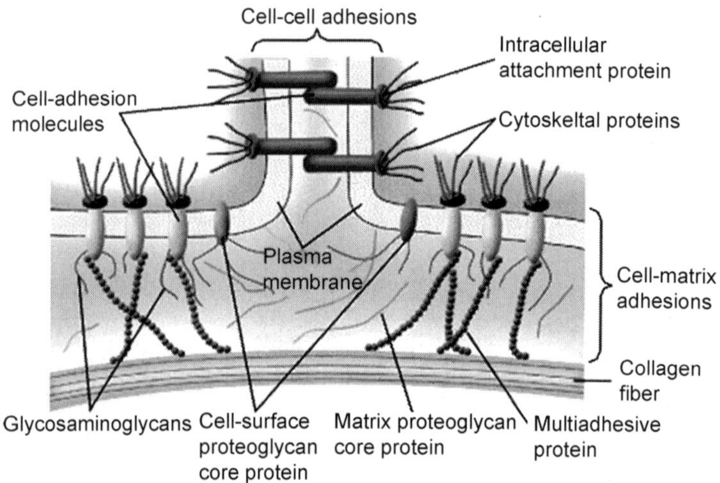

Fig.Integrating Cells into Tissues.

The extracellular matrix is not just an inert framework or cage that supports or surrounds cells. The matrix also communicates directly and indirectly with the intracellular signaling pathways that direct a cell to carry out specific functions. For example, the ability of hepatocytes—the principal cells in the liver—to express liver-specific proteins depends on their association with a matrix of appropriate composition. Specific ECM components can directly activate cytosolic signal-transduction pathways by binding to cell-adhesion protein receptors in the plasma membrane.

Alternatively, by binding growth factors and other hormones, the matrix can either sequester these signals from cells or, conversely, present them to cells, thereby indirectly inducing or inhibiting signaling pathways. Morphogenesis—the later stage of embryonic development during which form is achieved by cell movements and rearrangements—also is critically dependent on ECM components, which are constantly being remodeled, degraded, and resynthesized locally. Even in adults—in areas of wounding, for example degradation and resynthesis of ECM components occurs.

In this chapter, we focus on the structure of and interactions between ECM components, cell-adhesion molecules, and cell-adhesion junctions—the main structures that permit animal cells to form organized tissues. Plant cells are surrounded by a cell wall that is thicker and more rigid than the extracellular matrix. Although the plant cell wall and the extracellular matrix serve many of the same functions, they are structurally very different.

REGULATORS OF CELL CYCLE PROGRESSION

One of the most exciting developments of the last decade has been the elucidation of the molecular mechanisms that control the progression of eukaryotic cells through the division cycle. Our current understanding of cell cycle regulation has emerged from a convergence of results obtained through experiments on organisms as diverse as yeasts, sea urchins, frogs, and mammals. These studies have revealed that the cell cycle of all eukaryotes is controlled by a conserved set of protein kinases, which are responsible for triggering the major cell cycle transitions.

MPF

Three initially distinct experimental approaches contributed to identification of the key molecules responsible for cell cycle regulation. The first of these avenues of investigation originated with studies of frog oocytes. These oocytes are arrested in the G_2 phase of the cell cycle until hormonal stimulation triggers their entry into the M phase of meiosis. In 1971, two independent teams of researchers (Yoshio Masui and Clement Markert, as well as Dennis Smith and Robert Ecker) found that oocytes arrested in G_2 could be induced to enter M phase by microinjection of cytoplasm from oocytes that had been hormonally stimulated.

It thus appeared that a cytoplasmic factor present in hormone-treated oocytes was sufficient to trigger the transition from G_2 to M in oocytes that had not been exposed to hormone. Because the entry of oocytes into meiosis is frequently referred to as oocyte maturation, this cytoplasmic factor was called maturation promoting factor (MPF). Further studies showed, however, that the activity of MPF is not restricted to the entry of oocytes into meiosis. To the contrary, MPF is also present in somatic cells, where it induces entry into M phase of the mitotic cycle. Rather than being specific to oocytes, MPF thus appeared to act as a general regulator of the transition from G_2 to M. The second approach to understanding cell cycle regulation was the genetic analysis of yeasts, pioneered by Lee Hartwell and his colleagues in the early 1970s. Studying the budding yeast *Saccharomyces cerevisiae*, these investigators identified temperature-sensitive mutants that were defective in cell cycle progression.

The key characteristic of these mutants (called *cdc* for *c*ell *d*ivision *c*ycle mutants) was that they underwent growth arrest at specific points in the cell cycle. For example, a particularly important mutant designated *cdc28* caused the cell cycle to arrest at START, indicating that the Cdc28 protein is required for passage through this critical regulatory point in G_1. A similar collection of cell cycle mutants was isolated in the fission yeast *Schizosaccharomyces pombe* by Paul Nurse and his collaborators. These mutants included *cdc2*, which arrests the *S. pombe* cell cycle both in G_1 and at the G_2 to M transition (the major regulatory point in fission yeast). Further studies showed that *S.. cerevisiae cdc28* and *S. pombe cdc2* are functionally homologous genes, which are required for passage through START as well as for entry into mitosis in both species of yeasts. To avoid confusion resulting from the difference in genetic nomenclature between *S.. cerevisiae* and *S. pombe*, the protein encoded by both genes will be called Cdc2 in this text. Further studies of *cdc2* yielded two important insights. First, molecular cloning and nucleotide sequencing revealed that *cdc2* encodes a protein kinase—the first indication of the prominent role of protein phosphorylation in regulating the cell cycle. Second, a human gene related to *cdc2* was identified and shown to function in yeasts, providing a dramatic demonstration of the conserved activity of this cell cycle regulator.

The third line of investigation that eventually converged with the identification of MPF and yeast genetics emanated from studies of protein synthesis in early sea urchin embryos. Following fertilization, these embryos go through a series of rapid cell divisions. Intriguingly, studies with protein synthesis inhibitors had revealed that entry into M phase of these embryonic cell cycles requires new protein synthesis. In 1983, Tim Hunt and his colleagues identified two proteins that display a periodic pattern of accumulation and degradation in sea urchin and clam embryos. These proteins accumulate throughout interphase and are then rapidly degraded toward the end of each mitosis. Hunt called these proteins cyclins (the two proteins were designated cyclin A and cyclin B) and suggested that they might function to induce mitosis, with their periodic accumulation and destruction controlling entry and exit from M phase. Direct support for such a role of cyclins was provided in 1986, when It was shown that microinjection of cyclin A into frog oocytes is sufficient to trigger the G_2 to M transition.

These initially independent approaches converged dramatically in 1988, when MPF was purified from frog eggs in the laboratory of James Maller. Molecular characterization of MPF in several laboratories then showed that this conserved regulator of the cell cycle is composed of two key subunits: Cdc2 and cyclin B. Cyclin B is a regulatory subunit required for catalytic activity of the Cdc2 protein kinase, consistent with the notion that MPF activity is controlled by the periodic accumulation and degradation of cyclin B during cell cycle progression. A variety of further studies have confirmed this role of cyclin B, as well as demonstrating the regulation of MPF by phosphorylation and dephosphorylation of Cdc2.

In mammalian cells, cyclin B synthesis begins in S phase. Cyclin B then accumulates and forms complexes with Cdc2 throughout S and G_2. As these complexes form, Cdc2 is phosphorylated at two critical regulatory positions. One of these phosphorylations occurs on threonine-161 and is required for Cdc2 kinase activity. The second is a phosphorylation of tyrosine-15, and of the adjacent threonine-14 in vertebrates. Phosphorylation of tyrosine-15, catalysed by a protein kinase called Weel, inhibits Cdc2 activity and leads to the accumulation of inactive Cdc2/cyclin B complexes throughout S and G_2. The transition from G_2 to M is then brought about by activation of the Cdc2/cyclin B complex as a result of dephosphorylation of threonine-14 and tyrosine-15 by a protein phosphatase called Cdc25. Once activated, the Cdc2 protein kinase phosphorylates a variety of target proteins that initiate the events of M phase. In addition, Cdc2 activity triggers the degradation of cyclin B, which occurs as a result of ubiquitin-mediated proteolysis. This proteolytic destruction of cyclin B then inactivates Cdc2, leading the cell to exit mitosis, undergo cytokinesis, and return to interphase.

CYCLINS AND CYCLIN-DEPENDENT KINASES

The structure and function of MPF (Cdc2/cyclin B) provide not only a molecular basis for understanding entry and exit from M phase, but also the foundation for elucidating the regulation of other cell cycle transitions. The insights provided by characterization of the Cdc2/cyclin B complex have thus had a sweeping impact on understanding cell cycle regulation. In particular, further research has established that both Cdc2 and cyclin B are members of large families of related proteins, with different members of these families controlling progression through distinct phases of the cell cycle.

The Cdc2 controls passage through START as well as entry into mitosis in yeasts. It does so, however, in association with distinct cyclins. In particular, the G_2 to M transition is driven by Cdc2 in association with the mitotic B-type cyclins (Clb1, Clb2, Clb3, and Clb4). Passage through START, however, is controlled by Cdc2 in association with a distinct class of cyclins called G_1cyclins or Cln's. Cdc2 then associates with different B-type cyclins (Clb5 and Clb6), which are required for progression through S phase. These associations of Cdc2 with distinct B-type and G_1 cyclins direct Cdc2 to phosphorylate different substrate proteins, as required for progression through specific phases of the cell cycle. The cell cycles of higher eukaryotes are controlled not only by multiple cyclins, but also by multiple Cdc2-related protein kinases. These Cdc2-related kinases are known as Cdk's (for *c*yclin-*d*ependent *k*inases). As the original member of this family, Cdc2 is also known as Cdk1, with other currently identified family members being designated Cdk2 through Cdk8.

These multiple members of the Cdk family associate with specific cyclins to drive progression through the different stages of the cell cycle. For example, progression from G_1 to S is regulated principally by Cdk2 and Cdk4 (and in some cells Cdk6) in association with cyclins D and E. Complexes of Cdk4 and Cdk6 with the D-type cyclins (cyclin D1, D2, and D3) play a critical role in progression through the restriction point in G_1. Cyclin E is expressed later in G_1, and Cdk2/cyclin E complexes are required for the G_1 to S transition and initiation of DNA synthesis. Complexes of Cdk2 with cyclin A function in the progression of cells through S phase. The transition from G_2 to M is then driven by complexes of Cdc2 with cyclin B.

The activity of Cdk's during cell cycle progression is regulated by at least four molecular mechanisms. As already discussed for Cdc2, the first level of regulation involves the association of Cdk's with their cyclin partners. Thus, the formation of specific Cdk/cyclin complexes is controlled by cyclin synthesis and degradation. Second, activation of Cdk/ cyclin complexes requires phosphorylation of a conserved Cdk threonine residue around position 160. This activating phosphorylation of the Cdk's is catalysed by an enzyme called CAK (for *C*dk-*a*ctivating *k*inase), which may itself be composed of a Cdk (Cdk7) complexed with cyclin H. Complexes of Cdk7 and cyclin H are also associated with the transcription factor TFIIH, which is required for initiation of transcription by RNA polymerase II. It thus appears that this member of the Cdk family may participate in transcription as well as cell cycle regulation. In contrast to the activating phosphorylation by CAK, the third mechanism of Cdk regulation involves inhibitory phosphorylation of tyrosine residues near the Cdk amino terminus, catalysed by the Weel protein kinase. In particular, both Cdc2 and Cdk2 are inhibited by phosphorylation of tyrosine-15, and the adjacent threonine-14 in vertebrates. These Cdk's are then activated by dephosphorylation of these residues by members of the Cdc25 family of protein phosphatases.

In addition to regulation of the Cdk's by phosphorylation, their activities can also be controlled by the binding of inhibitory proteins (called Cdk inhibitors or CKIs) to Cdk/cyclin complexes. In mammalian cells, two families of Cdk inhibitors are responsible for regulating different Cdk/cyclin complexes. Members of the Cip/Kip family regulate all stages of progression through G_1 and S phase by inhibiting complexes of Cdk2, 4, and 6 with cyclins A, D, and E. In contrast, members of the Ink4 family are specific for complexes of Cdk4 and 6 with cyclin D, so the Ink4 CKIs only regulate progression through the restriction point in G_1. In yeast, different CKIs similarly regulate distinct stages of cell cycle progression by inhibiting specific Cdk/cyclin complexes. Control of Cdk inhibitors thus provides an additional mechanism for regulating Cdk activity. The combined effects of these multiple modes of Cdk regulation are responsible for controlling cell cycle progression in response both to checkpoint controls and to the variety of extracellular stimuli that regulate cell proliferation.

Growth Factors

The proliferation of animal cells is regulated largely by a variety of extracellular growth factors that control the progression of cells through the restriction point in late G_1. In the absence of growth factors, cells are unable to pass the restriction point and become quiescent, frequently entering the resting state known as G_0, from which they can reenter the cell cycle in response to growth factor stimulation. This control of cell cycle progression by extracellular growth factors implies that the intracellular signalling pathways stimulated downstream of growth factor receptors ultimately act to regulate components of the cell cycle machinery.

One critical link between growth factor signalling and cell cycle progression is provided by the D-type cyclins. Cyclin D synthesis is induced in response to growth factor stimulation as a result of signalling through the Ras/Raf/ERK pathway, and the D-type cyclins continue to be synthesized as long as growth factors are present. However, the D-type cyclins are also rapidly degraded, so their intracellular concentrations rapidly fall if growth factors are removed. Thus, as long as growth factors are present through G_1, complexes of Cdk4, 6/cyclin D drive cells through the restriction point. On the other hand, if growth factors are removed prior to this key regulatory point in the cell cycle, the levels of cyclin D rapidly fall and cells are unable to progress through G_1 to S, instead becoming quiescent and entering G_0. The inducibility and rapid turnover of D-type cyclins thus integrates growth factor signalling with the cell cycle machinery, allowing the availability of extracellular growth factors to control the progression of cells through G_1.

Since cyclin D is a critical target of growth factor signalling, it might be expected that defects in cyclin D regulation could contribute to the loss of growth regulation characteristic of cancer cells. Consistent with this expectation, many human cancers have been found to arise as a result of defects in cell cycle regulation, just as many others result from abnormalities in the intracellular signalling pathways activated by growth factor receptors. For example, mutations resulting in continual unregulated expression of cyclin D1 contribute to the development of a variety of human cancers, including lymphomas and breast cancers. Similarly, mutations that inactivate the Ink4 Cdk inhibitors (*e.g.*, p16) that bind to Cdk4, 6/cyclin D complexes are commonly found in human cancer cells.

The connection between cyclin D, growth control, and cancer is further fortified by the fact that a key substrate protein of Cdk4, 6/cyclin D complexes is itself frequently mutated in a wide array of human tumours. This protein, designated Rb, was first identified as the product of a gene responsible for retinoblastoma, a rare inherited childhood eye tumor. Further studies then showed that mutations resulting in the absence of functional Rb protein are not restricted to retinoblastoma but also contribute to a variety of common human cancers. Rb is the prototype of a tumor suppressor gene—a gene whose inactivation leads to tumor development. Whereas oncogene proteins such as Ras and cyclin D drive cell proliferation, the proteins encoded by tumor suppressor genes act as brakes that slow down cell cycle progression. Additional examples of cell cycle regulators encoded by tumor suppressor genes include the Ink4 Cdk inhibitors that bind Cdk4, 6/cyclin D complexes and the important growth regulator p53, which was discussed earlier in this chapter. Further studies of Rb have revealed that it plays a key role in coupling the cell cycle machinery to the expression of genes required for cell cycle progression and DNA synthesis.

The activity of Rb is regulated by changes in its phosphorylation as cells progress through the cycle. In particular, Rb becomes phosphorylated by Cdk4, 6/cyclin D complexes as cells pass through the restriction point in G_1. In its underphosphorylated form (present in G_0 or early G_1), Rb binds to members of the E2F family of transcription factors, which regulate expression of several genes involved in cell cycle progression, including the gene encoding cyclin E. E2F binds to its target sequences in either the presence or absence of Rb. However, Rb acts as a repressor, so the Rb/E2F complex suppresses transcription of E2F-regulated genes. Phosphorylation of Rb by Cdk4, 6/cyclin D complexes results in its dissociation from E2F, which then activates transcription of its target genes. Rb thus acts as a molecular switch that converts E2F from a repressor to an activator of genes required for cell cycle progression. The control of Rb via Cdk4, 6/cyclin D phosphorylation in turn couples this critical regulation of gene expression to the availability of growth factors in G_1.

INHIBITORS OF CELL CYCLE PROGRESSION

Cell proliferation is regulated not only by growth factors but also by a variety of signals that act to inhibit cell cycle progression. For example, agents that damage DNA result in cell cycle arrest, presumably to allow time for the cell to repair the damage. In addition, cell contacts and a variety of extracellular factors act to inhibit rather than stimulate proliferation of their target cells. The effects of such inhibitory signals are also mediated by regulators of the cell cycle machinery, frequently via the induction of Cdk inhibitors. A good example of the action of Cdk inhibitors is provided by cell cycle arrest in response to DNA damage, which is mediated by the protein p53. The p53 protein is a transcriptional regulator that functions, at least in part, to stimulate expression of the Cdk inhibitor p21.

The p21 protein inhibits several Cdk/cyclin complexes, and its induction by p53 appears to represent at least one mechanism of p53-dependent cell cycle arrest following DNA damage. In addition to inhibiting cell cycle progression via its interaction with Cdk's, p21 may directly inhibit DNA replication. In particular, p21 binds to proliferating cell nuclear antigen (PCNA) is a subunit of DNA polymerase d. Thus, p21 may play a dual role in cell cycle arrest induced by DNA damage, not only blocking cell cycle progression by inhibiting Cdk's but also directly inhibiting DNA replication in S phase cells. The best-characterized extracellular inhibitor of animal cell proliferation is TGF-b—a polypeptide factor that inhibits the proliferation of a variety of types of epithelial cells by arresting cell cycle progression in G_1. This action of TGF-b appears to be mediated by induction of the Cdk inhibitor p15, which binds to Cdk4, 6/cyclin D complexes. In the resulting absence of Cdk4 activity, Rb phosphorylation is blocked and the cell cycle is arrested in G_1. A different molecular mechanism is used to control cell cycle progression through the G_2 checkpoint, which prevents entry into mitosis in the presence of unreplicated or damaged DNA. Arrest of the cell cycle at this checkpoint is mediated by a protein kinase called Chk1, which is activated in response to DNA damage or incomplete replication. Chk1 phosphorylates the protein phosphatase Cdc25, thereby preventing Cdc25 from dephosphorylating and activating Cdc2. In the absence of Cdc2 activation, progression to mitosis is blocked and the cell remains arrested in G_2.

THE M PHASE

M phase is the most dramatic period of the cell cycle, involving a major reorganization of virtually all cell components. During mitosis (nuclear division), the chromosomes condense, the nuclear envelope of most cells breaks down, the cytoskeleton reorganizes to form the mitotic spindle, and the chromosomes move to opposite poles. Chromosome segregation is then usually followed by cell division (cytokinesis).

Stages of Mitosis

Although many of the details of mitosis vary among different organisms, the fundamental processes that ensure the faithful segregation of sister chromatids are conserved in all eukaryotes. These basic events of mitosis include chromosome condensation, formation of the mitotic spindle, and attachment of chromosomes to the spindle microtubules. Sister chromatids then separate from each other and move to opposite poles of the spindle, followed by the formation of daughter nuclei.

Mitosis is conventionally divided into four stages—prophase, metaphase, anaphase, and telophase. The beginning of prophase is marked by the appearance of condensed chromosomes, each of which consists of two sister chromatids (the daughter DNA molecules produced in S phase). These newly replicated DNA molecules remain intertwined throughout S and G_2, becoming untangled during the process of chromatin condensation. The condensed sister chromatids are then held together at the centromere, which is a DNA sequence to which proteins bind to form the kinetochore—the site of eventual attachment of the spindle microtubules. In addition to chromosome condensation, cytoplasmic changes leading to the development of the mitotic spindle initiate during prophase. The centrosomes (which had duplicated during interphase) separate and move to opposite sides of the nucleus. There they serve as the two poles of the mitotic spindle, which begins to form during late prophase.

In higher eukaryotes the end of prophase corresponds to the breakdown of the nuclear envelope. However, nuclear envelope breakdown is not a universal feature of mitosis. In particular, yeasts and many other unicellular eukaryotes undergo "closed mitosis," in which the nuclear envelope remains intact. In these cells the spindle pole bodies are embedded within the nuclear envelope, and the nucleus divides in two following migration of daughter chromosomes to opposite poles of the spindle. Following completion of prophase, the cell enters prometaphase—a transition period between prophase and metaphase. During prometaphase the microtubules of the mitotic spindle attach to the kinetochores of condensed chromosomes. The kinetochores of sister chromatids are oriented on opposite sides of the chromosome, so they attach to microtubules emanating from opposite poles of the spindle. The chromosomes shuffle back and forth until they eventually align on the metaphase plate in the centre of the spindle. At this stage, the cell has reached metaphase.

Most cells remain only briefly at metaphase before proceeding to anaphase. The transition from metaphase to anaphase is triggered by breakage of the link between sister chromatids, which then separate and move to opposite poles of the spindle. Mitosis ends with telophase, during which nuclei reform and the chromosomes decondense. Cytokinesis usually begins during late anaphase and is almost complete by the end of telophase, resulting in the formation of two interphase daughter cells.

MPF and Progression to Metaphase

Mitosis involves dramatic changes in multiple cellular components, leading to a major reorganization of the entire structure of the cell. These events are initiated by activation of the MPF protein kinase (Cdc2/cyclin B). It appears that MPF not only acts as a master regulator of the M phase transition, phosphorylating and activating other downstream protein kinases, but also acts directly by phosphorylating some of the structural proteins involved in this cellular reorganization.

The condensation of interphase chromatin to form the compact chromosomes of mitotic cells is a key event in mitosis, critical in enabling the chromosomes to move along the mitotic spindle without becoming broken or tangled with one another. The chromatin in interphase nuclei condenses nearly a thousand fold during the formation of metaphase chromosomes. Such highly condensed chromatin cannot be transcribed, so transcription ceases as chromatin condensation takes place. Despite the fundamental importance of this event, we do not fully understand either the structure of metaphase chromosomes or the molecular mechanism of chromatin condensation. However, protein complexes called condensins have recently been found to drive chromosome condensation by wrapping DNA around itself, compacting chromosomes into the condensed mitotic structure.

The condensins are phosphorylated directly by the Cdc2 protein kinase, which drives chromatin condensation by activating condensins as cells enter mitosis. One molecular alteration that generally accompanies chromosome condensation is phosphorylation of histone H1, so it is noteworthy that histone H1 is also a substrate for Cdc2. However, histone H1 phosphorylation is not required for mitotic chromosome condensation, so the significance of H1 phosphorylation by Cdc2 is unclear. In contrast, chromosome condensation has been shown to require phosphorylation of histone H3. Perhaps surprisingly, however, histone H3 is not phosphorylated by Cdc2 and the kinase responsible for H3 phosphorylation in mitotic cells remains to be identified. Breakdown of the nuclear envelope, which is one of the most dramatic events of mitosis, represents the most clearly defined target for MPF action. The Cdc2 phosphorylates the lamins, leading directly to depolymerization of the nuclear lamina. This is followed by fragmentation of the nuclear membrane into small vesicles, which eventually fuse to form new daughter nuclei at telophase. The endoplasmic reticulum and Golgi apparatus similarly fragment into small vesicles, which can then be distributed to daughter cells at cytokinesis. The breakdown of these membranes is also induced by MPF, and may in part be mediated by Cdc2 phosphorylation of the Golgi matrix protein GM130, which is required for the docking of COPI-coated vesicles to the Golgi membrane. Phosphorylation and inactivation of GM130 by Cdc2 inhibits vesicle docking and fusion, leading to fragmentation of the Golgi apparatus. However, additional targets of Cdc2 may also be involved, and the mechanisms by which MPF leads to membrane fragmentation remain to be fully elucidated.

The reorganization of the cytoskeleton that culminates in formation of the mitotic spindle results from the dynamic instability of microtubules. At the beginning of prophase, the centrosomes move to opposite sides of the nucleus. The rise in MPF activity then induces a dramatic change in the dynamic behaviour of microtubules. First, the rate of microtubule disassembly increases, resulting in depolymerization and shrinkage of the interphase microtubules. This disassembly is thought to result from phosphorylation of microtubule-associated proteins, either by MPF itself or by other MPF-activated protein kinases.

In addition, the number of microtubules emanating from the centrosomes increases, so the interphase microtubules are replaced by large numbers of short microtubules radiating from the centrosomes. The breakdown of the nuclear envelope then allows some of the spindle microtubules to attach to chromosomes at their kinetochores, initiating the process of chromosome movement that characterizes prometaphase. The proteins assembled at the kinetochore include microtubule motors that direct the movement of chromosomes toward the minus ends of the spindle microtubules, which are anchored in the centrosome. The action of these proteins, which draw chromosomes toward the centrosome, is opposed by the growth of the spindle microtubules, which pushes the chromosomes away from the spindle poles.

Consequently, the chromosomes in prometaphase shuffle back and forth between the centrosomes and the centre of the spindle.

Microtubules from opposite poles of the spindle eventually attach to the two kinetochores of sister chromatids (which are located on opposite sides of the chromosome), and the balance of forces acting on the chromosomes leads to their alignment on the metaphase plate in the centre of the spindle. The spindle consists of both kinetochore microtubules, which are attached to the chromosomes, and polar microtubules, which overlap with one another in the centre of the cell. In addition, short astral microtubules radiate outward from the centrosomes toward the cell periphery.

Proteolysis and the Inactivation of MPF

An important cell cycle checkpoint monitors the alignment of chromosomes on the metaphase spindle. Once this has been accomplished, the cell proceeds to initiate anaphase and complete mitosis. The progression from metaphase to anaphase results from ubiquitin-mediated proteolysis of key regulatory proteins, triggered by activation of a ubiquitin ligase called the anaphase-promoting complex. Activation of the anaphase-promoting complex is induced by MPF at the beginning of mitosis, so MPF ultimately triggers its own destruction. The anaphase-promoting complex remains inhibited, however, until the cell passes the metaphase checkpoint, after which activation of the ubiquitin degradation system brings about the transition from metaphase to anaphase and progression through the rest of mitosis.

Activation of the anaphase-promoting complex leads to the degradation of at least two key regulatory proteins. The onset of anaphase results from proteolytic degradation of a protein called Scc1, a component of a complex of proteins called cohesins that maintain the connection between sister chromatids while they are aligned on the metaphase plate. Degradation of Scc1 is not catalysed directly by the anaphase-promoting complex, which instead degrades a regulatory protein called Pds1. Degradation of Pds1 in turn activates another protein, called Esp1, which leads to proteolysis of the cohesin Scc1. Cleavage of Scc1 breaks the linkage between sister chromatids, allowing them to segregate by moving to opposite poles of the spindle. The separation of chromosomes during anaphase then proceeds as a result of the action of several types of motor proteins associated with the spindle microtubules.

The other key regulatory protein targeted for ubiquitination and degradation by the anaphase-promoting complex is cyclin B. Degradation of cyclin B leads to inactivation of MPF, which is required for the cell to exit mitosis and return to interphase. Many of the cellular changes involved in these transitions are simply the reversal of the events induced by MPF during entry into mitosis. For example, reassembly of the nuclear envelope, chromatin decondensation, and the return of microtubules to an interphase state probably result directly from loss of MPF activity and dephosphorylation of proteins that had been phosphorylated by MPF at the beginning of mitosis.

Cytokinesis

The completion of mitosis is usually accompanied by cytokinesis, giving rise to two daughter cells. Cytokinesis usually initiates in late anaphase and is triggered by the inactivation of MPF, thereby coordinating nuclear and cytoplasmic division of the cell. The cytokinesis of animal cells is mediated by a contractile ring of actin and myosin II filaments that forms beneath the plasma membrane. The location of this ring is determined by the position of the mitotic spindle, so the cell is eventually cleaved in a plane that passes through the metaphase plate perpendicular to the spindle. Cleavage proceeds as contraction of the actin-myosin filaments pulls the plasma membrane inward, eventually pinching the cell in half.

The mechanism of cytokinesis is different for higher plant cells, which are surrounded by rigid cell walls. Rather than being pinched in half by a contractile ring, these cells divide by forming new cell walls and plasma membranes inside the cell. In early telophase, vesicles carrying cell wall precursors from the Golgi apparatus associate with spindle microtubules and accumulate at the former site of the metaphase plate. These vesicles then fuse to form a large, membrane-enclosed, disclike structure, and their polysaccharide contents assemble to form the matrix of a new cell wall (called a cell plate). The cell plate expands outward, perpendicular to the spindle, until it reaches the plasma membrane. The membrane surrounding the cell plate then fuses with the parental plasma membrane, dividing the cell in two.

THE EUKARYOTIC CELLS

Recognition that modern-day organisms fall into three distinct evolutionary lineages came from comparisons of the nucleotide sequences of genes common to all organisms and of ribosomal RNAs, particularly the RNA species found in the small ribosomal subunit. Prokaryotes, which lack a defined nucleus and have a simple subcellular organization, form two of the lineages—the bacteria and archaea. The third lineage is the eukaryotes (eukarya), whose cells have a membrane-limited nucleus containing most of the cellular DNA, numerous specialized organelles, and a complex cytoskeleton.

Despite the differences in cellular organization between prokaryotes and eukaryotes, all cells share certain structural features and carry out DNA replication, protein synthesis, production of ATP, and many other complicated metabolic events in basically the same way. At the molecular level there is a close relationship between protein structure and function. Likewise, at the cellular level, function and structure are intimately related. Thus this chapter describes the basic structural components of cells and the general functions of each, focusing on eukaryotic cells. We begin by describing the capabilities and applications of various techniques of microscopy used to visualize cells and subcellular structures. We then consider the basic methods for purifying biological membranes and subcellular structures. The purified preparations produced by cell fractionation permit detailed studies on the structure and function of biomembranes and other subcellular structures. Following this overview of experimental techniques, we continue the discussion of biomembranes begun in earlier chapters. The surface membrane found on all cells and the membranes that line eukaryotic organelles all have the same basic architecture — a phospholipid bilayer; the unique function of each type of membrane is determined primarily by the proteins it contains. The final section briefly describes the structure and function of the main internal organelles and fibers of eukaryotic cells, as well as the extracellular substances that surround cells and give them shape and strength. This review sets the stage for later discussions of cell function at the molecular level.

MICROSCOPY AND CELL ARCHITECTURE

The modern, detailed understanding of cell architecture is based on several types of microscopy. Because there is no one "correct" view of a cell, it is essential to understand the characteristics of the key cell-viewing techniques, the types of images they produce, and their limitations. Schleiden and Schwann, using a primitive light microscope, first described individual cells as the fundamental unit of life, and light microscopy has continued to play a major role in biological research.

The development of electron microscopes greatly extended the ability to resolve subcellular particles and has yielded much new information on the organization of plant and animal tissues. The nature of the images depends on the type of light or electron microscope employed and on the way in which the cell or tissue has been prepared. Each technique is designed to emphasize particular structural features of the cell.

The epithelial cell lining the small intestine, appears when viewed by three different microscopic techniques. In this section, we focus on the most common application of light and electron microscopy—to visualize fixed, killed cells. Although this approach reveals much information, a critical question about such results is how true to life is the image of a biological specimen that has been fixed, stained, and dehydrated before examination? Thus we also consider some of the refinements that allow microscopy of unaltered or less altered specimens.

Light Microscopy

The *compound microscope,* the most common microscope in use today, contains several lenses that magnify the image of a specimen under study. The total magnification is a product of the magnification of the individual lenses: if the *objective lens* magnifies 100-fold (a 100X lens, the maximum usually employed) and the *eyepiece* magnifies 10-fold, the final magnification recorded by the human eye or on film will be 1000-fold.

However, the most important property of any microscope is not its magnification but its resolving power, or resolution — its ability to distinguish between two very closely positioned objects. Merely enlarging the image of a specimen accomplishes nothing if the image is blurry. The resolution of a microscope lens is numerically equivalent to *D,* the minimum distance between two distinguishable objects; the smaller the value of *D,* the better the resolution. *D* depends on three parameters, all of which must be considered in order to achieve the best possible resolution: the *angular aperture,* a, or half-angle of the cone of light entering the objective lens from the specimen; the *refractive index, N,* of the air or fluid medium between the specimen and the objective lens; and the *wavelength,* 1, of incident light: D = (0.611) ÷ (*N* × sin a). Decreasing the value of 1 or increasing either *N* or a will decrease the value of *D* and thus improve the resolution. Note that the magnification is not part of this equation.

The angular aperture, a, depends on the width of the objective lens and its distance from the specimen. Moving the objective lens closer to the specimen increases the angle a and thus sin a, and therefore reduces *D* (*i.e.*, increases the resolution).

Intuitively, one can recognize that increasing a allows a greater fraction of the light emanating from the specimen to enter the objective lens. The refractive index *N* is a measure of the degree to which a medium bends a light ray that passes through it; the refractive index of air is defined as 1.0. Use of immersion oil, which has a refractive index of 1.5, is a simple way to reduce *D* by 33 percent. An intuitive explanation for this improvement is that a medium with a higher refractive index than air, if placed between the specimen and the objective lens, will "bend" more of the light emanating from the specimen such that it goes into the lens. Finally, the shorter the wavelength of incident light, the lower will be the value of *D* and the better the resolution.

Due to limitations on the values of α, λ, and *N*, the *limit of resolution* of a light microscope using visible light is about 0.2 µm (200 nm). No matter how many times the image is magnified, the microscope can never resolve objects that are less than ≈0.2 µm apart or reveal details smaller than ≈0.2 µm in size. This is true because the maximum angular aperture for the best objective lenses is 70° (sin 70° = 0.94). With the visible light of shortest wavelength (blue, λ = 450 nm) and with an immersion oil (*N* = 1.5) above the sample, then or about 0.2 µm.

$$D = \frac{0.61 \times 450nm}{1.5 \times 0.94} = 194nm$$

Despite this limit of resolution, the light microscope can be used to track the location of a small bead of known size to a precision of only a few nanometers! If we know the precise size and shape of an object — say, a 5-nm sphere of gold—and if we use a video camera to record the microscopic image as a digital image, then a computer can calculate the position of the *centre* of the object to within a few nanometers.

This technique has been used, to nanometer resolution, for tracking the movement of gold particles attached via antibodies to specific proteins on the surface of living cells.

SAMPLES FOR LIGHT MICROSCOPY USUALLY ARE FIXED, SECTIONED, AND STAINED

Specimens for light microscopy are commonly fixed with a solution containing alcohol or formaldehyde, compounds that denature most proteins and nucleic acids. Formaldehyde also cross-links amino groups on adjacent molecules; these covalent bonds stabilize protein-protein and protein – nucleic acid interactions and render the molecules insoluble and stable for subsequent procedures. Usually the sample is then embedded in paraffin or plastic and cut into thin sections of one or a few micrometers thick. Alternatively, the sample can be frozen without prior fixation and then sectioned; this avoids the denaturation of enzymes by fixatives such as formaldehyde.

Since the resolution of the light microscope is 0.2 mm and mitochondria and chloroplasts are 1 mm long (about the size of bacteria), theoretically one should be able to see these organelles. However, most cellular constituents are not coloured and absorb about the same degree of visible light, so that they are hard to distinguish under a light microscope unless the specimen is stained. Thus the final step in preparing a specimen for light microscopy is to stain it, in order to visualize the main structural features of the cell or tissue. Many chemical stains bind to molecules that have specific features. For example, *hematoxylin* binds to basic amino acids (lysine and arginine) on many different kinds of proteins, whereas *eosin* binds to acidic molecules (such as DNA, and aspartate and glutamate side chains). Because of their different binding properties, these dyes stain various cell types sufficiently differently that they are distinguishable visually. Two other common dyes are *benzidine*, which binds to heme-containing proteins and nucleic acids, and *fuchsin,* which binds to DNA and is used in Fuelgen staining. If an enzyme catalyzes a reaction that produces a coloured or otherwise visible precipitate from a colourless precursor, the enzyme may be detected in cell sections by their coloured reaction products. This technique is called *cytochemical staining.*

FLUORESCENCE MICROSCOPY CAN LOCALIZE AND QUANTIFY SPECIFIC MOLECULES IN CELLS

Perhaps the most versatile and powerful technique for localizing proteins within a cell by light microscopy is fluorescent staining of cells and observation in the *fluorescence microscope.* A chemical is said to be *fluorescent* if it absorbs light at one wavelength (the *excitation wavelength*) and emits light (fluoresces) at a specific and longer wavelength.

Most fluorescent dyes emit visible light, but some (such as Cy5 and Cy7) emit infrared light. In modern fluorescence microscopes, only fluorescent light emitted by the sample is used to form an image; light of the exciting wavelength induces the fluorescence but is then not allowed to pass the filters placed between the objective lens and the eye or camera.

REVEALING SPECIFIC PROTEINS IN FIXED CELLS

Four very useful dyes for fluorescent staining are rhodamine and Texas red, which emit red light; Cy3, which emits orange light; and fluorescein, which emits green light. These dyes have a low, nonspecific affinity for biological molecules, but they can be chemically coupled to purified antibodies specific for almost any desired macromolecule. When a fluorescent dye – antibody complex is added to a permeabilized cell or tissue section, the complex will bind to the corresponding antigens, which then light up when illuminated by the exciting wavelength, a technique called *immunofluorescence microscopy*. By staining a specimen with two or three dyes that fluoresce at different wavelengths, multiple proteins can be localized within a cell.

REVEALING SPECIFIC PROTEINS IN LIVING CELLS

Fluorescence microscopy can also be applied to live cells. For example, purified actin may be chemically linked to a fluorescent dye. Careful biochemical studies have established that this "tagged" molecule is indistinguishable in function from its normal counterpart. If the tagged protein is *microinjected* into a cultured cell, the endogenous cellular and injected tagged actin monomers copolymerize into normal long actin fibers. This technique can also be used to study individual microtubules within a cell.

Another technique for detecting specific proteins within living cells takes advantage of *green fluorescent protein* (GFP), a naturally fluorescent protein found in the jellyfish *Aequorea victoria.* The bioluminescence of this organism, which radiates a green fluorescence, is due to GFP. This 238-aa protein contains serine, tyrosine, and glycine residues whose side chains have spontaneously reacted with one another to form a fluorescent chromophore. By recombinant DNA techniques the GFP gene can be introduced into living cultured cells or into specific cells of an entire animal. Because the introduced gene will express GFP, the cells will emit a green fluorescence when irradiated; this GFP fluorescence can be used to localize the cells within a tissue. Alternatively, the gene for GFP can be fused to the gene for another protein of interest, producing a recombinant DNA encoding one long chimeric protein that contains the entirety of both proteins. Cells in which this recombinant DNA has been introduced will synthesize this chimeric protein, whose green fluorescence will reveal the subcellular localization of the protein.

DETERMINING THE INTRACELLULAR CONCENTRATION OF CA^{2+} AND H^+ IONS

Changes in the cytosolic concentration of Ca^{2+} ions or pH frequently signal changes in cellular metabolism. The Ca^{2+} concentration in the cytosol of resting cells, for instance, is about 10^{-7} M. Many hormones or other stimuli cause a rise in cytosolic Ca^{2+} to 10^{-6} M; this, in turn, causes changes in cellular metabolism, such as contraction of muscle. The fluorescent properties of certain dyes, such as *fura-2*, facilitate measurement of the concentration of free Ca^{2+} in the cytosol. This dye contains five carboxylate groups that form ester linkages with ethanol. The resulting fura-2 ester is lipophilic and can diffuse from the medium across the plasma membrane into cells. Within the cytosol, esterases hydrolyze fura-2 ester yielding fura-2, whose free carboxylate groups render the molecule nonlipophilic, so it cannot cross cellular membranes and remains in the cytosol.

Each fura-2 molecule can bind a single Ca^{2+} ion but no other cellular cation, and the amount of fura-2 bound to Ca^{2+} is proportional, over a certain range, to the Ca^{2+} concentration. The fluorescence of fura-2 at one particular wavelength is enhanced when Ca^{2+} is bound, and the fluorescence is proportional to the Ca^{2+} concentration. At another wavelength the fluorescence of fura-2 is the same whether or not Ca^{2+} is bound and provides a measure of the total amount of fura-2 in the segment of the cell. By examining cells continuously in the fluorescence microscope and measuring rapid changes in the ratio of fura-2 fluorescence at these two wavelengths, one can quantify rapid changes in the fraction of fura-2 that has a bound Ca^{2+} ion and thus in the concentration of cytosolic Ca^{2+}. The fluorescence of other dyes is sensitive to the H^+ concentration and can be used in a similar way to monitor the cytosolic pH of living cells.

ORGANELLES OF THE EUKARYOTIC CELL

The various techniques described earlier have led to an appreciation of the highly organized internal structure of eukaryotic cells, marked by the presence of many different organelles. Here we present a brief overview of the major organelles. Unique proteins in the interior and membranes of each type of organelle largely determine its specific functional characteristics. Later chapters will examine the key roles that different organelles and the cytosol play in the functioning of eukaryotic cells.

ACIDIC ORGANELLES

Lysosomes provide an excellent example of the ability of intracellular membranes to form closed compartments in which the composition of the *lumen* (the aqueous interior of the compartment) differs substantially from that of the surrounding cytosol. Found in animal cells, lysosomes are bounded by a single membrane and are responsible for degrading certain components that have become obsolete for the cell or organism. In some cases, materials taken into a cell by endocytosis or phagocytosis also are degraded in lysosomes. Endocytosis refers to the process by which extracellular materials are taken up by invagination of a segment of the plasma membrane to form a small membrane-bounded vesicle (endosome).

In phagocytosis, relatively large particles are enveloped by the plasma membrane and internalized. Lysosomes contain a group of enzymes that degrade polymers into their monomeric subunits. For example, nucleases degrade RNA and DNA into their mononucleotide building blocks; proteases degrade a variety of proteins and peptides; phosphatases remove phosphate groups from mononucleotides, phospholipids, and other compounds; still other enzymes degrade complex polysaccharides and lipids into smaller units. *Tay-Sachs disease* is caused by a defect in one enzyme catalyzing a step in the lysosomal breakdown of certain glycolipids called gangliosides, which are abundant in nerve cells—with devastating consequences. The symptoms of this inherited disease usually are evident before the age of 1. Affected children commonly become demented and blind by age 2, and die before their third birthday. Nerve cells from such children are greatly enlarged with swollen lipid-filled lysosomes.

All the lysosomal enzymes work most efficiently at acid pH values and collectively are termed *acid hydrolases.* A hydrogen ion pump and a Cl^- channel protein in the lysosomal membrane maintain the pH of the interior at ≈4.8. The pump hydrolyzes ATP and uses the released free energy to pump H^+ ions from the cytosol into the lumen of the lysosome; the Cl^- channel allows Xλ ions to enter. Together they transport HCl. The acid pH helps to denature proteins, making them accessible to the action of the lysosomal hydrolases, which themselves are resistant to acid denaturation. Lysosomal enzymes are poorly active at the neutral pH values of cells and most extracellular fluids. Thus if a lysosome releases its contents into the cytosol, where the pH is between 7.0 and 7.3, little degradation of cytosolic components takes place.

Lysosomes vary in size and shape, and several hundred may be present in a typical animal cell. In effect, they function as sites where various materials to be degraded collect. *Primary lysosomes* are roughly spherical and do not contain obvious particulate or membrane debris. *Secondary lysosomes,* which are larger and irregularly shaped, appear to result from the fusion of primary lysosomes with other membrane organelles; they contain particles or membranes in the process of being digested. The process by which an aged organelle is degraded in a lysosome is called *autophagy* ("eating oneself").

PLANT VACUOLES

Most plant cells contain at least one membrane-limited internal vacuole. The number and size of vacuoles depend on both the type of cell and its stage of development; a single vacuole may occupy as much as 80 percent of a mature plant cell. Plant cells store water, ions, and nutrients such as sucrose and amino acids within these vacuoles. We will see how such materials are accumulated in vacuoles. Vacuoles also act as receptacles for waste products and excess salts taken up by the plant and may function similarly to lysosomes in animal cells. Like lysosomes, vacuoles have an acidic pH, maintained by a proton pump and a Cl^- channel protein in the vacuole membrane, and contain a battery of degradative enzymes. Similar storage vacuoles are found in green algae and many microorganisms such as yeast. Like most cellular membranes, the vacuolar membrane is permeable to water but is poorly permeable to the small molecules stored within it. Because the solute concentration is much higher in the vacuole lumen than in the cytosol or extracellular fluids, water tends to move by osmotic flow into vacuoles, just as it moves into cells placed in a hypotonic medium.

This influx of water causes both the vacuole to expand and water to move into the cell from the wall, creating hydrostatic pressure, or *turgor,* inside the cell. This pressure is balanced by the mechanical resistance of the cellulose-containing cell wall that surrounds plant cells. Most plant cells have a turgor of 5 – 20 atmospheres (atm); their cell walls must be strong enough to react to this pressure in a controlled way. Unlike animal cells, plant cells can elongate extremely rapidly—at rates of 20–75 μm/h. This elongation, which usually accompanies plant growth, occurs when a segment of the somewhat elastic cell wall stretches under the pressure created by water taken into the vacuole.

PEROXISOMES DEGRADE FATTY ACIDS AND TOXIC COMPOUNDS

All animal cells (except erythrocytes) and many plant cells contain peroxisomes, a class of small organelles (≈0.2 – 1 μm in diameter) bounded by a single membrane. (*Glyoxisomes* are similar organelles found in plant seeds that oxidize stored lipids as a source of carbon and energy for growth. They contain many of the same types of enzymes as peroxisomes as well as additional ones used to convert fatty acids to glucose precursors.) Peroxisomes contain several *oxidases* — enzymes that use molecular oxygen to oxidize organic substances, in the process forming hydrogen peroxide (H_2O_2), a corrosive substance. Peroxisomes also contain copious amounts of the enzyme *catalase,* which degrades hydrogen peroxide to yield water and oxygen:

$$2H_2O_2 \xrightarrow{catalase} 2\,H_2O + O_2$$

In contrast to oxidation of fatty acids in mitochondria, which produces CO_2 and is coupled to generation of ATP, peroxisomal oxidation of fatty acids yields acetyl groups and is not linked to ATP formation. The energy released during peroxisomal oxidation is converted to heat, and the acetyl groups are transported into the cytosol, where they are used in the synthesis of cholesterol and other metabolites. In most eukaryotic cells, the peroxisome is the principal organelle in which fatty acids are oxidized, thereby generating precursors for important biosynthetic pathways. Particularly in liver and kidney cells, various toxic molecules that enter the bloodstream also are degraded in peroxisomes, producing harmless products.

In the human genetic disease *X-linked adrenoleukodystrophy* (ADL), peroxisomal oxidation of very long chain fatty acids is defective. The *ADL* gene encodes the peroxisomal membrane protein that transports into peroxisomes an enzyme required for oxidation of these fatty acids. Individuals with the severe form of ADL are unaffected until mid-childhood, when severe neurological disorders appear, followed by death within a few years.

Mitochondria Are the Principal Sites

Most eukaryotic cells contain many mitochondria, which occupy up to 25 percent of the volume of the cytoplasm. These complex organelles, the main sites of ATP production during aerobic metabolism, are among the largest organelles, generally exceeded in size only by the nucleus, vacuoles, and chloroplasts. Mitochondria contain two very different membranes, an outer one and an inner one, separated by the intermembrane space. The outer membrane, composed of about half lipid and half protein, contains proteins that render the membrane permeable to molecules having molecular weights as high as 10,000. In this respect, the outer membrane is similar to the outer membrane of gram-negative bacteria.

The inner membrane, which is much less permeable, is about 20 percent lipid and 80 percent protein — a higher proportion of protein than occurs in other cellular membranes. The surface area of the inner membrane is greatly increased by a large number of infoldings, or *cristae,* that protrude into the *matrix,* or central space. In nonphotosynthetic cells, the principal fuels for ATP synthesis are fatty acids and glucose. The complete aerobic degradation of glucose to CO_2 and H_2O is coupled to synthesis of as many as 36 molecules of ATP. In eukaryotic cells, the initial stages of glucose degradation occur in the cytosol, where two ATP molecules per glucose molecule are generated. The terminal stages, including those involving phosphorylation coupled to final oxidation by oxygen, are carried out by enzymes in the mitochondrial matrix and cristae.

As many as 34 ATP molecules per glucose molecule are generated in mitochondria, although this value can vary because much of the energy released in mitochondrial oxidation can be used for other purposes (*e.g.*, heat generation and the transport of molecules into or out of the mitochondrion), making less energy available for ATP synthesis. Similarly, virtually all the ATP formed during the oxidation of fatty acids to CO_2 is generated in the mitochondrion. Thus the mitochondrion can be regarded as the "power plant" of the cell.

CHLOROPLASTS

Except for vacuoles, chloroplasts are the largest and most characteristic organelles in the cells of plants and green algae. They can be as long as 10 μm and are typically 0.5 – 2 μm thick, but they vary in size and shape in different cells, especially among the algae. Like the mitochondrion, the chloroplast is surrounded by an outer and an inner membrane. Chloroplasts also contain an extensive internal system of interconnected membrane-limited sacs called thylakoids, which are flattened to form disks; these often are grouped in stacks called *grana* and embedded in a matrix, the *stroma*. The thylakoid membranes contain green pigments (chlorophylls) and other pigments and enzymes that absorb light and generate ATP during photosynthesis. Part of this ATP is used by enzymes located in the stroma to convert CO_2 into three-carbon intermediates; these are then exported to the cytosol and converted to sugars. Perhaps surprisingly, the molecular mechanisms by which ATP is formed in mitochondria and chloroplasts are very similar. Chloroplasts and mitochondria share other features: Both often migrate from place to place within cells and also contain their own DNA, which encodes some of the key organellar proteins. The proteins encoded by mitochondrial or chloroplast DNA are synthesized on ribosomes within the organelles. However, most of the proteins in each organelle are encoded in nuclear DNA and are synthesized in the cytosol; these proteins then are incorporated into the organelles.

THE ENDOPLASMIC RETICULUM

Generally, the largest membrane in a eukaryotic cell encloses the endoplasmic reticulum (ER) — a compartment comprising a network of interconnected, closed, membrane-bounded vesicles. The endoplasmic reticulum has a number of functions in the cell but is particularly important in the synthesis of many membrane lipids and proteins. The *smooth endoplasmic reticulum* is smooth because it lacks ribosomes; regions of the *rough endoplasmic reticulum* are studded with ribosomes.

DIFFERENT BETWEEN EUKARYOTES AND PROKARYOTES

- Eukaryotes contain organelles that perform various specific functions for the cell.
- A highly organized cytoskeleton is present in eukaryotes.

Eukaryotes are typically more complex than prokaryotes and appear more organized when examined in the microscopic. This organization into different intracellular compartments likely reflects the demands of a more complex cellular system. It may also be that the formation of these separate organelles allowed the subsequent evolution of more elaborate cells. Figure of the various structures found in the typical eukaryotic cells. Note that chloroplasts are only found in photosynthetic organisms.

The organelles are the nucleus, the mitochondria, the endoplasmic reticulum and the golgi apparatus. Internally eukaryotic cells have a cytoskeleton that determines cell structure and in plants this is supplemented by cell walls. We discuss these structures and their functions in eukaryotic cells in the following sections.

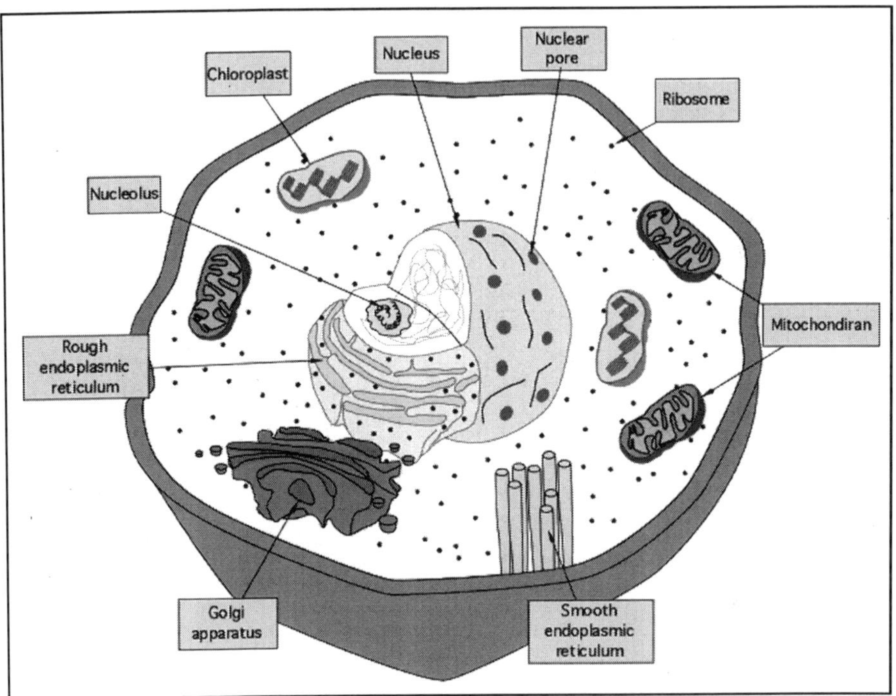

Fig. Eukaryotic Cell Structure

UNIQUE STRUCTURES IN EUKARYOTES

- Microfilaments and microtubules form a network of fibers that determine cell shape, participate in cell division, move organelles around the cell and form motility structures.
- The endoplasmic reticulum is a network of tube and vesicles involved in membrane synthesis and in export of protein and other substances from the cell.
- The Golgi apparatus is a double membrane structure mainly concerned with the maturation of proteins synthesized in the endoplasmic reticulum and the formation of lysosomes.

- Lysosomes are membrane encircled structures that contain digestive enzymes. Their exact role in the cell depends upon the cell type.

Microfilaments and Microtubules

The cytoskeleton is a network of filaments and fibers found in the cytoplasm of many eukaryotic cells. It serves four known roles in cells.

- The cytoskeleton helps to determine the shape of the cell.
- It is involved in cell division, allowing the separation of the chromosomes into the daughter cells.
- The cytoskeleton moves organelles around the cell.
- It is involved in motility either through the use of flagella or by amoeboid movement.

The cytoskeleton components can be divided into three classes based upon the size, distribution and function of the filaments. Microfilaments are the smallest at 4 to 6 nm in diameter and are made of actin. These lie beneath the surface of the cell membrane and are anchored to it, forming a web inside the cell. They dictate the cell's shape and can also be involved in motility by contraction or expansion of the filament.

Filaments may also tether organelles to the membrane and help move them around the cell. This movement can be important for modulation of organelle function. Intermediate filaments are 10 nm in diameter and are made of keratin, which is the same protein found in hair and fingernails. These filaments take different forms and are found in many types of cells, but their exact function is unknown. They may play a structural role similar to that of some microfilaments. Figure shows some examples of cytoskeletal elements.

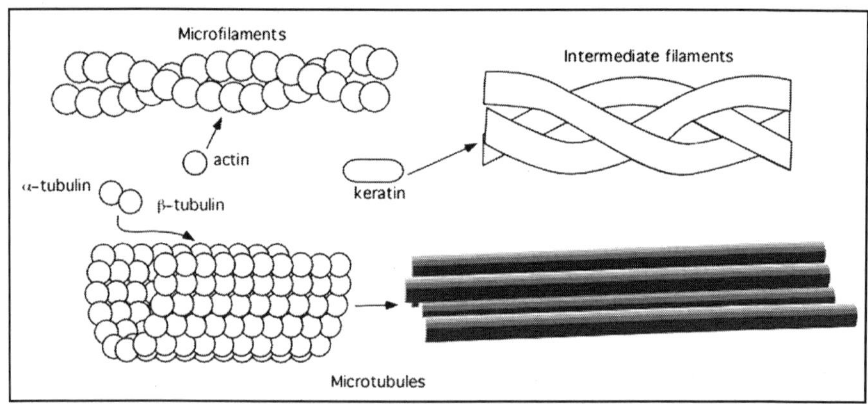

Fig. Cytoskeletal Elements

Eukaryotic cells have several different types of scaffolding proteins to help them keep their shape. Microfilaments, microtubules, and centrioles are all important structural elements. In addition to the above, there are more complex fibers and structures. Microtubules are hollow cylindrical structures that are

20-25 nm in diameter containing tubulin as the major structural protein. Tubulin polymerizes into a helical cylindrical structure and thirteen of these protofilaments then combine to make a microtubule. Microtubules can form the basis of a number of different structures. Many of these structures form to perform a necessary function and are then disassembled afterwards. Often microtubules form centrioles that contain nine sets of microtubules arranged in a circular matrix.

These are 400 nm long and 150 nm wide and are usually found in pairs at right angles to each other. Centrioles are important in proper chromosome segregation and cell division as discussed in the section on the nucleus below. Microtubules are also part of the basal bodies of flagella and cilia. Cilia and flagella are examples of more permanent structures that contain microtubules. Both cilia and flagella have a similar structure of nine pairs of microtubules arranged in a circular fashion around a tenth pair that runs down the centre. Both are attached to the membrane and project into the environment. In a process that requires energy, the microtubules in the outer ring are moved with respect to each other, causing the cilia or flagella to bend and snap back in a whip-like fashion.

This bending causes the movement of liquid near the structures such that spent liquid with few nutrients and waste products is moved away from the cell and is replaced by fresh liquid containing nutrients and oxygen. The beating of cilia and flagella can also push the cell through its environment. Cilia are 2 to 10 μm long and 0.5 μm wide and are shorter and typically more numerous than flagella with hundreds of them on some types of cells. An example of a ciliated organism is the unicellular protist Paramecium, which can be found in fresh water ponds. The microbe is a predator of bacteria and motility is vital in this life-style, both for chasing down prey and moving away from danger.

Cilia cover the surface of Paramecium and move the organism through the environment by beating in a coordinated fashion. Figure shows one example of a protozoan. Flagella are 50-100 μm in length and there are typically only one or two per cell. Eukaryotic flagella are larger than those found on bacteria or archaea and have a more complex structure. Flagella are found in many unicellular creatures with one example being the dinoflagellates and their primary role is cell motility.

These aquatic creatures contain two flagella; one encircling the body of the organism while the other is attached in a perpendicular fashion to the first. Dinoflagellates are often photosynthetic and important as primary producers in the oceans.

Endoplasmic Reticulum

The endoplasmic reticulum (ER) is a finely divided system of interconnected membranes, consisting of tubules and vesicles that loop through the cell and are contiguous with the nuclear membrane. A drawing of the ER is shown in Figure. It functions in the synthesis of membranes and membrane proteins and is also involved in protein secretion. Not surprisingly, the ER is especially prominent in cells doing a large amount of protein secretion.

The ER works very closely with the Golgi apparatus to carry out these functions. There is no structure in bacterial cells that is analogous to the ER, but many of the same functions are carried out on the inside surface of the cellular membrane in bacteria. ER comes in two types: rough ER and smooth ER.

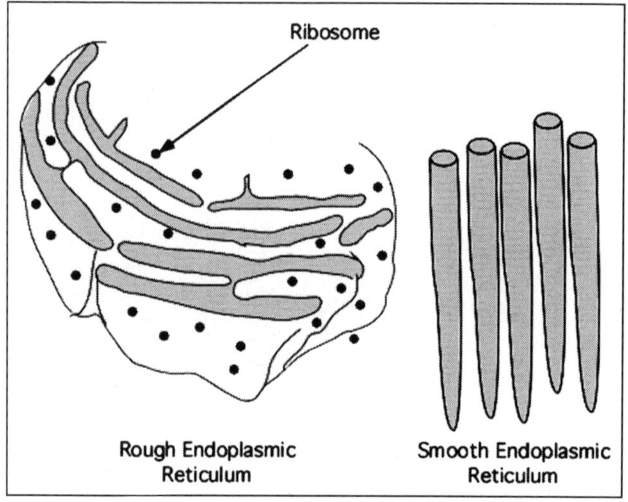

Fig. The Endosplasmic Reticulum

Eukaryotic cells contain a network of passages that connect various elements of the cell and are also important in secretion. Rough ER gets its appearance from the presence of ribosomes on its surface as seen in the electron microscope, and its function is the production, processing and export of proteins. During translation an appropriate signal guides the ribosome to the ER membrane and causes the protein to be synthesized directly across the membrane into the lumen of the ER.

There proteins may be processed or modified by the addition of carbohydrates to form glycoproteins. After processing proteins move slowly through the ER and are packaged into vesicles of ER membrane called transition vesicles. These release from the ends of the ER and move by elements of the cytoskeleton either to the Golgi apparatus or to the plasma membrane. Once contact is made between the transition vesicle and the Golgi or the plasma membrane, the two fuse and release the contents of the vesicle into the target compartment. Smooth ER does not contain ribosomes and the lumen and membrane of smooth ER contain a variety of enzymes that perform many functions including modification of toxins and synthesis of steroids.

Golgi Apparatus

The Golgi apparatus is an organelle containing a double membrane and it is mainly devoted to the processing of proteins synthesized in the ER. A drawing of the Golgi apparatus is shown in Figure. It is found in many eukaryotic cells, but it lacks a well-formed structure in many fungi and ciliate protozoa. It consists of regions of stacked contiguous membranes containing no ribosomes. Each membrane sac is 15 to 20 nm thick and separated from the next stack by about 30 nm. A complex network of tubes and vesicles extend from the edges of these sacs into the surrounding cytoplasm. The stack of membranes has a definite polarity with those near the ER (the cis face) having a different shape and enzyme content than those at the opposite end (the trans or maturing face). Studies of the Golgi apparatus appear to show material flowing into the cis face from vesicles, through the apparatus and then exiting at the trans face.

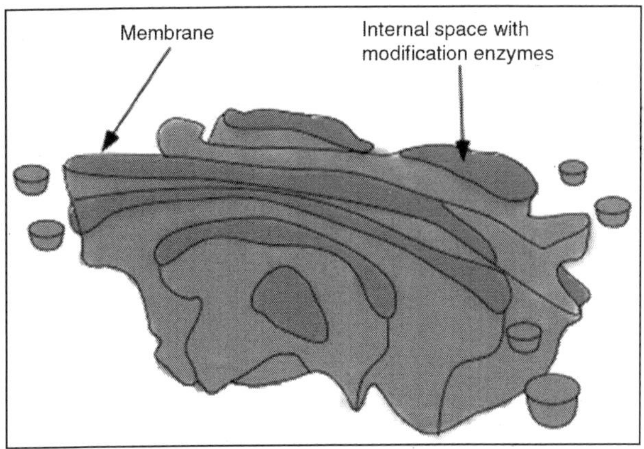

Fig. The Golgi Apparatus

The Golgi apparatus is involved in the glycosylation and proteolytic processing of proteins that are to be secreted into various cellular organelles or to the outside environment. Proteins often pass through the Golgi apparatus as part of their maturation. The role of the Golgi apparatus is to package material for export, but its exact function varies depending upon the organism. For example, Giardia and Entamoeba utilize Golgi apparatus to form cell walls during cyst formation.

It often participates in the synthesis of cell membranes and the final processing of proteins before export. In many cases the Golgi apparatus contains glycosylation enzymes that add sugars to proteins as they move through its lumen. The type of glycosylation that takes place is dependent upon signals contained within the protein sequence. The Golgi also processes enzymes using proteases, which clip at specific amino acids sequences to form mature proteins and hormones. These mature proteins then move to their final destinations, which may be in the membrane, in lysosomes or secreted into the environment.

Lysosomes

One of the most important functions of the Golgi apparatus is the synthesis of lysosomes, an organelle that is found in a variety of eukaryotic cells. Lysosomes are spherical structures enclosed in a single membrane that can vary in size from 50 nm to several μm.

They are involved in intracellular digestion and contain enzymes (called hydrolases) that digest many types of macromolecules. Hydrolases function best under acidic conditions (pH 3.5 to 5.0) and the lysosome maintains this pH by membrane proteins that pump protons into its interior. Enzymes bound for lysosomes are synthesized on ribosomes that deposit them in the rough ER. They then move though the smooth ER and Golgi apparatus before being package into lysosomes. In some cases lysosomes can also bud off from the smooth ER. Lysosomes serve a variety of functions depending upon the cell type. In unicellular eukaryotes they are often digestive structures that take bacteria or other substances from the outside environment and degrade them into usable nutrients. In mammals, lysosomes serve to eliminate unwanted particles, either cell structures that are no longer needed or foreign macromolecules (from viruses or bacteria) that have invaded the cell. In any case the hydrolytic enzymes and low pH typically inactivate and then degrade any particle that enters the lysosome

EUKARYOTIC CELLS ABSORB THINGS BY ENDOCYTOSIS

- The major pathway into eukaryotic cells is by endocytosis. This process takes two forms, phagocytosis and pinocytosis.
- The Golgi apparatus, ER, lysosomes and endocytosis work as a coordinated whole to move particles into and out of the cell.

Lysosomes are an important part of endocytosis, a process that eukaryotic cells use to take up particles from the environment. Endocytosis is illustrated in Figure. The process creates membrane-bound cavities filled with fluid and solid materials. Larger membrane-enclosed cavities are called vacuoles while smaller ones are called vesicles. Endocytosis comes in two forms, phagocytosis and pinocytosis.

Phagocytosis involves the engulfment of large particles, even microorganisms, into membrane-bound compartments. It is a process used most often in the immune system and is described in detail in the chapter on infection and immunity. Pinocytosis involves the recognition of specific particles in the environment as described below. The process is used by unicellular eukaryotic microbes to ingest food and by multicellular organisms to take in certain macromolecules traveling from other parts of the organism. Eukaryotic cells absorb materials from the outside by using endocytosis. This is a standard pathway through which most material enters the cell. Pinocytosis begins when protein receptors on the cell surface bind to the target molecule to be ingested.

The bound particle migrates to an area of the membrane that is rich in a protein called clathrin.

This protein forms a matrix and causes an indentation in the membrane called a clathrin-coated pit. The entrance of a receptor with a bound particle begins a process of invagination at the pit that results in the internalization of the pit and the engulfment of the incoming particle inside a membrane vesicle.

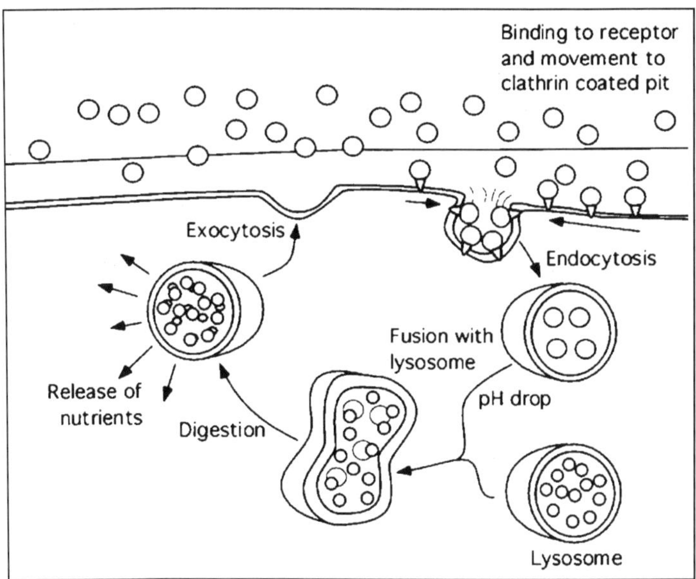

Fig. The Process of Endocytosis

This vesicle is called an endosome and once inside the cell, proteins in the endosome membrane begin to pump protons inside, dropping the internal pH of the endosome. The clathrin on the membrane of the endosome then migrates back to the plasma membrane to repeat the cycle. The endosome fuses with a lysosome to form an endolysosome, which causes the ingested particle to be degraded.

The valuable breakdown products are transported out of the endolysosome into the cell. The spent endolysosome is called a residual body and sometimes fuses with the plasma membrane to release remaining compounds into the environment. The ER, Golgi apparatus, lysosomes and endosomes seem to operate as a coordinated whole, functioning in the import and export of materials.

It is probably correct to think of these structures as a functional unit and the term vacuome has been coined to describe them. Ribosomes in the ER manufacture proteins and these are modified in the ER and Golgi. The mature proteins eventually find their way to the plasma membrane for export or become part of lysosomes.

Lysosomes serve in the endocytotic pathway to take up materials and process them for cell use. They also digest spent cell constituents. All of these processes occur within membrane structures and this carefully controls the import and export of materials from the cell. Considering the extent of these structures in the cell it is remarkable that the membranes of these structures, especially lysosomes, never rupture. If they did it would be catastrophic to the cell and would rapidly lead to cell death.

THE NUCLEUS HOLDS THE CELLS GENETIC MATERIAL IN EUKARYOTES

- The genome of eukaryotes is sequestered to a membrane bound organelle called the nucleus.
- The nucleus is the site of replication and transcription.
- Most eukaryotes have more than one chromosome in their nucleus and replication of these chromosomes proceeds through a sequence of steps that are visible with a light microscope.
- The nucleus contains visible spots called nucleoli, which are the location or ribosome synthesis.

While the bacterial cell does seem to sequester its chromosome to a portion of the cytoplasm, there is no demarcation that divides the nucleoid from the rest of the cell. In eukaryotes the nuclear membrane separates the cell's DNA from the cytoplasm.

The nucleus is the largest and most clearly visible organelle of eukaryotic cells. It contains almost all the cell's DNA and is the site of chromosome replication and transcription. It has two layers of membrane encircling it called the nuclear envelope, with the outer layer being contiguous with the ER. Scattered throughout this nuclear envelope are circular openings known as nuclear pores. These pores are highly discriminatory, allowing easy movement in and out of the nucleus of only appropriate macromolecules such as proteins with specific sequences.

In eukaryotes, the chromosome is not a single circular piece of DNA as in most prokaryotes. Rather, it is split into a number of linear chromosomes with each cell containing at least two copies of each chromosome. The exceptions are those cells that specialize in reproduction and only contain one copy of the cell's chromosomes. Each piece of DNA is complexed with special basic structural proteins called histones that seem to be important in keeping the DNA organized.

The DNA is also bound by other proteins involved in its maintenance and the entire set of DNA and associated proteins is called chromatin. For much of the cell cycle chromatin consists of long DNA strands formed into beads by association of histones along it length. Figure shows a nucleus in the midst of mitosis, with the chromosomes visible. Eukaryotic DNA replication takes place during the cell phase called mitosis. At this time, protein synthesis is halted and the chromosomes condense. The sister chromatids meet at the middle of the cell and then migrate to two separate poles. This movement is coordinated by centrosomes, kinetochores and microtubules. In actively growing cells the DNA is replicated from numerous sites, rather than the single bi-directional origin in prokaryotes. This is necessary due to the much larger amount of DNA found in most eukaryotic cells. During division in prokaryotes, the cell appears to simply split in two with each daughter cell receiving a chromosome. In contrast, eukaryotic cells go through a morphologically distinct phase, mitosis, to achieve separation of the chromosomes.

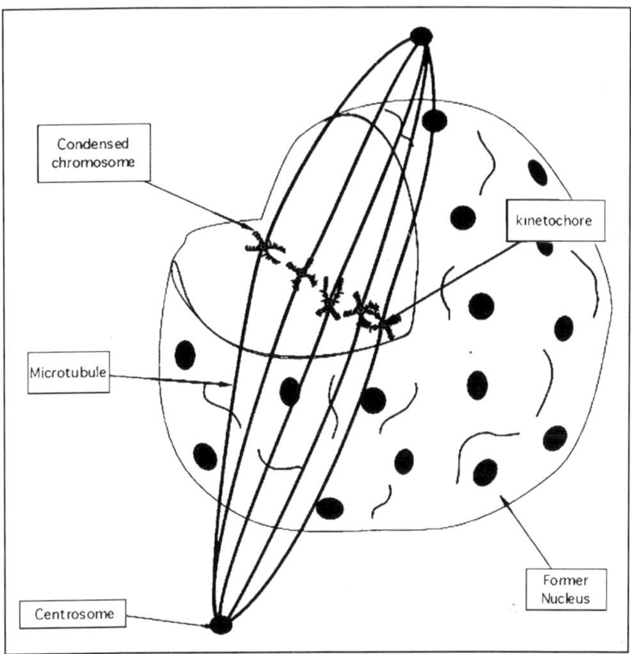

Fig. A eukaryotic cell in the middle of mitosis

One of the more important events of mitosis is the binding of additional histones and the contraction of the chromatin into compact structures that were called chromosomes due to their staining properties. (The original meaning of the term chromosomes is a colored body, but is now synonymous with a cell's DNA). The two daughter chromosomes formed during replication are physically separated into the separate daughter cells by the filaments called microtubules.

These attach at one end to the chromosome at a region termed the kinetochores and at the other end they attach to one of two regions of the cell called a centrosome. By depolymerization of the microtubules at each centrosome, each daughter chromosome is pulled away from its partner and toward a region that eventually reforms as a new nucleus. There are also a number of important differences in transcription between eukaryotes and prokaryotes.

In eukaryotes, mRNA transcription takes place in the nucleus and the finished mRNA moves through the nuclear pores and into the cytoplasm for translation by the ribosomes. The genes of eukaryotic cells also contain regions of largely unimportant DNA, termed introns, that do not code for protein.

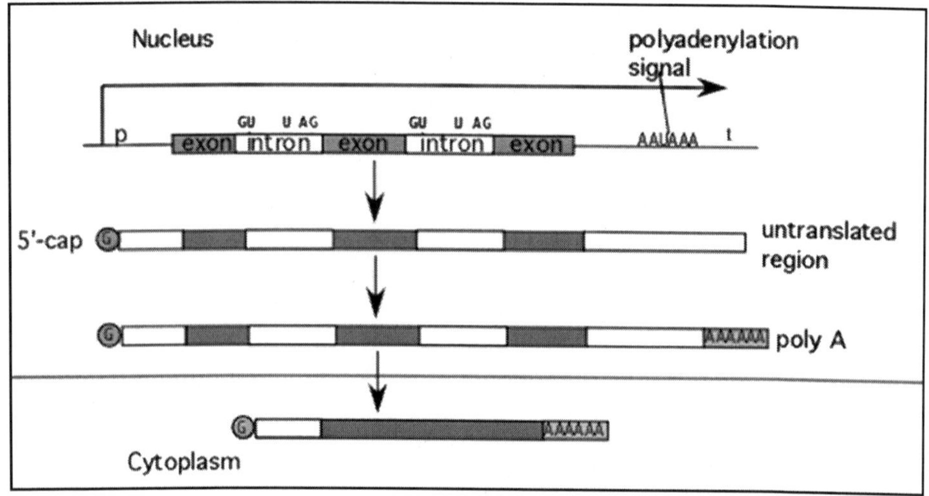

Fig. The Steps of Ggene Expression in Eukaryotes

After a gene is transcribed into mRNA these introns are removed before translation. One set of nuclear proteins removes these sequences and splices the actual coding sequences (exons) together to make the finished mRNA. The finished mRNA then travels from the nucleus to ribosomes in the cytoplasm. The mRNA of eukaryotic cells is also decorated with modifications at each end that affect the stability of the mRNA.

At the front end is usually a 5'-cap made of 7-methylguanine attached to the mRNA by a triphosphate linkage. At the 3' end of the mRNA is a long stretch of A bases (a poly-A tail) that have a role in mRNA stability as mentioned earlier in this chapter. Finally, while it is quite common in bacteria to have a number of genes on each mRNA transcript, the vast majority of mRNAs in eukaryotes code for only a single protein product. Figure shows the steps in gene expression in eukaryotes.

Processing genetic information into protein is more complex in eukaryotes than in prokaryotes. In part this is due to the existence of introns in most genes of "higher eukaryotes" (though introns are rare in yeast). These have to be removed, a poly-A tail added to the 3' end and a cap added to the 5' end of the mRNA before it reaches the cytoplasm for translation.

Eukaryotic cells also contain one or more dark-staining structures within the nucleus called nucleoli. Although they are not enclosed within a separate membrane the nucleoli are complexes with separate granular and fibrillar regions.

They are present in non-dividing cells, but frequently disappear during mitosis and again reappear after cell division is over. The nucleolus is the initial site of ribosome synthesis. This structure contains the DNA that codes for the ribosomal RNA genes. The ribosomal RNAs are synthesized and then processed to form the final rRNA molecules. These are then combined with several ribosomal proteins synthesized in the cytoplasm to form an initial ribosomal complex. The entire complex then migrates out of the nucleus into the cytoplasm where it combines with other proteins to form a ribosome.

MITOCHONDRIA AND PLASTIDS ARE ORGANELLES OF ENERGY GENERATION IN EUKARYOTIC CELLS

- Mitochondria are found in almost all eukaryotic cells and convert high-energy electrons into ATP.
- Plastids are factories for photosynthesis, converting light energy into high-energy electrons and ATP.
- Both of these structures trace their ancestry back to free-living prokaryotes. Mitochondria are involved in energy generation through respiration. Mitochondria have no fixed shape, but often look like short rods in transmission electron micrographs when viewed along their long axis. Each mitochondrion contains two membranes. The outer membrane is smooth and serves as a selective barrier. The inner membrane is highly convoluted and folded and contains high numbers of membrane complexes. Nutrients are oxidized inside the mitochondria by catabolic enzymes and the high-energy electrons extracted are donated to a respiratory chain in the inner membrane. These enzymes then create a proton gradient and this gradient is then used to synthesize ATP. ATP leaves the mitochondria and it serves as a source of energy for the rest of the cell's machinery.

Mitochondria are rod-shaped structures that resemble the shape of common bacteria. They contain two membranes, similar to what is found in gram-negative bacteria, and 70S ribosomes. Energy generation occurs at the inner membrane.

Plastids are specialized organelles involved in metabolism that are unique to plants and come in several forms. Amyloplasts are starch storage containers found in some plants. Chloroplasts are oval-shaped structures inside of plant and algal cells that contain an outer and inner membrane as shown in Figure.

The outer membrane serves a similar function to the outer membrane of mitochondria, while the inner membrane consists of a network of stacks of membranous disks, called thykaloids, which are attached together by narrow tubes of membrane.

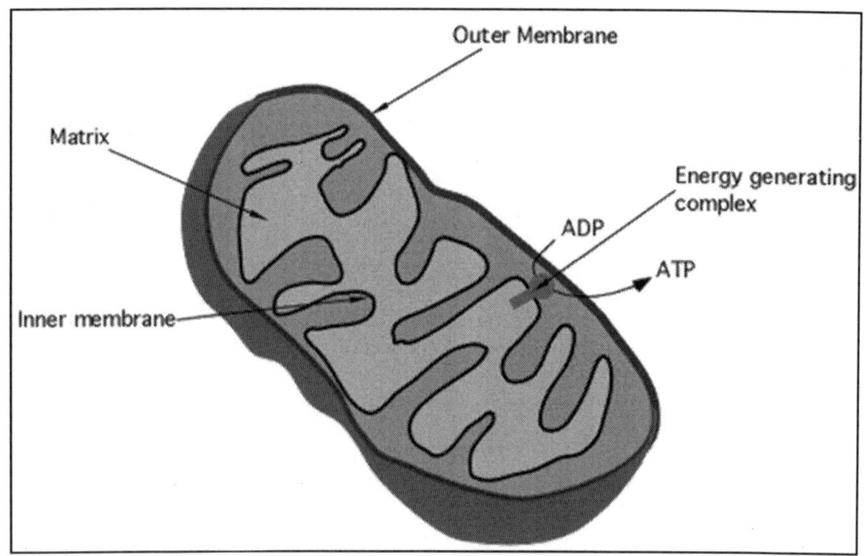

Fig. Mitochondria structure

The thykaloid membranes are the centers of photosynthesis in eukaryotes. They contain enzyme complexes that capture light and produce ATP and high-energy electrons that are used to form sugar from carbon dioxide. Chloroplasts are the site of the light reactions of photosynthesis.

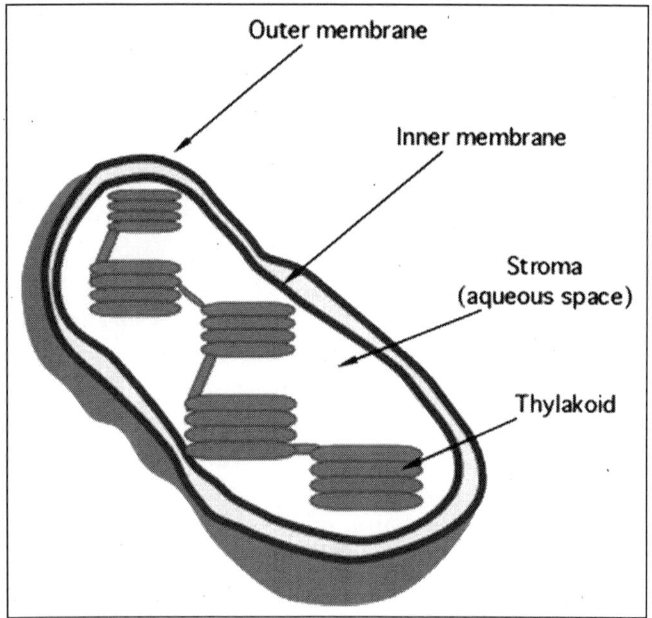

Fig. Chloroplast Structure

Light striking the chloroplast is converted into a proton motive force and this is used to generate energy. Chloroplasts contain two membranes, again similar to gram-negative bacteria, and the chloroplast itself is a relative of cyanobacteria. Mitochondria and chloroplasts each have a single circular chromosome and large numbers of ribosomes that are of bacterial (70S) and not eukaryotic (80S) form. The presence of two lipid bilayers, a circular chromosome and 70S ribosomes is consistent with the evolutionary hypothesis that we have already explained near the beginning of this chapter.

THE IMPLICATIONS OF EUKARYOTIC STRUCTURES ON THEIR GROWTH

- Eukaryotic cells are generally bigger than prokaryotic cells. This allows eukaryotes to have organelles, have slower turnover rates of macromolecules, and requires the presence of mechanisms to move things around the cell.
- The division of the chromosome from the rest of the cell and the larger size of eukaryotes, slows the rate at which they can replicate.
- The presence of organelles, which are sensitive to harsh conditions, might limit the extreme environments in which eukaryotes can grow.

Eukaryotic cells are generally much larger than prokaryotic ones and this difference in volume has several implications. First bigger cells can afford to have more things stored in the cytoplasm. This means it is not as costly to a eukaryotic cell to have structures taking up space. In a prokaryote, space is at a premium and anything not being used is pretty rapidly degraded. This may be one reason that organelles are possible.

Second, larger cells have a lower surface-to-volume ratio than do smaller cells and therefore prokaryotes effectively have more contact with their environment. This greater exposure can mean a more rapid response to changing environmental conditions.

Finally bigger cells have more of a challenge moving molecules within themselves. Prokaryotes can often depend on simple diffusion to move molecules around the cell, but this process might be too slow and inefficient in much larger cells. Eukaryotes overcome this by having specific transport mechanisms (*i.e.* microtubules) inside the cell. Size constrains eukaryotes in two important ways: how fast they can grow and what environments they can tolerate. The compartmentalization of the genome inside the nucleus limits the rate at which eukaryotic cells can divide. The complete cell division cycle in a multicellular eukaryotic organism depends upon the cell type, but even in rapidly dividing skin cells it takes at least 8 hours.

In unicellular yeast cultures, the shortest cell cycle is about 1.7 hours under ideal conditions. Due to their smaller genomes, lack of a nucleus and the ability to couple transcription and translation, bacteria can grow much faster. Clostridium perfringens has been shown to go through a complete division cycle in at little as 6.6 minutes at 43°C in beef cubes!

There are environments that have one or more physical characteristics that prevent the growth of most organisms. These extreme environments might be too hot, cold, acidic or alkaline for typical organisms to grow. However, a small subset of prokaryotes has evolved to take advantage of these environments and thrive, and prokaryotes always define the extremes of where life can exist. In part this is probably due to the fact that simpler cells have fewer "body parts" that must be changed in order for growth under very different conditions. In eukaryotes, they would need to modify not only their cytoplasmic contents to tolerate the extreme environment, but also the makeup of their organelles.

6

Structure and Function of Ribosomal

Ribosomes are cytoplasmic granules composed of RNA and protein, at which protein synthesis takes place. They were first observed by Palade in the electron microscope as dense particles or granules. Upon isolation, they were shown to contain approximately equal amounts of RNA and protein Label. To function actively in protein synthesis, they must be bound into complete ribosomes.

We know that ribosomes are but one of the required components necessary for the synthesis of protein. The others are messenger RNA, which carries the genetic message; soluble RNA, which carries amino acids to be synthesized; and guanosine triphosphate, which is the source of energy. A number of ribosomes may be attached to the same messenger, each manufacturing its own chain of polypeptides, called a polysome. Ribosomes are also found in the mitochondria and chloroplasts of eukaryotic cells.

They are always smaller than the 80S cytoplasmic ribosomes, and are comparable to prokaryotic ribosomes in both size and sensitivity to antibiotics, although the sedimentation values vary somewhat in different phyla. Prokaryotic and eukaryotic ribosomes do not differ in any fundamental way; both preform the same functions by the same set of chemical reactions.

The genetic code is the same in all living organism, and it has been demonstrated that eukaryotic ribosomes are able to translate bacterial mRNAs correctly. Eukaryotic ribosomes are much larger that prokaryotic ones and most of their proteins are different.

Antibiotics such as chloramphenicol inhibit bacterial but not eukaryotic ribosomes. Protein synthesis by eukaryotic ribosomes in inhibited by cycloheximide. Mitochondrial and chloroplast ribosomes resemble those in bacteria.

They are inhibited by chloramphenicol, and hybrid ribosomes containing one bacterial and one chloroplast ribosome subunit, for example, are fully active in protein synthesis. Hybrid eukaryotic ribosomes containing subunits from both plants and mammals are also active in protein synthesis, but are inactive if one of the subunits is derived from bacteria.

Some structural resemblance must exist, however, since reconstruction experiments have shown that two proteins from the large subunit of E. coli can replace the homologous proteins in mammalian ribosomes. In summary, there is little structural but considerably functional homology between prokaryotic and eukaryotic ribosomes.

Cells devote considerable effort to the production of these essential organells. For example, an E. coli cell contains approximately 15,000 ribosomes, each one with a molecular weight of about three million daltons. Ribosomes therefore represent twenty-five per cent of the total mass of these bacterial cellsLabel.

STRUCTURE

Ribosomes are tiny particles, about 200 A. It is composed of both proteins and RNA; in fact it has approximately 37 - 62% RNA, and rest are made up of proteins Label. The RNA present in ribosomes are obviously called ribosomal RNA, and they are produced in the nucleolus, which is a prominent globular structure in the nucleus.

Thus, the proteins are gene products of themselves, and one ribosome is made up of dozens of genes. The ribosomes fall into two categories: Those that are free to roam in the cytoplasm, and those that are bound to gigantic, cobwebby organelles made up of membranes, called the endoplasmic reticulum; thus, causing a rough surface. Although, the two kinds of ribosomes play similar roles in translating mRNA to produce proteins, they are very distinct in where its product is located.

The ribosomes in the cytoplasm allows its protein to roam about freely, while the bound ribosomes transfer their functional protein into the endoplasmic reticulum. In addition, ribosomes are also located within the mitochondria, and the chloroplast, but are only few in content.

This spherical particle of 23nm, is composed of two subunits; a large and small Label. In Eukaryotes, the co-efficient of ribosomes are 80s, of which is divided into 60s for the large, and 40s for the small subunit.

The 60s contain 28s rRNA, with a small fragment that is attached noncovalently and can be released upon heating; a 5.8s, and a very small - 120 nucleated of 5sRNA. Whereas, the 40s subunit has only a single 18s rRNA Label. In prokaryotes, however, the large and small subunits are split into 50s and 30s, making a total of 70s respectively. The 50s has two types of rRNA - a 23s and a 5s. It also has 32 different proteins. On the other hand, the 30s contains a single 16s rRNA, 21 different types of proteins.

To help better understand what the s stands for in rRNA, let us use the prokaryotes as an example. The 50s and 30s refers to the sedimentation coefficient of the two subunits. This coefficient is a measure of the speed with which the particles sediment through a solution when spun in an ultra centrifuge.

Thus, the particles with larger coefficient would centrifuge and settle much faster since it is has more mass than the particle with the smaller coefficient. 50s + 30s $\rightarrow$ 70s Note that the two subunits above make up the entire ribosomal molecule which is 70s. The reason the coefficients do not add up is because they are not proportional to the particle weight. During protein synthesis, ribosomes line up along the mRNA and form a polysome, also called the polyribosome. The mRNA is aligned in the gap between the 2 ribosomal subunits. It is possible that the nascent peptide chain grows through a channel or groove in the large ribosomal subunit. This is predicted to be the case since ribosomes protect a segment of 30-40 amino acids from degradation.

Speaking of amino acids, up to 30 ribosomes can attach on one strand of mRNA to form amino acid chains thus leading to protein formation. Ribosomes act as the backbone for many molecules during translation. It provides room for many structures to situate itself thus enhancing protein synthesis. For example, mRNA inserts itself between the two subunits; the peptidyl transferase complex - the enzyme that allows for the tRNA to break apart from the amino acid on P-site; this enzyme lays across the molecule, between the subunits. It contains the P and the A-site for tRNA binding.

Last but not least, the ribosome molecule allows the growing polypeptide chain, to emerge from the back of the structure, thus it is situated perpendicular to the mRNA chain. Ribosomes have a tertiary structure. Ribosomes make up a large part of cells in many species, which leads to protein manufacturing. For example, in E.Coli (bacteria), they make up about 1/4 of the total cell mass. They are intensely basiphilic (having high affinity for bases). Due to its complex structures, with many proteins and different kinds of RNA, researchers have found it very difficult to study the macro molecular structure of ribosomes, especially for the fact it is quite impossible to observe its crystal using an x-ray diffraction. Thus, scientists have been forced to use other means of study to map the proteins and RNA components in ribosome. Some of these are the cross-linking, immunoelectron microscopy, and low-angle neutron scattering methods.

The cross-linking shows the protein arrangement and the types of bonds it forms within itself. The neutron scattering experiments forms horizontal lines that show the entire structure of ribosome, with its two subunits, and shows where the proteins are arranged in the molecule. The empty regions around the proteins is where the rRNA is located. The immunoelectron microscopy, shows the proposed location of the 16s rRNA molecule of the small subunit, in prokaryotes.

FUNCTION

The ribosomes plays a very important role in protein synthesis, which is the process by which proteins are made from individual amino acids. Without the ribosomes the message would not be read, thus proteins could not be produced. Therefore, ribosomes play a very important role in role in protein synthesis. The primary agent in the process of translating the mRNA into a specific amino acid chain is the ribosome, which consists of two subunits. These subunits are made up of a third and extremely abundant type of RNA, ribosomal RNA (rRNA), and together contain up to eighty-two specific proteins assembled in a precise sequence Label. The ribosomes constituents must be put together in an extremely precise position and sequence.

This assembled ribosome displays a series of small groves, tunnels, and platforms, where the action of protein synthesis occurs Label. There are the active sites, each dedicated to one of the tasks required for translation of mRNA into protein. Proteins being synthesized for export out of the cell, are made by ribosomes attached to the rough endoplasmic reticulum. In contrast, proteins for use by the cell are generally made in the cytoplasm by free ribosomes. Several of these free ribosomes may attach to a single mRNA molecule, giving rise to the polyribosome or polysome Label. Protein synthesis takes place on polyribosomes (or polysomes) where 80S ribosomes associate with an mRNA coding for a given protein. The number of ribosomes associated in the polysomal chains depends on the size of the mRNA.

This is also associated with the size of the protein that is being synthesized. Outside the polyribosome, the ribosomes are dissociated and form a pool of free subunits. Transfer RNAs are also bound to the ribosome. There are quite a few factors involved in the formation of the initiation complex. These include: GTP, methionine tRNA, an initiation codon in mRNA, 80S ribosomes, and three protein factors Label. The process of protein synthesis begins with the capture of the tRNA, which is carrying an amino acid, by an initiation factor. This binds to a small ribosomal subunit, which occupies one of the active sites in the ribosomes, the P (protein) site. This initiation complex recognized and binds to the 5' end of an mRNA molecule and slides down to the initiation codon, which is always an AUG sequence of amino acids. The large subunit of the ribosome now joins the complex. A second tRNA is now brought into the ribosome by the elongation factor. If the anticodon of the tRNA pairs with the next codon of the message, the tRNA occupies the A (acceptor) site on the ribosome. This positions the second amino acid adjacent to the initiation methionine.

Then an enzyme, peptidyl transferase, which is part of the large ribosomal subunit mediates the separation of the first amino acid from its tRNA and the formation of a peptide bond between the initial methionine and the amino acid is formed. The P site is now occupied by an uncharged tRNA molecule label. The ribosome will now move down the mRNA by one codon, a process known as translocation. This movement shifts the growing polypeptide chain to the P position, and results in an empty A site, where a new charged tRNA can enter and pair, by forming a hydrogen bond between the codon and the anticodon. This holds the tRNA into place long enough for an even more stable binding to occur Label.

The uncharged tRNA that previously occupied the P site is booted out of the ribosome and will be recharged and recycled by the cell. The energy needed for this process is supplied by the hydrolysis of guanosine triphosphate (GTP).

The process then continues along the length of the mRNA, until the first stop codon is encountered. At that point the action of a termination factor releases the completed protein from the last tRNA and the ribosome dissociates into its component parts. Another function of the ribosomes occurs in the relation to the neuron and axons.

The cell body of a typical large neuron contains vast numbers of ribosomes. Although dendrites often contain some ribosomes, there are no ribosomes in the axon, and its protein must therefore be provided by the many ribosomes in the cell body.

STRUCTURE LINKAGE

Antibiotics are drugs produced by bacteria and fungi. These molecules function as drugs used in the chemotherapy of infectious disease, and follow the principle of selective toxicity. Selective toxicity follows the principle of using drugs that kill the harmful microorganism without damaging the host. As a result of its toxicity, antibiotics can affect the ribosomal structure, inhibiting protein synthesis. For instance, let us take the 70s ribosome of prokaryotes; antibiotics can target this structure and can adverse the effects on the cells of the host.

Among the antibiotics that interfere with protein synthesis are chloramphenicol, erythromyocin, streptomycin, and the tetracycline. This paper will be focused on these four antibioticsand its role played in the effect of the ribosome structure, thus leading to the change in protein synthesis. For instance, the chloramphenicol reacts with the 50s structure of the 70s prokaryote ribosome, by inhibiting the formation of the peptide bonds in the growing polypetide chain. Erythromyocin, the second antibiotic, also reacts with the same structure as chloromphenicol.

However, it has a very narrow range of activity, since it affects mostly the gram-positive bacteria. The other two antibiotics attract the 30s structure of the 70s prokaryotic ribosome. The tetracycline interferes with the attachment of the tRNA, which carries the amino acids, to the ribosome, thus preventing the addition of amino acids to the growing polypeptide chain.

One unique aspect of tetracycline is that it cannot penetrate well into the mammalian cells, therefore, it does not interfere with the mammalian ribosomes. Aminoglycoside antibiotics, a type of streptomycin, changes the shape of the 30s structure of the 70s prokaryotic ribosome, thus interfering with the initial stage of protein synthesis. This in turn, causes the misreading of the genetic code on the mRNA.

FUNCTION LINKAGE

RNA polymerase is the enzyme that directs transcription, which is the process by which the mRNA copy of a gene is synthesized. Transcription follows the same rules of base pairing as DNA replication. This base pattern ensure that an RNA transcript is a faithful copy of the gene. There are three stages of transcription: initiation, elongation, and termination. During initiation, the enzyme recognizes a promoter region, which lies upstream from the gene.

The polymerase binds tightly to the promoter and causes localized melting, or separation of the two DNA strands within the promoter. Then the polymerase starts building the RNA chain. Ribonucleoside triphosphates such as ATP, GTP, CTP and UTP are the building blocks the polymerase uses for this job Label. After the first nucleotide is in place, the polymerase binds the second nucleotide, joining it to the first. This forms the initial phosphodiester bond in the RNA chain.

The second stage is elongation, where the RNA polymerase directs the sequential binding of ribonucleotides to the RNA chain. As it does this, it moves along the DNA template and the melted DNA moves with it. This melted region exposes the bases of the template DNA one-by-one so that they can pair with the bases of the incoming ribonucleotides.

As soon as the transcription machinery passes, the two DNA strands wind around each other again, reforming the double helix. Only enough separation will occur so that the polymerase can read the DNA template Label. The final stage is termination, which allows the termination of transcription.

These work in conjunction with RNA polymerase, and is sometimes aided by another protein, to loosen the association between RNA product and DNA template. So the RNA dissociates from the RNA polymerase and DNA, thus terminating transcription. Transcription is very important in that it is the only step in expression of the genes for rRNAs.

It is also important to ribosomes, because it sets up the RNA in a 5' to 3' sequence, which allows the ribosomes to read the message 5' to 3'. Without transcription of a gene, the ribosome would not be able to translate an mRNA, thus not allowing it to be separated into its component parts. If this does not occur, then translation will also not occur, thereby affecting the function and role of ribosomes in translation.

Regulation linkage: Ribosomes are used by all living cells to synthesize proteins. This synthesis can be inhibited by antibiotics, which can have an effect on the organism. Some antibiotics target specific subunits of the ribosome or may target the entire ribosome completely. Streptomycin, for example, can target the 70s ribosome in some prokaryotes and cause adverse effects on the cell of the host. An antibiotic that affects the 30s ribosome is tetracycline. It can prevent the addition of amino acids to the growing polypeptide chain by interfering with the attachment of the tRNA onto the ribosome. The 50s ribosome may also be targeted by erythromycin, which blocks the translocation reaction on ribosomes. Other antibiotics that interfere with the ribosome to synthesize proteins include chloramphenicol, rifamycin, puromycin, cycloheximide, and anisomycin.

MESSENGER RNA

Messenger ribonucleic acid (mRNA) is a molecule of RNA encoding a chemical "blueprint" for a protein product. mRNA is transcribed from a DNA template, and carries coding information to the sites of protein synthesis: the ribosomes.

Here, the nucleic acid polymer is translated into a polymer of amino acids: a protein. In mRNA as in DNA, genetic information is encoded in the sequence of four nucleotides arranged into codons of three bases each. Each codon encodes for a specific amino acid, except the stop codons that terminate protein synthesis.

This process requires two other types of RNA: transfer RNA (tRNA) mediates recognition of the codon and provides the corresponding amino acid, while ribosomal RNA (rRNA) is the central component of the ribosome's protein manufacturing machinery.

MRNA

The brief existence of an mRNA molecule begins with transcription and ultimately ends in degradation. During its life, an mRNA molecule may also be processed, edited, and transported prior to translation. Eukaryotic mRNA molecules often require extensive processing and transport, while prokaryotic molecules do not.

Transcription

During transcription, RNA polymerase makes a copy of a gene from the DNA to mRNA as needed. This process is similar in eukaryotes and prokaryotes. One notable difference, however, is that eukaryotic RNA polymerase associates with mRNA processing enzymes during transcription so that processing can proceed quickly after the start of transcription. The short-lived, unprocessed or partially processed, product is termed *pre-mRNA*; once completely processed, it is termed *mature mRNA*.

Eukaryotic pre-mRNA Processing

Processing of mRNA differs greatly among eukaryotes, bacteria and archea. Non-eukaryotic mRNA is essentially mature upon transcription and requires no processing, except in rare cases. Eukaryotic pre-mRNA, however, requires extensive processing.

5' Cap Addition

A *5' cap* (also termed an RNA cap, an RNA 7-methylguanosine cap or an RNA m^7G cap) is a modified guanine nucleotide that has been added to the "front" or 5' end of a eukaryotic messenger RNA shortly after the start of transcription. The 5' cap consists of a terminal 7-methylguanosine residue which is linked through a 5'-5'-triphosphate bond to the first transcribed nucleotide. Its presence is critical for recognition by the ribosome and protection from RNases.

Cap addition is coupled to transcription, and occurs co-transcriptionally, such that each influences the other. Shortly after the start of transcription, the 5' end of the mRNA being synthesized is bound by a cap-synthesizing complex associated with RNA polymerase. This enzymatic complex catalyzes the chemical reactions that are required for mRNA capping. Synthesis proceeds as a multi-step biochemical reaction.

Splicing

Splicing is the process by which pre-mRNA is modified to remove certain stretches of non-coding sequences called introns; the stretches that remain include protein-coding sequences and are called exons. Sometimes pre-mRNA messages may be spliced in several different ways, allowing a single gene to encode multiple proteins. This process is called alternative splicing. Splicing is usually performed by an RNA-protein complex called the spliceosome, but some RNA molecules are also capable of catalyzing their own splicing.

Editing

In some instances, an mRNA will be edited, changing the nucleotide composition of that mRNA. An example in humans is the apolipoprotein B mRNA, which is edited in some tissues, but not others. The editing creates an early stop codon, which upon translation, produces a shorter protein.

POLYADENYLATION

Polyadenylation is the covalent linkage of a polyadenylyl moiety to a messenger RNA molecule. In eukaryotic organisms, most messenger RNA (mRNA) molecules are polyadenylated at the 3' end. The poly(A) tail and the protein bound to it aid in protecting mRNA from degradation by exonucleases. Polyadenylation is also important for transcription termination, export of the mRNA from the nucleus, and translation. mRNA can also be polyadenylated in prokaryotic organisms, where poly(A) tails act to facilitate, rather than impede, exonucleolytic degradation. Polyadenylation occurs during and immediately after transcription of DNA into RNA. After transcription has been terminated, the mRNA chain is cleaved through the action of an endonuclease complex associated with RNA polymerase. The cleavage site is characterized by the presence of the base sequence AAUAAA near the cleavage site. After the mRNA has been cleaved, 80 to 250 adenosine residues are added to the free 3' end at the cleavage site. This reaction is catalyzed by polyadenylate polymerase. Just as in alternative splicing, there can be more than one polyadenylation variant of a mRNA.

TRANSPORT

Another difference between eukaryotes and prokaryotes is mRNA transport. Because eukaryotic transcription and translation is compartmentally separated, eukaryotic mRNAs must be exported from the nucleus to the cytoplasm. Mature mRNAs are recognized by their processed modifications and then exported through the nuclear pore.

TRANSLATION

Because prokaryotic mRNA does not need to be processed or transported, translation by the ribosome can begin immediately after the end of transcription. Therefore, it can be said that prokaryotic translation is *coupled* to transcription and occurs *co-transcriptionally*.

Eukaryotic mRNA that has been processed and transported to the cytoplasm (i.e. mature mRNA) can then be translated by the ribosome. Translation may occur at ribosomes free-floating in the cytoplasm, or directed to the endoplasmic reticulum by the signal recognition particle. Therefore, unlike prokaryotes, eukaryotic translation *is not* directly coupled to transcription.

DEGRADATION

After a certain amount of time, the message is degraded by RNases. The limited lifetime of mRNA enables a cell to alter protein synthesis rapidly in response to its changing needs. Different mRNAs within the same cell have distinct lifetimes (stabilities).

In bacterial cells, individual mRNAs can survive from seconds to more than an hour; in mammalian cells, mRNA lifetimes range from several minutes to days. The greater the stability of an mRNA, the more protein may be produced from that mRNA.

The presence of AU-rich elements in some mammalian mRNAs tends to destabilize those transcripts through the action of cellular proteins that bind these motifs. Rapid mRNA degradation via AU-rich elements is a critical mechanism for preventing the overproduction of potent cytokines such as tumor necrosis factor (TNF) and granulocyte-macrophage colony stimulating factor (GM-CSF). Base pairing with a small interfering RNA (siRNA) or microRNA (miRNA) can also accelerate mRNA degradation.

mRNA STRUCTURE

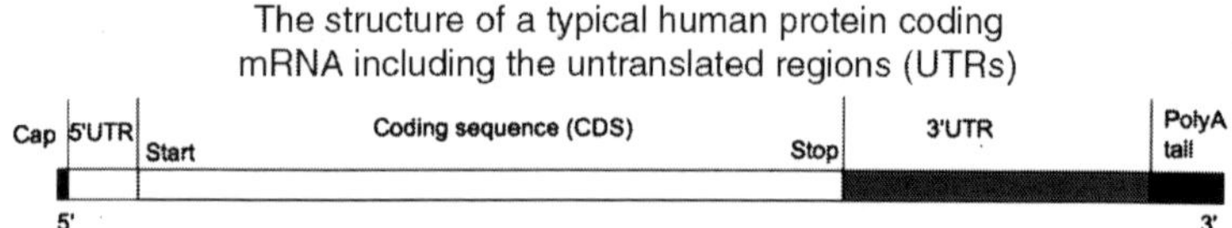

Fig. The Structure of a Mature Eukaryotic mRNA. A Fully Processed mRNA Includes a 5' cap, 5' UTR, Coding Region, 3' UTR, and Poly(A) tail.

5' Cap

The *5' cap* is a modified guanine nucleotide added to the "front" (5' end) of the pre-mRNA using a 5',5-Triphosphate linkage. This modification is critical for recognition and proper attachment of mRNA to the ribosome, as well as protection from 5' exonucleases. It may also be important for other essential processes, such as splicing and transport.

Coding Regions

Coding regions are composed of codons, which are decoded and translated into one (mostly eukaryotes) or several (mostly prokaryotes) proteins by the ribosome. Coding regions begin with the start codon and end with the one of three possible stop codons. In addition to protein-coding, portions of coding regions may also serve as regulatory sequences in the pre-mRNA as exonic splicing enhancers or exonic splicing silencers. Start codons are indicated by a AUG triplet. Stop codons are indicated by a UAA, UAG, or UGA.

Untranslated Regions

Untranslated regions (UTRs) are sections of the mRNA before the start codon and after the stop codon that are not translated, termed the five prime untranslated region (5' UTR) and three prime untranslated region (3' UTR), respectively. These regions are transcribed with the coding region and thus are exonic as they are present in the mature mRNA. Several roles in gene expression have been attributed to the untranslated regions, including mRNA stability, mRNA localization, and translational efficiency. The ability of a UTR to perform these functions depends on the sequence of the UTR and can differ between mRNAs.

The stability of mRNAs may be controlled by the 5' UTR and/or 3' UTR due to varying affinity for RNA degrading enzymes called ribonucleases and for ancillary proteins that can promote or inhibit RNA degradation. Translational efficiency, including sometimes the complete inhibition of translation, can be controlled by UTRs. Proteins that bind to either the 3' or 5' UTR may affect translation by influencing the ribosome's ability to bind to the mRNA.

MicroRNAs bound to the 3' UTR also may affect translational efficiency or mRNA stability. Cytoplasmic localization of mRNA is thought to be a function of the 3' UTR. Proteins that are needed in a particular region of the cell can actually be translated there; in such a case, the 3' UTR may contain sequences that allow the transcript to be localized to this region for translation.

Some of the elements contained in untranslated regions form a characteristic secondary structure when transcribed into RNA. These structural mRNA elements are involved in regulating the mRNA. Some, such as the SECIS element, are targets for proteins to bind. One class of mRNA element, the riboswitches, directly bind small molecules, changing their fold to modify levels of transcription or translation. In these cases, the mRNA regulates itself.

3' Poly(A) Tail

The 3' poly(A) tail is a long sequence of adenine nucleotides (often several hundred) added to the "tail" or 3' end of the pre-mRNA through the action of an enzyme, polyadenylate polymerase. In higher eukaryotes, the poly(A) tail is added onto transcripts that contain a specific sequence, the AAUAAA signal. The importance of the AAUAAA signal is demonstrated by a mutation in the human alpha 2-globin gene that changes the original sequence AATAAA into AATAAG, which can lead to hemoglobin deficiencies.

Monocistronic Versus Polycistronic Mrna

An mRNA molecule is said to be monocistronic when it contains the genetic information to translate only a single protein. This is the case for most of the eukaryotic mRNAs. On the other hand, polycistronic mRNA carries the information of several proteins, which are translated into several proteins. Most of the mRNA found in bacteria and archea are polycistronic. Dicistronic is the term used to describe a mRNA that encodes only two proteins.

TRANSFER RNA

Transfer RNA (abbreviated tRNA) is a small RNA (usually about 74-95 nucleotides) that transfers a specific amino acid to a growing polypeptide chain at the ribosomal site of protein synthesis during translation. It has a 3' terminal site for amino acid attachment. This covalent linkage is catalyzed by an aminoacyl tRNA synthetase.

It also contains a three base region called the anticodon that can base pair to the corresponding three base codon region on mRNA. Each type of tRNA molecule can be attached to only one type of amino acid, but because the genetic code contains multiple codons that specify the same amino acid, tRNA molecules bearing different anticodons may also carry the same amino acid.

STRUCTURE

tRNA has primary structure, secondary structure (usually visualized as the *cloverleaf structure*), and tertiary structure (all tRNAs have a similar L-shaped 3D structure that allows them to fit into the P and A sites of the ribosome).

- The 5'-terminal phosphate group.
- The acceptor stem is a 7-bp stem made by the base pairing of the 5'-terminal nucleotide with the 3'-terminal nucleotide (which contains the CCA 3'-terminal group used to attach the amino acid). The acceptor stem may contain non-Watson-Crick base pairs.
- The CCA tail is a CCA sequence at the 3' end of the tRNA molecule. This sequence is important for the recognition of tRNA by enzymes critical in translation. In prokaryotes, the CCA sequence is transcribed. In eukaryotes, the CCA sequence is added during processing and therefore does not appear in the tRNA gene.
- The D arm is a 4 bp stem ending in a loop that often contains dihydrouridine.
- The anticodon arm is a 5-bp stem whose loop contains the anticodon.
- The T arm is a 5 bp stem containing the sequence TØC where Ø is a pseudouridine.
- Bases that have been modified, especially by methylation, occur in several positions outside the anticodon. The first anticodon base is sometimes modified to inosine (derived from adenine) or pseudouridine (derived from uracil).

ANTICODON

An anticodon is a unit made up of three nucleotides that correspond to the three bases of the codon on the mRNA. Each tRNA contains a specific anticodon triplet sequence that can base-pair to one or more codons for an amino acid.

For example, one codon for lysine is AAA; the anticodon of a lysine tRNA might be UUU. Some anticodons can pair with more than one codon due to a phenomenon known as wobble base pairing. Frequently, the first nucleotide of the anticodon is one of two not found on mRNA: inosine and pseudouridine, which can hydrogen bond to more than one base in the corresponding codon position. In the genetic code, it is common for a single amino acid to be specified by all four third-position possibilities; for example, the amino acid glycine is coded for by the codon sequences GGU, GGC, GGA, and GGG.

To provide a one-to-one correspondence between tRNA molecules and codons that specify amino acids, 61 tRNA molecules would be required per cell. However, many cells contain fewer than 61 types of tRNAs because the wobble base is capable of binding to several, though not necessarily all, of the codons that specify a particular amino acid.

AMINOACYLATION

Aminoacylation is the process of adding an aminoacyl group to a compound. It produces tRNA molecules with their CCA 3' ends covalently linked to an amino acid. Each tRNA is aminoacylated (or *charged*) with a specific amino acid by an aminoacyl tRNA synthetase. There is normally a single aminoacyl tRNA synthetase for each amino acid, despite the fact that there can be more than one tRNA, and more than one anticodon, for an amino acid. Recognition of the appropriate tRNA by the synthetases is not mediated solely by the anticodon, and the acceptor stem often plays a prominent role.

Reaction:

- Amino acid + ATP → aminoacyl-AMP + PPi
- Aminoacyl-AMP + tRNA → aminoacyl-tRNA + AMP

BINDING TO RIBOSOME

The ribosome has three binding sites for tRNA molecules: the A, P and E sites. During translation the A site binds an incoming aminoacyl-tRNA as directed by the codon currently occupying this site. This codon specifies the next amino acid to be added to the growing peptide chain. The A site only works after the first aminoacyl-tRNA has attached to the P site. The P-site codon is occupied by peptdyl-tRNA that is a tRNA with multiple amino acids attached as a long chain.

The P site is actually the first to bind to aminoacyl tRNA. This tRNA in the P site carries the chain of amino acids that has already been synthesized. The E site is occupied by the empty tRNA as it is about to exit the ribosome.

tRNA GENES

Organisms vary in the number of tRNA genes in their genome. The nematode worm *C. elegans*, a commonly used model organism in genetics studies, has 29,647 genes in its nuclear genome, of which 620 code for tRNA. The budding yeast *Saccharomyces cerevisiae* has 275 tRNA genes in its genome. In the human genome, which according to current estimates has about 27,161 genes in total, there are about 4,421 non-coding RNA genes, which include tRNA genes. There are 22 mitochondrial tRNA genes; 497 nuclear genes encoding cytoplasmic tRNA molecules and there are 324 tRNA-derived putative pseudogenes.

Cytoplasmic tRNA genes can be grouped into 49 families according to their anticodon features. These genes are found on all chromosomes, except 22 and Y chromosome. High clustering on 6p is observed (140 tRNA genes), as well on 1 chromosome. tRNA molecules are transcribed (in eukaryotic cells) by RNA polymerase III, unlike messenger RNA which is transcribed by RNA polymerase II. pre-tRNAs contain introns; in bacteria these self-splice, whereas in eukaryotes and archaea they are removed by tRNA splicing endonuclease.

The existence of tRNA was first hypothesized by Francis Crick, based on the assumption that there must exist an adapter molecule capable of mediating the translation of the RNA alphabet into the protein alphabet. Significant research on structure was conducted in the early 1960s by Alex Rich and Don Caspar, two researchers in Boston, the Jacques Fresco group in Princeton

University and a United Kingdom group at King's College London. A later publication reported the primary structure in 1965 by Robert W. Holley. The secondary and tertiary structures were derived from X-ray crystallography studies reported independently in 1974 by American and British research groups headed, respectively, by Alexander Rich and Aaron Klug.

FUNCTIONAL ASPECTS OF MODIFIED BASES IN TRNA

Transfer RNA molecules (tRNAs) are typically about 75 nucleotides long and fold into stable tertiary structures not unlike polypeptides. While they have no independent chemical function, their structures are finely tuned to suit a number of steps in translation. They must be specifically recognized by the enzymes that attach amino acids to them, they must bind efficiently to the catalytic sites in the ribosome and they must form productive interactions with mRNA transcripts.

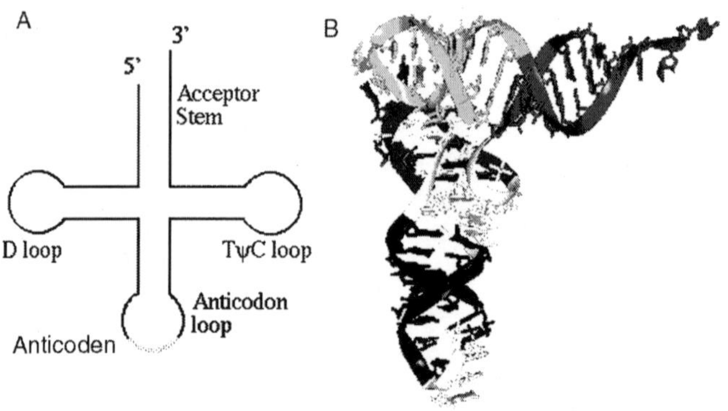

Fig. Two Representations of the Structure of tRNA

The focus of this exercise will be on the adaptation of tRNA structure that permits efficient and specific interactions with mRNA. The region of the tRNA molecule that interacts with the messenger is known as the anticodon.

It is comprised of three nucleotides that form part of the anticodon loop. These three bases are free to hydrogen bond with other nucleotides, particularly a set of three base sequences (codons) on mRNA that correspond to the amino acid for which the tRNA molecule is specific. In principle, all three bases of the anticodon could interact via normal Watson- Crick base-pairing to form a three base pair duplex with a single codon sequence.

In fact, the first nucleotide of the anticodon, position 34 in tRNA(Phe), which pairs with the third base of the codon, is known as the "wobble" base. The degeneracy of the genetic code means that several codons can specify a single amino acid. To reduce the number of tRNAs necessary to read the codons, the wobble position is often intended to base pair with two or three different bases at the 3' end of the codon. This reduces the specificity of interaction between the tRNA molecule and the transcript, and has the potential to lead to significant errors in translation if only two base pairs provide the thermodynamic stability to determine the interaction.

It appears that one approach taken by nature to enhance the specific base pairing interactions between tRNA and mRNA is to use modified bases at positions flanking the anticodon in the tRNA molecule. tRNA is notable for the number and variety of modified bases in its structure.

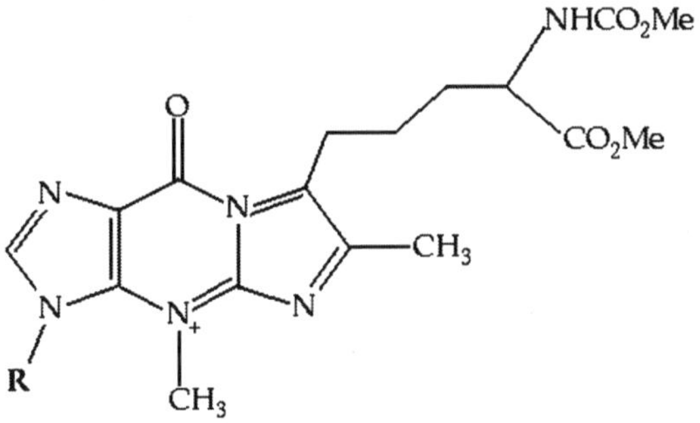

Fig. The Structure of Wybutosine, a Highly Modified Guanine Analog that Appears at Position 37, 3' to the Anticodon, in Yeast tRNA(Phe).

These bases are typically modified forms of the usual A, U, C and G's. Some are methylated analogues (for example 5-methyl uracil - otherwise known as thymine - is found in tRNA), and others are more dramatic, such as wybutosine.

These modifications are post-transcriptional. They are made after the tRNA molecule is transcribed from its gene. The purpose of this exercise is to investigate how modifications made to tRNA(Phe) from yeast, including the creation of wybutosine adjacent to the anticodon, aid in the function of that molecule.

tRNA(Phe) has an anticodon sequence GAA, that binds to sequences UUC and UUU on the messenger (both codons for phenylalanine). Because of the wobble at position 3 of the codon, and the U:A base pairs at positions 1 and 2, the interaction between tRNA(Phe) and mRNA is weaker than most. Here, we will examine how this potential problem is overcome.

THE MODEL

Three models are required for this exercise, though a number of others will need to be created in the course of the work. The first comes from coordinates for tRNA(Phe) deposited to the PDB by Kim and co-workers.*(1)* 6tna.pdb is a slightly modified version of the actual PDB entry, with a few changes made for ease of manipulation in Midas.

The The second model (arna.pdb is a undecamer duplex of RNA, derived from idealized coordinates of RNA in the A conformation, containing the sequence UUC at the 5' end of one of the strands. brna.pdb is a decamer of duplex RNA, like rna.pdb, but in the B conformation.

RIBOSOMAL RNA

Ribosomal RNA (rRNA) is the central component of the ribosome, the protein manufacturing machinery of all living cells. The function of the rRNA is to provide a mechanism for decoding mRNA into amino acids and to interact with the tRNAs during translation by providing peptidyl transferase activity.

INSIDE THE RIBOSOME

The ribosome is composed of two subunits, named for how rapidly they sediment when subject to centrifugation. tRNA is sandwiched between the small and large subunits and the ribosome catalyzes the formation of a peptide bond between the 2 amino acids that are contained in the tRNA.

The ribosome also has 3 binding sites called A, P, and E.

- The A site in the ribosome binds to an aminoacyl-tRNA (a tRNA bound to an amino acid).
- The NH2 group of the aminoacyl-tRNA which contains the new amino acid, attacks the carboxyl group of peptidyl-tRNA (contained within the P site) which contains the last amino acid of the growing chain called peptidyl transferase reaction.
- The tRNA that was holding on the last amino acid is moved to the E site, and what used to be the aminoacyl-tRNA is now the peptidyl-tRNA.

A single mRNA can be translated simultaneously by multiple ribosomes.

PROKARYOTES VS. EUKARYOTES

Both prokaryotic and eukaryotic can be broken down into two subunits (the S in 16S represents Svedberg units):

Type	Size	Large Subunit	Small Subunit
prokaryotic	70S	50S (5S, 23S)	30S (16S)
eukaryotic	80S	60S (5S, 5.8S, 28S)	40S (18S)

Note that the S units of the subunits cannot simply be added because they represent measures of sedimentation rate rather than of mass. The sedimentation rate of each subunit is affected by its shape, as well as by its mass.

Prokaryotes

In prokaryotes a small 30S ribosomal subunit contains the 16S rRNA. The large 50S ribosomal subunit contains two rRNA species (the 5S and 23S rRNAs). Bacterial 16S, 23S, and 5S rRNA genes are typically organized as a co-transcribed operon. There may be one or more copies of the operon dispersed in the genome (for example, *Escherichia coli* has seven). Archaea contains either a single rDNA operon or multiple copies of the operon. The 3' end of the 16S rRNA (in a ribosome) binds to a sequence on the 5' end of mRNA called the Shine-Dalgarno sequence.

Eukaryotes

In contrast, eukaryotes generally have many copies of the rRNA genes organized in tandem repeats; in humans approximately 300–400 rDNA repeats are present in five clusters (on chromosomes 13, 14, 15, 21 and 22).

The 18S rRNA in most eukaryotes is in the small ribosomal subunit, and the large subunit contains three rRNA species (the 5S, 5.8S and 28S rRNAs).

Mammalian cells have 2 mitochondrial (12S and 16S) rRNA molecules and 4 types of cytoplasmic rRNA (28S, 5.8S, 5S (large ribosome subunit) and 18S (small subunit). 28S, 5.8S, and 18S rRNAs are encoded by a single transcription unit (45S) separated by 2 Internally transcribed spacer (ITS). The 45S rDNA organized into 5 clusters (each has 30-40 repeats) on chromosomes 13, 14, 15, 21, and 22. These are transcribed by RNA polymerase I. 5S occurs in tandem arrays (~200-300 true 5S genes and many dispersed pseudogenes), the largest one on the chromosome 1q41-42. 5S rRNA is transcribed by RNA polymerase III. The tertiary structure of the small subunit ribosomal RNA (SSU rRNA) has been resolved by X-ray crystallography. The secondary structure of SSU rRNA contains 4 distinct domains — the 5', central, 3' major and 3' minor domains. A model of the secondary structure for the 5' domain (500-800 nucleotides) is shown.

Translation

Translation is the net effect of proteins being synthesized by ribosomes, from a copy (mRNA) of the DNA template in the nucleus. One of the components of the ribosome (16s rRNA) base pairs complementary to a sequence upstream of the start codon in mRNA.

IMPORTANCE OF rRNA

Ribosomal RNA characteristics are important in medicine and in evolution.

- rRNA is the target of several clinically relevant antibiotics: chloramphenicol, erythromycin, kasugamycin, micrococcin, paromomycin, ricin, sarcin, spectinomycin, streptomycin, and thiostrepton.
- rRNA is the most conserved (least variable) gene in all cells. For this reason, genes that encode the rRNA (rDNA) are sequenced to identify an organism's taxonomic group, calculate related groups, and estimate rates of species divergence. For this reason many thousands of rRNA sequences are known and stored in specialized databases such as RDP-II and the European SSU database.

SYNTHESIS OF RIBOSOMAL RNA - rRNA

The 70S ribosome of prokaryotes consists of a 308 subunit and a 50S subunit. The former contains 16S rRNA and the latter 238 and 58 rRNAs. In bacterial genes the sequences specifying 168, 238 and some times also5S RNA are arranged in a series mRNA is transcribed from DNA as a 30S unit, which bas been called p308.

Experimental evidence indicates that the 308 unit has the 16S component at the 5' end and the 238 component at the 3' end, with spacer units between the two components. In some prokaryotes 5S rRNA is transcribed at or near the 3' end.

During processing the p30S transcriptional unit is cleaved within the spacer segment by RNase III into 25S and l8S segments. These are then reduced to p23S and pl6S segments respectively, also by RNase III. Secondary trimming of these intermediates yields the final size, 23S and 168, respectively.

The enzyme involved in secondary trimming is probably a ribonuclease, designated as RNase M or maturase.In E. coli trimming of the pl6s component involves removal of a total of about 200 nucleotides from both 3' and 5' sides. The final cleavages occur when the rRNAs are associated with structural proteins of ribosomes in the 'preribosomal particles'.

Some nucleoside modifications take place in the p30S component, while others occur only after cleavage. Methylaiions found in 23S rRNA occur early at the p30S stage, whereas modifications found in 168 rRNA take place at a late stage of processing, in some cases after association with specific ribosomal proteins. In E. coli the precursors of 58 rRNA are only a few nucleotides larger than mature SS rRNA. In Bacillus there are two types of larger precursors. One, designated as p5A, is 180 nucleotides long, while the other, P5B' is 140-150 nucleotides long. Both are apparently transcribed on different 5S genes.

Cleavage of p5A by the enzyme RNase M5 splits it into three components, a 5' terminal piece of about 20 nucleotides, mature 5S rRNA of 118 nucleotides and a 3' terminal piece of about 40 nucleotides. The 5' and 3' terminal pieces are broken down to their monouculeotides by an exonuclease, which thus serves a scavenging function.

Mature 5S rRNA is resistant to this enzyme. The precursor p5B is also split by RNase M5 into a 22 nucleotide 5' terminal piece, a mature 58 mRNA and some smaller 3' terminal pieces. In eukaryotes the 808 ribosome consists of 40S and 60S subunits. The 408 subunits contain 16-18S rRNA, and the 60S subunits 25-28S rRNA, S.88 RNAand.5S rRNA. The genes coding for 16-188 rRNA and 25S-288 rRNA arc arranged in clusters of hundreds to thousands of copies. In eukaryotes the transcribed rRNA is 45S rRNA in mammals and 36-38S rRNA in lower eukaryotes, with molecular weights of 4.5 million and 2.6-2.8 million, respectively. There are currently two schemes for the possible arrangement of 28S and 18S segments within the 458 rRNA precursor; and the processing of the precursor rRNA.

- According to one scheme 458 rRNA has the general form: 5'P-28s rRNA - spacer - 18SRNA-spacer-3' OH. This transcriptional RNA has a life time of about 15 minutes during which methylation of the ribose moiety in the 288 and 188 regions takes place. An endonuclease cleaves 458 rRNA into 418 and 208components.

The cleavage takes place towards the 5' side of the 208 rRNA components. 41S rNA is degraded through 36S and 328 stages to mature 28S rRNA. During the three steps cleavage takes place in the non-methylated spacer segment by exonucleases. Similarly degration of the non methylated spacer region of the 20S component yields 18S rRNA.

- According to the second scheme the transcribed spacer sequence is situated at the 5'P end of 45S rRNA and the 28S com-ponent at the 3'OH end. The bulk of the experimental evidence supports this scheme which would also unify transcription in prokaryotes and eukaryotes. Processing occurs almost entirely within the nucleolus in eukaryote cells. The number of nucleoli within a cell depends on the extent to which the rRNA genes within a chromosome are dispersed.

45S transcriptional rRNA undergoes cleavage at four sites. Cleavage at site I removes the transcribe4 spacer sequence at the 5'P end. In primitive eukaryotes it may occur before transcription is completed. The order of cleavage at sites 2 and 3 varies in different species. Cleavage at site 2 separates 18S rRNA.

A segment of 140 nucleotides, with a sedimentation coefficient of 5.8S (formerly 78), situated between sites 3 and 4, remains covalently linked to the segment formed by.cleavage 3. In yeast tRNAs about 10% of 5.88 RNA is extended at the 5' end by 6-7 nucleotides. Cleavage 4 results in final trimming of the large rRNA segment. 58 rRNA is transcribed separately.

Analyses of the terminal sequences of 16-18S rRNA from many organisms reveal that the 3' end is almost always A-OH and the 5' end is pU. As many as 8 nucleotides at the 3' end are apparently uniform. They may be involved in the binding of mature rRNA to mRNA. Similarly 6 or more nucleotides may be uniform at the 5' end.

Nucleoside modifications take place on the initial transcript 45S rRNA in regions that give rise to mature 285 and 18S rRNAS. A few base methylations (6 in 18S rRNA) occur at later stages. This late methylation is characteristic of prokaryotes. Mefhy1ation seems to be essential for cleavage.

CLOSURE OF RRNA

Bacterial genome projects usually comprise three well-defined stages. The first one is the construction of high quality random shotgun libraries with varying insert sizes. Following this, the next task is typically to generate a deep coverage of the genome, with random shotgun sequences. This step generates the bulk of the sequences that are used to compile the finished genome sequence, and it is in many ways the heart of the project. Shotgun sequences are then assembled into sequence contigs, each representing a separate, non-overlapping portion of the genome.

The resulting assembly is then converted into a continuous genome sequence through a more laborious and time-consuming finishing phase. Due to the small size and low complexity of bacterial genomes, a genome sequence is only considered finished when all bases meet an acceptable quality level and the individual sequences are assembled into a single continuous consensus sequence. During the finishing phase of a bacterial genome project, the overall sequence quality is improved and sequence data are generated to close gaps between existing contigs.

Finishing can take a variety of forms, but it almost always involves the generation of sequences from large insert clones that have been demonstrated to span virtual gaps in the shotgun assembly, and by the use of alternative strategies to close real gaps when the corresponding DNA fragment has not appeared in any of the libraries generated for the sequencing project. The occurrence of real gaps is often associated with statistical cloning fluctuation of the shotgun model, repetitive regions in the genome (for example ribosomal operons), and with sequences refractory to the cloning system used. Strategies for real gap closure that have been used so far include the use of alternative cloning vectors, combinatory PCR, with primers derived from contig ends, subtractive hybridization, physical mapping, and direct sequencing of genomic DNA.

The time required for finishing depends on the nature of the genome and most particularly on the structure and frequency of repetitive sequences, but this phase easily outlasts that of the generation of the shotgun sequences and is indeed one of the key rate-limiting steps in bacterial genome sequencing. Here we describe the assembly process and finishing phase of the *Chromobacterium violaceum* genome project. *C. violaceum* is a free-living, gram-negative bacteria that is highly abundant in the water and banks of the Negro River in the Brazilian Amazon, and produces a violacein pigment with antimicrobial activity against some important pathogens as well as antiviral and anticancer activity. The sequencing and analysis of the *C. violaceum* genome were entirely executed by the Brazilian National Genome Sequencing Consortium.

A random shotgun strategy was used, and the resulting sequences were assembled into 57 contigs. These were then organized into 19 scaffolds, using the information from shotgun and cosmid clones; sequences of which were located in different contigs. Forty-seven virtual gaps within the scaffolds were closed by whole insert sequencing of the corresponding shotgun/cosmid clones. Eighteen real gaps were, for the most part, closed by applying the PCR-assisted contig extension (PACE), followed by specific confirmatory and oriented combinatory PCR. PACE involves the generation of stepwise extensions from the ends of contigs by PCR, until the closure of individual gaps is achieved. This methodology has proven to be especially useful for extending the multicopy ribosomal operons, and has greatly accelerated the finishing phase of the *C. violaceum* genome project.

METHODS

Pace Methodology

PACE was developed as a two-step PCR strategy, with nested primers derived from contig ends, as described by Carraro et al., 2003. Briefly, a set of 96 primers was randomly selected, with no reference to their precise sequence. Pairs of outward-facing specific nested primers were then designed, approximately 140 bp from contig ends and 40 bp apart from each other. Specific primers were checked for alignment to a single position of the genome with the FASTA programme. Secondary structures and dimer formation were verified using the Oligo Tech software. PCR was then performed in 96-well plates, using 80 ng genomic DNA as templates for the first reaction and 1 ìl of a 1:100 dilution of the first reaction as templates in a subsequent nested reaction.

Specific PCR

All contig extensions were confirmed by specific PCR, followed by direct sequencing of the resulting PCR products. Reaction mixtures for specific PCR contained 80 ng genomic DNA, 250 ìM dNTP, 1.5 mM MgC12, 1 U Platinum High Fidelity *Taq* DNA polymerase (Invitrogen, Carlsbad, CA, USA), and 10 ìM of each specific primer in a final volume of 20 ìl. Reactions were carried out at 94°C for 30 s, 60°C for 30 s, and 68°C for 2 min from 35 cycles. An initial cycle of 94°C for 2 min, and a final extension at 68°C for 6 min was used.

Combinatory PCR

For ribosomal operon orientation, primers were designed in the flanking regions of the 16S and 5S rRNA genes (outside of the operon repeat unit) and PCRs were undertaken in a combinatory way. The expected fragment sizes were between 5.2 and 5.8 kb, depending on the annealing position of the primers in the flanking region. PCRs were performed in a final volume of 20 ìl, containing 1.5 mM MgC12, 300 ìM dNTPs, 2 U Platinum High Fidelity *Taq* DNA polymerase (Invitrogen) and 15 ìM of each flanking primer. Reactions were carried out at 94°C for 30 s, 60°C for 30 s and 68°C for 5 min, for 35 cycles. Initial denaturation step at 94°C for 2 min, and final extension step at 68°C for 6 min, was used. Two almost identical versions of the rRNA operon were identified during the assembly phase. The difference between the two copies resides in a 100-bp insertion in the intergenic region between the 16S and ILE genes and a 74-bp insertion between the ILE and ALA genes (copy A - contains 100 bp-16S/ILE, 74 bp- ILE/ALA and copy B - does not contain 100 bp-16S/ ILE, 74 bp-ILE/ALA). In order to position correctly the two versions in the genome, a primer located 150 bp downstream of the 100-bp insertion (in the ILE gene) was used in a combinatory way in amplification reactions, together with one of the eight specific primers corresponding to the 16S rRNA flanking region.

PCR reactions were carried out as described above, except for modifications in the cycling parameters (94°C for 30 s, 60°C for 30 s and 68°C for 2 min). Sequence specificity was checked by BLASTN analysis of the two fragment extremities against the two available ribosomal operon copies. A fragment was considered specific to one of the two copies if the high quality sequence portion aligned with the corresponding copy with at least 95% identity.

Product Analysis

Three to five microliters from each PCR were loaded onto a 1% ethidium bromidestained agarose gel. Single PACE products, or fragments of expected size in the case of confirmatory or combinatory PCRs, were purified with the QIAquickTM PCR Purification Kit and sequenced directly using the same specific primer used for amplification on an ABI PrismR 3100 DNA sequencer. High-quality sequences with more than 300 bp, and Phred quality greater than 20, were analysed for specificity. Sequence specificity was checked by BLASTN analysis against the available genome assembly, and a fragment was considered specific if at least 25 bp of the high quality sequence aligned with at least 95% identity with the end of the corresponding contig.

Chromobacterium Violaceum Genome Assembly

The sequencing and analysis of the *C. violaceum* genome were entirely executed by the Brazilian National Genome Sequencing Consortium, comprising 25 sequencing laboratories, one bioinformatics centre, and three coordination laboratories, distributed throughout Brazil. In the initial phase of the project, random shotgun sequencing produced approximately 80,000 high quality reads, generated from both ends of pUC18 clones, with insert sizes raging from 2.0 to 4.0 kb. Additionally, both ends of 3,350 cosmid clones, with an average insert size of 40 kb, were also sequenced, providing a validation check of the final assembly.

The shotgun sequences, corresponding to approximately 13-fold genome coverage, were assembled into 57 contigs. These shotgun contigs were then organized into 19 scaffolds, using the information from shotgun and cosmid clones, the end sequences of which were located in different contigs. Forty-seven virtual gaps within the 19 scaffolds were closed by whole insert sequencing of the corresponding shotgun/cosmid clones. Points of genome assembly instability and regions of low quality sequences were identified using the Autofinisher Programme, and were resolved by re-sequencing of the selected clones. Real gaps that did not involve the rRNA gene (n = 18) were mainly closed by applying PACE, as detailed in Carraro et al. Here, we relate the methodology used to close rRNA-related gaps and to correctly position the eight ribosomal operon units in the *C. violaceum* genome assembly.

Closure of Real Gaps by Pace

The PACE technique depends on rare mismatched primer-template interactions that occur between arbitrary primers and template DNA with an unknown sequence, even under highly stringent conditions. These are captured through elevated PCR-cycle repetition, and through the use of specific anchoring primers, corresponding to adjacent regions of known sequence (contig ends). Thus, PACE allows the generation of stepwise extensions from the ends of contigs by PCR, until the closure of individual gaps is achieved. When we started using PACE to close the *C. violaceum* genome assembly, seven of the existing 38 contig ends ended with the 5S rRNA sequence and three ended with the 16S rRNA sequence, suggesting the existence of at least seven identical copies of the rRNA operon.

Twenty-two PACE reactions were initially applied to extend contig ends that did not contain rRNA-derived sequences, resulting in 137 specific sequences, allowing the immediate closure of 15 real gaps in the *C. violaceum* assembly due to their relatively small size. Of these, six apparently linked the contig to an rRNA gene (one to a 5S rRNA gene and five to 16S rRNA genes), suggesting a total of eight copies of the ribosomal operon. In two cases (gap 191-221 of 1691 bp, gap 202-199L of 1790 bp), additional rounds of PACE were undertaken until the complete closure of each gap was achieved. All contig extensions and gap closures were confirmed by specific PCR, followed by direct sequencing of the PCR products.

Closure of Gaps Related to rRNA Sequences

We used a single pair of nested primers specific for the 16S rRNA gene to check for the extension of contigs ending in 16S rRNA. Thirty-five PACE fragments were selected for sequencing and high quality sequences were further analysed using BLASTN to check for their specificity. Of these, 34 were found to be specific, and they confirmed five novel flanking regions for the 16S rRNA gene (contigs 194, 205, 222, 230 and 183), in addition to the three known, giving a total of eight rRNA operons. All 16S and 5S rRNA contig extensions were confirmed by specific PCR, followed by direct sequencing of the PCR products. Seventy-two specific PCR reactions, including real gaps and rRNA-related gaps, were performed, corresponding to the four possible combinations between the external and nested primers from both contig ends. At least one fragment from each junction was submitted for sequencing to check the specificity.

Positioning of the Eight Ribosomal Operons in the Chromobacterium Violaceum Genome Assembly

After the 16S PACE protocol, the genome assembly was composed of eight unoriented contigs, ending either with 16S rRNA or 5S rRNA sequences. To assemble the contigs in the correct order, we used a combinatory PCR strategy, with primers derived from the non-repetitive genomic region flanking the 16S and 5S rRNA gene.

PCR fragments of expected size were sequenced from both extremities, and high quality sequences were aligned to the genome assembly. The orientation of the contigs was confirmed if at least 100 bp of the sequenced fragment aligned with the end of the corresponding contigs, outside the common ribosomal operon sequence. The whole genome assembly revealed a slight difference between the sequences, corresponding to the ribosomal operons. Two almost identical versions were identified.

The difference between the two versions resided in a 100-bp insertion between the 16S and ILE genes, and 74 bp between the ILE and ALA genes. In order to correctly place the two versions (A and B), in the eight already-oriented ribosomal operon copies, a primer located 150 bp downstream of the 100-bp insertion was used in a combinatory way, together with one of the eight specific primers corresponding to the 16S rRNA-flanking region. Using this strategy, we found out that most (6/8) of the rRNA operon copies contained the 100-bp and 74-bp insertion (copy A). These results were independently confirmed by the finding of a larger number of reads in the genome assembly in which the 100-bp insertion was not present.

After the initial assembly of shotgun reads, the *C. violaceum* genome sequence was organized into 19 scaffolds. The genome sequence assembly was mainly made difficult by the eight almost identical rRNA operon copies dispersed throughout the genome. Using PACE, we were able to generate contig extensions with an average of 1 kb in length from all contigs, which closed the majority of gaps in a single round of experimentation. In addition, the PACE methodology proved to be extremely useful for extending the multicopy ribosomal operons.

The surprisingly high success rate of our approach can be attributed to deep shotgun coverage and the small size of the gaps in the *C. violaceum* genome assembly. However, deep shotgun coverage is not a pre-requisite for using PACE. Analyses made by our group demonstrated that the methodology can be applied at early stages of the bacterial genome assembly, drastically reducing the time and cost of the finishing phase.

Usually the finishing phase of a genome project takes between 50 to 60% of the total time of the project. In the case of the *C. violaceum* genome project we used approximately 30% of the total time required to conclude the project, corresponding to a reduction of at least 40% of the estimated time. The importance of PACE is that, at a pre-determined point in the shotgun sequencing, primers can be generated and contig extensions obtained with no requirement for a one-by-one analysis of the gaps. Using successive PACE reactions, it is possible to close gaps in bacterial genomes in a stepwise fashion, regardless of their size. Based on the experience accumulated in the *C. violaceum* genome project we strongly recommend the use of the PACE methodology in the finishing phase of bacterial genome projects.

7

Sequence of the Human Genome

A Person who has had the measles once is extremely unlikely to get a second attack of the same disease; he is, as we say, immune to it. Many other infectious diseases confer immunity upon recovery, although not always for such a long period. Immunity is the result of a defence mechanism of the body against the invading microorganism.

Antibodies are formed which combine with the microorganisms and often destroy them by, for example, clumping or dissolving them. Lasting immunity is found when the antibodies outlast the infection.

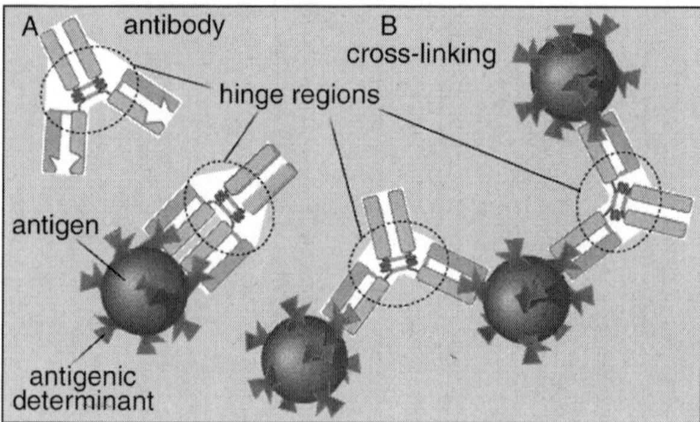

Fig. Antigen and Antibody.

Antibodies are highly specific in their reactions. Measles are a virus, but a person who has acquired immunity against measles is not, for that reason, less likely to contract German measles or poliomyelitis or any other virus disease.

The specificity of antibodies is a consequence of the manner in which they arise. When a microorganism gets inside a warm blooded animal, its protein acts as an antigen, that is, it stimulates the production of antibodies by the infected individual.

Proteins are highly complex molecules; they consist of thousands of atoms which are arranged into specific configurations. An antigen functions like a mold for shaping a host protein into its own mirror image. When the antibody has become free of this mold it still "fits" the exact type of antigen by which it was produced, but no others. Not only the protein of microorganisms but any foreign protein acts as antigen. This is the reason why no successful skin grafts can be made between different persons, not even between mother and child. No two individuals have exactly the same proteins in their skin cells, and when a graft is made it is eventually destroyed by the very antibodies whose formation it induces.

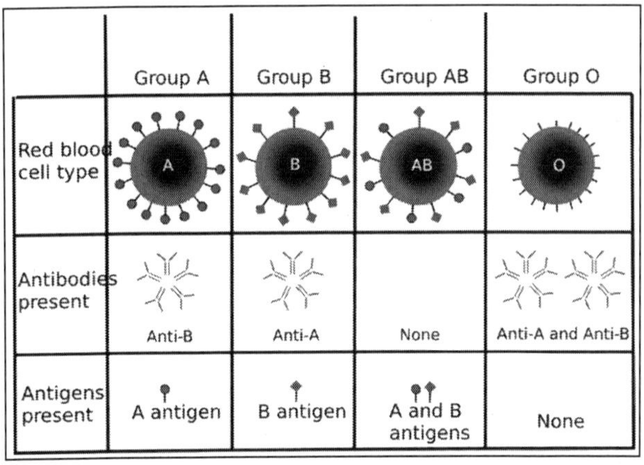

	Group A	Group B	Group AB	Group O
Red blood cell type	A	B	AB	O
Antibodies present	Anti-B	Anti-A	None	Anti-A and Anti-B
Antigens present	A antigen	B antigen	A and B antigens	None

Fig. Antigens and Antibodies in the A-B-O Human Blood Groups.

Antigens are also carried by the protein of the human blood cells. Fortunately, individual specificity is here less pronounced than in skin proteins; otherwise no transfusion of a patient with whole blood would be possible.

Yet not every combination between donor and patient is compatible, and in the early days of blood transfusion fatal accidents arose from clots of transfused blood in the veins of a patient. The reason for this is now well understood, and such accidents are avoided by the right choice of donor. A number of different antigens are carried by human red blood cells. Most of them can be detected only by the fact that they induce antibodies when the blood is injected into a warm blooded animal, for example, a rabbit.

But there is one group of antigens for which ready made antibodies occur in human serum, the fluid in which the blood cells are suspended. The main antigens of this group are called *A* and *B*, and human beings can be divided into four groups according to whether their red blood cells carry one of these antigens (*A* or *B*), both (*AB*), or neither (*O*).

In addition, the serum of each individual contains preformed antibodies to those antigens that are not present in the blood cells. Thus, in blood group *A* the serum contains anti-B, in blood group *B* it contains anti-*A*, in blood group *O* it contains anti-*A* and anti-*B*. and in blood group *AB* it contains neither. For successful transfusion the donor has to be chosen so that his blood is not clotted by the antibodies of the patient. It is probable that all antigenic specificities are gene controlled. In many cases this has been shown to be true. The *A-B-O* blood groups are determined by the action of three alleles of one gene; if, for the sake of simplicity, we waive the rule that alleles of the same gene should be denoted by the same letter, we may call these alleles *A*, *B* and *O*.

We were dealing with *pairs* of alleles: round and wrinkled, red and white, tasting and nontasting. There is, however, no reason why any given gene should exist in only two alternative forms. Indeed we know of many cases of so called "multiple alleles" in which a gene has mutated to a series of different alleles.

It shows a wild rabbit and three mutant varieties, produced by the action of multiple alleles of a gene controlling coat colour. The chinchilla rabbit has silvery fur; the Himalayan rabbit is white apart from black ears, feet, nose and tip of tail, and it has pink eyes; the albino rabbit is wholly white with pink eyes.

The alleles producing these effects are usually denoted by the letters *C* for full (wild) colour; c^{ch} for chinchilla; c^H for Himalayan; and *c* for albino. Any individual can, of course, carry only two of these genes and can be either homozygous for one of them, like the rabbits in the diagram, or heterozygous for two, for example, c^Hc.

In heterozygotes, coat colour depends on the dominance relationship between the two alleles; since c^H is dominant over *c*, c^Hc rabbits look Himalayan.

Fig. Multiple Alleles in the Rabbit.

The possible genotypes for A, B, and O in man and the blood groups that correspond to them are set out in the following table:

	homozygous			heterozygous		
Genotype	OO	AA	BB	AO	BO	AB
Blood group	O	A	B	A	B	AB

You will notice that each gene produces its characteristic antigen independent of its partner gene. As a result O, which does not produce an antigen, cannot be recognized in the presence of either A or B. In other words,

O is recessive to A and B, and persons of blood group A or B can be of two different genotypes according to whether or not they are heterozygous for O. This plays an important role in the medicolegal application of the A-B-O blood groups.

Newborn infants in maternity wards are labeled with the name of the mother. Very occasionally doubt may arise as to whether a mother has been given the right baby, and blood typing may then help to decide this question. In a well known case a woman, let us call her Mrs. Smith, found on her return from the hospital that her baby carried the label "Brown." There had been a Mrs. Brown in the same ward as Mrs. Smith, and she had, indeed, been given an infant labeled "Smith."

A glance at the table will show you that between them they carried nothing but O genes and could not have a child with any blood group but O. On the other hand, the Smiths could *not* have a child of blood group O; for, while Mrs. Smith belonged to blood group O, Mr. Smith belonged to AB. He was, therefore heterozygous for A and B, and had to transmit one of these alleles to his child. When it was found that baby "Brown" was O and Baby "Smith" was A, it was evident that the labels were correct and that the two women had been given the wrong infants.

Not all cases can be decided so easily and unambiguously. Suppose that both Smith and Brown parents had belonged to blood group A. Then either couple might have produced baby "Smith" with blood group A. Moreover, either couple might have produced baby "Brown" with blood group O; for A individuals may carry O genes and can transmit them to their children. Fortunately, many more blood groups, controlled by different series of alleles, have been discovered in recent years and can be used for deciding difficult cases.

This refers also to the use of blood typing in cases of disputed paternity, many of which cannot be decided on the basis of the A-B-O system alone. Mr. Smith, as we just saw, could not have fathered a child of blood group O; similarly, Mr. Brown could not have fathered a child belonging to blood group A, B, or AB. But the father of an A baby from an A mother might belong to any of the four major blood groups. In such a dilemma, recourse to other blood groups may help narrow the issue. In any case, blood group tests in cases of disputed paternity fulfill only a negative function; they can exclude certain men from possible fatherhood, but they cannot indict any particular man. Suppose blood group O were alleged to be the father of an illegitimate child of blood group O by a woman of blood group B. On purely genetical grounds, there is no objection to this assumption, but neither is there an objection to any other man of blood group O or against any men of blood groups A and B.

Only where the baby shares a very rare blood group with the alleged father, but not with the mother, will the geneticist feel reasonably convinced of the correctness of the allegation, although the court may not accept his judgment.

In 1921 a Norwegian judge made medicolegal history by convicting a man of paternity on account of a rare dominant abnormality foreshortened fingers, or "brachydactyly" which was absent in the mother but present in both the infant and the alleged father.

DANGER TO INFANTS

Like the alleles which determine ability or inability to taste PTC, the A, B and O blood group alleles occur not only in human beings but also in the large anthropoid apes. A new and highly important human blood group was discovered in 1940 through tests on lower monkeys. When rabbits were injected with blood from a Rhesus monkey, their serum was found to contain an antibody which combined with the red blood cells of about 85 per cent of individuals in a White population; the remaining 15 per cent gave no such reaction.

Persons whose blood reacts with the antibody against Rhesus blood are called Rh-positive; those who do not react are called Rh-negative. Both types of individual occur in all human races, but the proportions of positives and negatives differ. Evidently the red blood cells of Rh-positive individuals carry an antigen which is so similar to the antigen on the Rhesus blood cells that the antibody does not distinguish between them. The blood cells of Rh-negative persons lacks this antigen.

When an Rh-negative person is given a blood transfusion from an Rh-positive donor, his body reacts like that of a rabbit into which Rhesus antigen is injected: it makes antibodies against the foreign antigen.

Since it takes some time before a sufficient amount of antibody is produced, a first transfusion of this kind is usually perfectly harmless; but the antibodies persist in the serum of the transfused person, and should he receive a second transfusion with Rh-positive blood there may be a severe reaction. This explains why accidents had happened occasionally in blood transfusions between individuals of the same A-B-O blood group. No antibodies are formed by an Rh-positive individual who is given a transfusion with Rh-positive blood: for such a patient the Rh-antigen is not a foreign antigen.

Of much greater importance is the fact that the new blood group furnished the explanation for a severe and hitherto rather mysterious disease, called "hemolytic disease of the newborn." Before we go into this, we shall have to consider the inheritance of the Rh character. At first sight it is very simple. Rh-positive children invariably have at least one Rh-positive parent. Rh-negative children, on the other hand, may issue from any combination of parents: both Rh-positive, or one Rh-positive, the other Rh-negative. Thus Rh-positive parents may carry the Rh-negative gene and transmit it to their children, while Rh-negative parents carry only Rhnegative genes.

Expressed in Mendelian terms, this means that we are dealing with a dominant allele, Rh, and a recessive allele, rh. Rh-negative individuals must be homozygous (rh rh), but Rhpositive individuals may be homozygous (Rh Rh) or heterozygous (Rh rh). In reality, more alleles are known and the situation is far from simple; but for most practical purposes these complications can be disregarded.

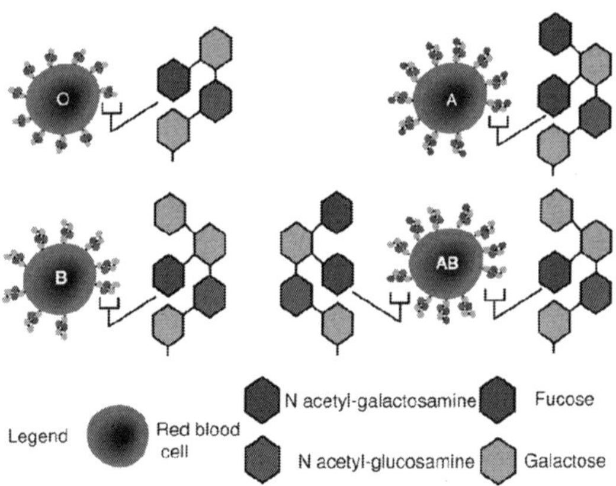

Fig. Blood Group System

With this simple genetical picture in mind, let us return to hemolytic disease of the newborn. About once in several hundred pregnancies among White women, a newborn child suffers from a severe anemia with jaundice which in the worst cases may be fatal if blood transfusion cannot be carried out in time. The disease has a tendency to affect several children in the same family, and this seemed to point to a hereditary basis; but there are also special features which do not fit into any simple Mendelian explanation. Thus first born children hardly ever are affected; but once a couple have had one hemolytic child, more of their children are likely to be affected. In some families all children following the birth of the first affected one suffer from the disease. The explanation came when it was found that in almost all cases of hemolytic disease the mother was Rh-negative and the child Rh-positive and that, moreover, the blood of the mother contained anti-Rh antibody.

On the contrary, no such antibody is found in the blood of Rh-negative women who never have been pregnant or who have given birth only to Rh-negative children. This shows that an Rh-negative woman reacts to an Rh-positive fetus as she would to transfusion with Rhpositive blood: she makes antibodies against the Rh-antigen. As after blood transfusion, the level of antibodies takes time to build up and usually no reaction occurs with the fetus that stimulated antibody production. This explains why first born children rarely suffer from hemolytic disease

But when an Rh-negative woman is pregnant with a second Rh-positive child her antibodies may get into the circulation of the fetus and destroy its blood cells. When an Rh-negative woman bears an Rh-positive child, the *Rh* gene in the fetus must have come from the father's side.

The father, therefore is either homozygous, *Rh Rh*, or heterozygous, *Rh rh*. In the former case, he will transit an *Rh* gene to every child; this is the type of marriage where, once a sufficiently high level of antibodies has been established in the mother, every child will be affected. In the second case, half the children will inherit an *rh* gene from both father and mother and will be Rh-negative; these children are in no danger from the antibodies in the mother's blood.

Fortunately, a large proportion of pregnancies which, from the blood groups of father and mother, might lead to hemolytic disease of the newborn do not, in fact, do so, partly because of the low level of antibodies in early pregnancies, partly because of differences in the ease with which antigens and antibodies are exchanged between mother and child, partly perhaps for still other reasons.

All the same, many hospitals make sure that pregnancies of an Rhnegative mother and an Rh-positive father get special attention. In severe cases, an affected child may be saved by the replacement of all its blood with blood that is free from antibodies.

Similar gene controlled incompatibilities between the blood groups of mother and offspring are known in a number of animal species, for example, rabbits, dogs, and horses. Not always, however, do the antibodies harm the fetus before birth. In the horses, for example, antibodies are produced when a stallion, carrying on its blood cells an antigen *R* sires a foal with this antigen by a mare lacking it. The mare produces antibodies against *R*, but it takes at least three pregnancies by an *R* stallion before the danger limit has been reached. In a subsequent pregnancy by an *R* stallion, the antibodies in the mother's milk may kill the foal; but the foal may be saved by fostering it out to another mare.

GENES AND CHARACTERS

There is hardly any character in any organism that is not to some extent under the influence of the genotype. Genes control the colors and shapes of animals and plants, the ability of animals to see, hear and smell, the resistance of a man to an infecting bacterium, and the ability of the bacterium to infect the man. They influence the rate at which a calf will grow on a given diet, the degree to which a child can profit from education, and the likelihood that a person's balance of mind will give way under mental stress.

They determine whether or not a Chlamydomonas has flagellae for swimming about, and whether or not a yeast culture can grow on galactose. Genes influence the most superficial as well as the most fundamental processes of life; they act at every stage of development. Indeed, the very chance of survival at any period from conception to extreme old age is to some degree under the control of the genotype. In chickens, rats and mice, mutant genes have been shown to kill the embryo by severely disturbing the development of such vital organs as the skeleton, the brain, the kidneys.

Doubtless, the normal alleles of these genes play a role in controlling the normal development of these organs. Equally doubtless, similar genes will play this role in the human embryo. At the other end of the life span we find that inbred lines of mice differ in the average length of life of the individual. There is little reason to doubt that also in man longevity is in part genedetermined, although the very great genetical and environmental variability among human individuals makes it difficult to provide conclusive evidence for it.

It seems that the great majority of genes are indispensable for the survival of the organism. In experimental animals or plants, very small pieces of chromosome can be destroyed by, for instance, X rays. Attempts to breed individuals that lack the same piece in both partner chromosomes rarely are successful; for no matter which piece is missing, it almost invariably seems to contain genes without which survival is impossible.

In Drosophila, it could be shown that even genes controlling such a seemingly irrelevant character as eye colour or shape of wing margin need to be present in at least one chromosome for the embryo to develop into a living fly. If such genes had no other effects than their easily observed ones, this would be difficult to understand. There is, however, ample evidence that most, if not all, genes produce a variety of effects.

This evidence comes from a study of mutant genes. The very existence of a normal gene can be inferred only after it has mutated at least once, and its effects must be deduced from a comparison between normal and mutated individuals. Many mutant genes have several obvious effects. In Drosophila, the mutant gene "lozenge" makes the eyes look sticky like boiled sweets, removes the claws and pads from the feet so that the flies can no longer walk up the glass wall of their culture bottle, and makes the females infertile.

Japanese waltzing mice are deaf; Frizzle hens have immature ovaries and lay few eggs. It is safe to conclude that the normal allelomorphs of these genes are necessary for the normal development of eyes, feet, and genetical organs in Drosophila; of ears and nervous system in mice; of plumage and ovaries in fowl. Whenever a mutant gene is studied thoroughly, it is found to produce a whole array of effects.

Most mutant genes reduce viability; it is therefore not surprising that their normal alleles are necessary for life. The ability of a gene to produce several distinct effects is called "pleiotropy," that is, "many directedness." From what has been said just now it is clear that pleiotropy is the rule and not the exception.

Just as every gene influences many characters, so is every character influenced by many genes; this is known as "gene interaction." In a way, gene interaction and pleiotropy are complementary to each other; for to state that many genes act pleiotropically on viability is only a different way of saying that viability is influenced by many genes. Similarly, many mutant genes and, by inference, their normal alleles affect fertility and, conversely, fertility depends on the action of all these genes as well as on that of others which do not advertise their presence by striking effects on superficial characters. Even very specific characters, like the colour of the iris in the human eye, depend on the action of many genes.

In order to avoid details, the "characters" have been chosen as broadly as possible and might better be called "organs" or "functions." The choice of more narrowly defined characters, such as eye colour or structure of the tail, would have yielded an essentially similar picture.

A line connecting a gene with a character signifies that the gene influences the character. The diagram has been kept simple by restricting it to a few genes and characters. Had we entered more characters on the right, more lines would have started in most genes on the left. Had we entered more genes on the left, more lines would have come together in most characters on the right. The irregular way in which the lines cross each other shows that the constellation of characters controlled by a single gene varies between genes.

Thus the genes *fi*, *st*, *Va*, and *v* all cause a similar nervous disorder resulting in circling and shaking movements; but fidget mice can hear, while the other three mutant types are deaf; only the gene *Va* produces an irregular pattern on the coat, only *fi* leads to sores in the eyes, and only *st* shortens the tail. Again, it should be remembered that these interactions between mutant genes and characters indicate some of the ways in which the normal alleles interact in normal development.

PLEIOTROPY: ONE GENE AFFECTS SEVERAL CHARACTERS

Sickle Cell Anemia

In 1910 a young West Indian Negro consulted an American physician about a feverish cold. Since he complained of not having felt well for quite a long time, the doctor undertook a thorough examination, in the course of which he found that the young man suffered from severe anemia of a hitherto not recorded type.

Under the microscope his blood presented a striking picture. While in normal blood the red cells are uniformly round disks, the blood of the patient contained a high proportion of curiously sickle shaped red cells. Several years later the same abnormality was found in the blood of a father and his son, both suffering from anemia. The condition was called "sickle cell anemia"; it is quite distinct from the more usual forms of anemia.

Once attention had been directed to it, more cases were reported, at first exclusively from Negroes. It soon became obvious that the disease was in some way hereditary, but its accurate genet ical basis was determined only when a technique had been developed that ensured reliable detection of the sickling phenomenon.

a

b

Fig. (*a*) Blood Corpuscles of a Patient with Sickle cell Anemia. In (*b*) the Sickling has been Emphasized by Keeping the Drop of Blood Sealed off from the Air.

This technique consists in keeping the sample of blood for a day or more sealed off from the air, when the cells of patients assume bizarre shapes. With this technique it was established that, in addition to the typical blood picture found in patients suffering from sickle cell anemia, a mild, but distinctly recognizable form of sickling occurs in the blood of about 9% American Negroes who do not suffer from anemia.

This milder, harmless form of sickling is especially frequent among the relatives of patients suffering from sickle cell anemia; in particular, it is practically always found in both parents of an anemic child. This led to the interpretation, fully borne out by further investigations, that the condition is due to a gene *S*, which in heterozygotes *S*+ causes the mild form of sickling, unaccompanied by anemia, while homozygotes *SS* are anemic.

Laboratory investigations of the abnormal blood revealed that the gene *S* affects the hemoglobin, that is, the pigmented, ironcontaining protein of the red blood cells. In an electric field like the one the hemoglobin molecules of *SS* individuals move more slowly than the hemoglobin molecules of normal individuals. The blood of heterozygotes *S*+ contains both kinds of hemoglobin in approximately equal proportions; apparently one kind is formed under the influence of the *S* gene, the other under the influence of its normal allele.

This, incidentally, is a good illustration of the fact that the terms "dominant" and "recessive" are relative, and that a gene may be dominant at one level of observation and recessive at another. At the level of hemoglobin chemistry, the gene *S* and its normal allelomorph act independently of each other, similar to the blood group genes *A* and *B* at the level of the microscopic blood picture, *S* is incompletely dominant over its normal allele; finally, at the level of the manifest disease, *S* is recessive to its normal allele.

A search for the presence of the gene *S* in many countries gave curious results.

At first it seemed as though the gene was restricted to Negroes, in Africa and in the two Americas. Later, individuals whose blood could be made to sickle were found also among Italians, Greeks, and other peoples around the Mediterranean, and among South Indian tribes.

The distribution of such individuals, however, was exceedingly patchy. While large numbers of them occurred in certain parts of Africa, they were practically absent from others. Even within the same territory there often were striking differences in the proportion of individuals with sickle cells. In Uganda, for example, this proportion was nearly one half in one tribe and nearly zero in another. The most puzzling aspect of this survey was that sicklers (S^+) have such high frequencies in certain populations. Homozygotes for *S* rarely reach adulthood and have children; nature effectively sterilizes them. Although, sterilization of homozygotes for a recessive gene is a very inefficient way of getting rid of the gene, it will keep its frequency low.

In fact, other genes that result in "natural sterilization" of the homozygotes, such as the gene for amauritic idiocy, are rare even in heterozygotes. It therefore seemed that in certain populations the childless death of the homozygotes was overbalanced by the fact that the heterozygotes had some natural advantage which gave them a better chance than the rest of the population to have children, half of whom would inherit the gene *S*.

A possible clue to the nature of this advantage was provided by the discovery that sicklers are frequent in malaria ridden regions and are rare or absent in malaria free regions. This suggested that the gene *S* in some way protects its carriers against the malaria parasite. Investigations to test this idea lent support to it.

Out of 288 young children in a malaria district of Africa, 43 were heterozygous for *S* and of these 12, that is, 28%, were found to carry the malaria parasite in their blood; of the remaining 245 children, all without the gene *S*, 113, that is, 46 per cent, carried the parasite. In one African tribe 30 volunteers offered themselves for infection with malaria; 15 of them were S^+ and of these only 2 contracted the disease; the other 15 were ++, and of these only 1 remained free of it.

Altogether, there is good, although not yet wholly conclusive, evidence for a pleiotropic effect of the gene *S*. The same altered hemoglobin that in homozygotes results in anemia makes the blood of heterozygotes an unsuitable environment for the malaria parasite. Since malaria affects physical and mental development in various ways, the final piciotropic effects of a gene controlling susceptibility to this disease will be correspondingly far reaching.

If this explanation of the curious distribution of the gene S is correct, we are dealing with a case in which a heterozygote for two alleles (S^+) is superior to the homozygotes for either (SS or $++$). This situation has been mentioned as a possible contributory cause of hybrid vigour.

Hereditary Dwarfism in the Mouse

They did not breed, but when their littermates were bred together, some of the litters contained about one quarter dwarf mice. The character thus was attributed to the action of a recessive gene. Mice homozygous for this gene stop growing at the end of the second week or a little later and usually reach a weight of only 6-8 grams as compared with a weight of 20 grams or more for normal mice. An adult dwarf mouse together with its normal littermate. The dwarf gene has many pleiotropic effects. Early on in development, even before growth has stopped, homozygotes can be distinguished from their normal littermates by blunt snouts and relatively shorter ears and tails. At the stage when normal young mice are lively and easily excited, dwarf mice are timid and quiet.

They are very sensitive to temperature changes; but when they are kept at the right temperature they tolerate starvation better than normal mice. They have a shortened life span and are sterile in both sexes.

Examination of their inner organs revealed that most of their hormone producing glands, such as the thyroid, the pituitary, and the gonads, were undersized and abnormal in structure. Now, it is well known that the pituitary, a small gland at the base of the skull, occupies a key position in the control of hormone production; for this reason it is often referred to as "the master gland." It therefore occurred to some scientists to attempt to cure the hereditary dwarfism by treatment with pituitary.

Pieces of fresh rat pituitary were implanted daily under the skin of dwarf mice which had stopped growing. The success of the treatment was spectacular. The treated animals attained normal weight and body proportions, they became lively and vigorous and behaved in every way like normal mice. The males became fertile, but the females remained sterile, although their ovaries regained normal size and structure. Similarly, all other hormone producing glands, with one notable exception, became normal. The excep tional gland, which did not respond to the treatment, was the pituitary. This is exactly what one would expect if the basic defect, from which all the others are derived, is an abnormality of the master gland; for, while implanted pituitary can supply all the hormones by which a normal pituitary controls its subordinate glands, it cannot cure the source of the trouble in the pituitary itself.

GENE INTERACTION: ONE CHARACTER DEPENDS ON MANY GENES

Chlorophyll Formation

In the cells of green plants the green pigments, together with some yellow ones, are carried inside numerous small bodies, the so called "chloroplasts".

Primitive plants, for instance, algae, may have one large chloroplast instead. Before the invention of the electron microscope, not much detail could be discerned in a chloroplast. Nowadays electron micrographs reveal a highly complicated structure.

Combined with genetical studies observations with the electron microscope have become an important tool for understanding the role of the genes in the development of the chloroplast, and although this work is still at its beginning, a rough picture of gene interaction has started to emerge. Electron microscopic studies have shown that the normal chloroplast (V) develops by stages II-IV from a small colorless granule (I), which in the course of development not only changes its size and shape but also forms a complex interior network of lamellae between which some globules are suspended.

At certain stages of this structural process, the chloroplast forms pigments, first yellow and then green ones. In the diagram, the source of the pigments seems to lie outside the chloro0plast; this has been done for the sake of clarity: in reality, the pigments form inside the chloroplast. It is certain that a vast number of genes are involved in the control of these processes, for many different mutations have been shown to interfere with chlorophyll and chloroplast formation.

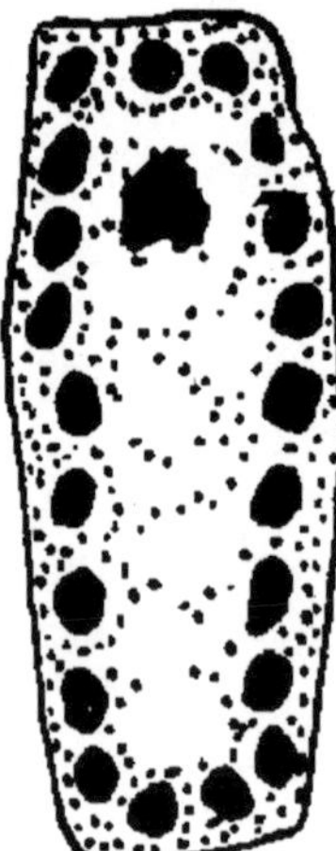

Fig. Plant Cell with Nucleus and Chloroplasts

More than a hundred such genes have proclaimed their existence by mutation; many more may have remained undetected because they have not yet given rise to observed mutations. It is also known that different genes control different steps of chlorophyll formation; for mutations may arrest chloroplast development at different stages and in different ways.

Genes Acting in Series: an Illustration from Human Life

Genes, like *A, B, D, G* or C_1, C_2, C_3 or F_1, F_2, F_3, that act in series resemble workers at a conveyer belt; nonfunctioning genes of such a series resemble absentee workers. Let us see what happens in a factory production line when one of the workers is absent. The assembly of a doll by four girls working at a conveyer belt. Ann has a supply of dolls' heads and fastens them to the bodies which are sent to her on the belt.

She sends her product to Beryl, who adds the arms. Beryl sends her product to Cathie, who adds the legs. Cathie sends her product to Dot, who dresses the doll and puts the finished product into a box. Let us suppose that no worker can carry out any other manipulation than her own and, moreover, that she can do this only when the previous manipulations have been completed.

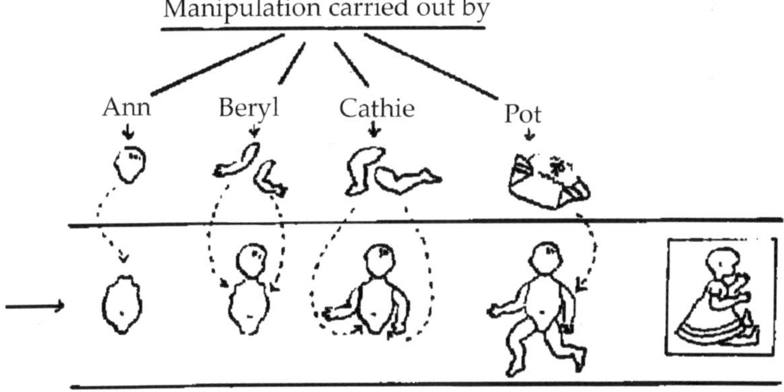

Fig. Assembly of a Doll on a Conveyor Belt.

This might, for instance, happen if the various parts of the body were attached to each other by strings with hooks which could be used in only one order. Now consider what happens when Ann is absent. If the conveyer belt is not stopped, headless and limbless bodies pass Ann's place, and since they cannot be manipulated by the other workers they accumulate at the end. If, instead, Beryl is absent, Ann will be able to carry out her manipulation but Cathie and Dot will not, and limbless bodies with heads will accumulate at the end of the line.

Cathie's absence will result in an accumulation of legless bodies, and Dot's absence in an accumulation of naked but otherwise complete dolls. In general, when a worker is absent there will be an accumulation of the unfinished product furnished by the preceding worker.

Suppose now that the foreman of the production line is anxious to fulfill his quota of finished dolls and that he has the chance of obtaining either the finished product or the intermediate ones from a parallel production line which has a surplus of all of them.

The finished product will, of course, help him in any case; but in order to keep the remainder of the workers occupied he may prefer to ask for an unfinished one. When Cathie is absent, Dot can be occupied by a supply of naked dolls, but Ann and Beryl will have to remain idle. When Beryl is absent, Dot can again be occupied by a supply of naked dolls; but now it would be preferable to obtain the previous product, legless dolls, for this would keep both Cathie and Dot busy.

Finally, when the absentee is Ann, all remaining girls can be occupied if Beryl is supplied with limbless bodies; failing this, Cathie and Dot can be kept busy by a supply of legless bodies, or Dot alone by a supply of naked but otherwise complete dolls. In general, then, the choice between usable outside supplies is the greater the earlier along the line the absence occurred; when the last worker is absent only the finished product will do; when the first worker is absent all unfinished ones with the exception of the first will serve the purpose.

Synthesis of an Amino Acid by a Bacterium

One of the main constituents of living matter is protein. A protein molecule is a long chain of simpler organic molecules, called "amino acids." About twenty different amino acids are known to occur in proteins, and most organisms require all or most of them for life.

Higher organisms like man obtain their amino acids through the protein in their diet. Many microorganisms, for instance most bacteria, can make or "synthesize" at least some of the required amino acids from simple inorganic ingredients of their food. Where genetical methods could be employed in the study of amino acid synthesis that even bacteria can be studied genetically—it was found that amino acid synthesis proceeds in steps, each step being controlled by a different gene.

A, B, C, D = normal genes controlling successive steps in the synthesis of *AA* = an amino acid P_1, P_2, P_3 = intermediate products; precursors of *AA* = block in the reaction series where one of the genes is not functioning These genes, then, act in series like the girls at the conveyer belt. Our analysis of the "synthesis" of a doll will help us understand what happens when one of these genes mutates to an inactive state. Synthesizes the first step; gene *B*, comparable to Beryl, synthesizes the second; gene *C*, comparable to Cathie, synthesizes the third, and *D*, comparable to Dot, the fourth. *AA*, comparable to the dressed doll, is the finished amino acid; P_1, P_2 and P_3 are three intermediate products, or "precursors," comparable to the three unfinished stages of the doll. Bacteria in which *D* is not functioning can survive only when they are supplied with the finished amino acid; for *D*, like Dot, is the only one that can carry out the final step. Bacteria with an inactive gene *C* can survive on either the amino acid or its precursor P_3. Bacteria with an inactive gene *B* can use not only the amino acid and P_3 but also P_2.

Finally, bacteria in which A is not functioning can use all three precursors as well as the amino acid. Often such mutant strains accumulate precursors which precede the break in the production line; thus a strain with an inactive gene B may accumulate P_1, or a strain with an inactive gene *D*, P_3.

The study of chemical reactions in the living organism is called "biochemistry." The study of genes controlling biochemical reactions is called "biochemical genetics." The biochemical genetics of microorganisms has cleared up the manner in which many important compounds are synthesized under the control of genes acting in series.

There is a remarkable similarity in the biochemical pathways by which cells of widely differing organisms synthesize essential compounds such as amino acids and vitamins, and the biochemical genetics of microorganisms has contributed at least as much to biochemical as to genetical knowledge.

A gene-controlled Biochemical Sequence in Man

Lastly we shall consider an example from human genetics where genes have been shown to work in series not on the synthesis but on the breakdown of an amino acid. In 1908 an English physician, Garrod,called Inborn Errors of Metabolism.

In it he dealt with a number of inherited disorders of human metabolism, that is, of those biochemical processes by which the living organism synthesizes new compounds or breaks down existing ones. One of the conditions described by Garrod was alkaptonuria, an abnormality characterized by a blackening of the urine and a blackening and hardening of the cartilage (gristle).

The disease is inherited as a recessive autosomal condition and must therefore be due to the malfunctioning of a single gene. When the urine of patients was analysed it was found to contain a substance called alkapton, which is absent from the urine of normal persons. Now alkapton is one of the breakdown products of protein, or rather of one particular amino acid, called "phenylalanine."

When patients were given extra quantities of this amino acid in their diet, they excreted correspondingly more alkapton in their urine, while normal persons did not do so. In metabolism, phenylalanine is in the last resort broken down to carbon dioxide (CO_2) and water (H_2O). It appears that alkapton is an intermediate product in this chain of breakdown reactions, and that the gene for alkaptonuria interrupts the chain just after this stage has been reached.

Subsequently, a second gene—likewise autosomal and recessive —was found to intervene in the same chain of metabolic reactions, but at an earlier step. In this case, the mutation that led to the discovery of the gene has very serious effects; for homozygous individuals suffer from an extreme type of mental deficiency. The urine of these individuals contains large amounts of a substance called "phenylpyruvic acid," and the disease is called "phenylpyruvic idiocy" or "phenylketonuria".

In the metabolism of phenylalanine, phenylpyruvic acid is the first breakdown product. In patients it is, therefore, the first step in the metabolic reaction chain that is blocked by the mutated gene; in normal individuals, it is the first step that is mediated by the normal allele.

DNA

DNA, Like Protein, forms giant molecules which consist of smaller and simpler components. The components of protein are called "amino acids", the components of nucleic acid are called "nucleotides." Each nucleotide, in turn, consists of three parts: a sugar molecule, a phosphate group, and a so called purine or pyrimidine base. Purines and pyrimidines are organic molecules of a kind found in many natural products; caffeine, for instance, is a purine.

All nucleotides of DNA contain the same sugar molecule, called "desoxyribose"; the nucleotides of RNA contain, instead, a sugar called "ribose," which differs from desoxyribose by having one additional oxygen atom. The bases in the different nucleotides of DNA are not all the same; there are four bases, two purines and two pyrimidines, and each nucleotide carries one of them. Rarely, a fifth base may occur.

To give the formulae of the purines and pyrimidines of DNA would take us too far into chemistry; but it will simplify the presentation of what follows if we know their chemical names. The two purines are called "adenine" (*A*) and "guanine" (*G*): the two pyrimidines are called "thymine" (*T*) and "cytosine" (*C*).

Originally, it was thought that a giant molecule of DNA always contained equal numbers of *A, G, T* and *C*; that, in fact, each such giant molecule could be subdivided into groups of four nucleotides, carrying between them just one *A*, one *G* and one *T* and one *C*. If this were correct, the potentialities of DNA as carrier of a genetical code would be rather limited; for—as can easily be veri fied—each group of four nucleotides could at best yield only $2 \times 3 \times 4 = 24$ different "letters," and many such groups of four would be required for the spelling of even a simple "word," although thousands of "words," that is, gene-controlled processes, have to be specified for the development of even the simplest organism. Until recently, these considerations inclined most geneticists to regard the protein component of the chromosome as the essential carrier of genetic information.

This concept of the structure of DNA has now been abandoned. The proportion between the four bases in a molecule of DNA differs between species, and it is not always 1: 1: 1: 1. There is no longer any reason to assume that the bases in a DNA molecule are arranged in groups of four; on the contrary, it is probable that they can be arranged in every possible sequence, such as *AATGCGA* or *TGTGCCA* and so on.

This at once does away with the limitations of coding. It is evident that even very short sequences of, say, three neighboring nucleotides can "spell" large varieties of different "letters" if all possible combinations and permutations among the four bases are allowed. Three nucleotides may, for instance, spell *AAA* or *CCC* or *AGA* or *TCA* and so on. For bacteria and viruses, where DNA has been shown to carry the genetic information, it is reasonable to assume that the code is spelled by the sequence of nucleotides, and it is at least probable that this is true also for higher organisms.

There remains, however, one peculiar numerical relationship among the four bases of DNA: the number of purines is always equal to the number of pyrimidines ($A+G = T+C$). Closer analysis shows that this matching between purines and pyrimidines goes even further, and that the number of adenines equals that of thymines ($A = T$), while the number of guanines equals that of cytosines ($G = C$). Thus, if in the DNA of a given species there is, for example, an excess of adenine over guanine, there is a corresponding excess of thymine over cytosine.

This chemical property of DNA remained unexplained until in 1953 two young scientists at Cambridge, the Englishman Crick and the American Watson, put forward a new idea for the structure of DNA. Their model not only explained the curious quantitative relationship among the bases, it also gave a simple interpretation of one of the most puz zling biological properties of the genetic material, its ability for accurate replication.

Before we consider the Watson-Crick model of DNA, let us look at the difficulties that confronted geneticists when they tried to understand gene replication. It is difficult to think of similar processes in nature which might provide a key to the mystery. When a crystal of, say, ordinary household salt is kept in a strong solution of the same salt it "grows" by the addition of salt molecules, and the new molecules are added on in such a way that the characteristic shape of the crystal—in this particular case a cube—is maintained. The resulting large crystal can easily be split again into smaller cubes.

There is a certain resemblance between gene replication and this process of growth and subdivision according to a fixed geometrical pattern, but the differences are much more striking than the similarities. A crystal has a simple molecular structure; its atoms are arranged in regular patterns. Genes—whatever their particular chemical structure—have such highly specific functions that they must differ from each other in complex and subtle ways. It is these complexities and subtleties that are replicated faithfully every time a gene multiplies, and this cannot be achieved by the simple physical forces that govern the growth of a crystal. A better comparison can be made with antibody formation. Antigens resemble genes in their high degree of specificity, and they transfer specificity to the antibodies that they provoke.

The hypothesis that the procedure by which a cell makes "copies" of its genes is similar to that by which an antibody copies the specificity of the antigen is an attractive idea, but it has its difficulties. The antibody is not a copy of the antigen; their relationship to each other is that of the mold, or "template," for the stamping of a coin.

In order to replicate the antigen molecule itself, the antibody would have to serve as template for making an antigen, which then indeed would be exactly like the original one. Thus, theories of gene replication that are based on a comparison with antibody formation have to assume that the gene is the template for some intermediate product, which in turn is the template for the new gene.

The outstanding feature of the Watson Crick model is that it can explain gene replication without recourse to a hypothetical intermediate carrier of specificity. Watson and Crick based their model on the X-ray diffraction picture of DNA.

When X rays pass through a crystal they are scattered by its atoms, and from the pattern that the scattered beams produce on the X-ray plate an experienced scientist can infer the relative positions of the atoms to each other.

This technique has been taken over for substances that are not true crystals but whose molecules, like those of protein or nucleic acid, consist of regularly arranged and spaced subunits. X-ray diffraction work on nucleic acid has been carried out for many years.

The DNA molecule, only part of which is shown, consists of two threads that are wound spirally round each other. Since the diameter of the windings remains the same throughout the molecule, it is more correct to speak of helical winding and to call the whole structure a double helix.

In reality, of course, the strands are not solid fibers but consist of individual atoms, held together by strong chemical bonds. Weaker bonds connect the two strands crosswise and hold them together.

Chemically, the strands are formed by alternating units of phosphate and sugar; the bases are attached to the sugar, one base to each sugar unit, and they reach out into the hollow cylinder formed by the two winding strands.

Opposite bases are held together by weak chemical bonds. If we imagine the strands unwound and the whole structure flattened out it would look somewhat like this: Now, of the two kinds of base, the purines are considerably larger than the pyrimidines. If a pyrimidine on one strand were paired with a pyrimidine on the other, the diameter of the whole structure in this region would be narrower than in a region where two purines were paired.

If one wants to preserve the same width throughout the whole DNA molecule, one has to arrange the bases in such a way that at every crosslink a purine is paired with a pyrimidine. Since, as the X-ray diffraction picture shows, the diameter of DNA does not change from region to region, this must be the way the bases are arranged in nature.

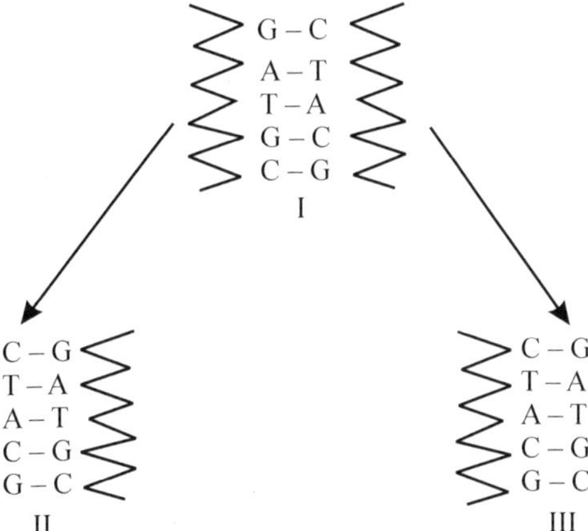

You will realise that this conclusion automatically results in equal proportions of purines and pyrimidines and explains one of the hitherto puzzling numerical relations among the bases. Moreover, when the chemical configurations of the individual bases are taken into account it turns out that, in order to fit into the available space, adenine has to pair with thymine, and guanine with cytosine.

This explains why there are always as many adenines as thymines and as many guanines as cytosines in a molecule of DNA. The base sequences in the two strands are thus complementary to each other: if we choose one of them arbitrarily, the other one is no longer open to choice. If, for instance, the sequence in a region of one strand is AGGTTCTGAC then the opposite sequence necessarily is TCCAAGACTG

It remains to discuss the greatest attraction of the Watson Crick model for the geneticist: its ability to provide a simple mechanism for gene replication. According to present day theory, DNA duplicates itself in the following way: First, the two strands unwind and separate; how this is achieved is still somewhat of a puzzle.

Once the strands are free, the base sequences, which according to our present view carry the genetical code, are exposed to the surrounding fluid, which contains all kinds of chemical building blocks, among them phosphate, sugar, and the four purine and pyrimidine bases required for synthesis of DNA.

From this store of building material each base of an original strand attaches to itself a new base, and these new bases then become linked with each other by a new sugar phosphate strand, which proceeds to wind round the old one and form a double helix with it. The most important point about the whole process is that the old bases are not free in the choice of the new ones.

Unless each base chooses its complementary one, that is, unless A chooses T, T chooses A, G chooses C, and C chooses G, there will be a bulge or a constriction in the new double helix wherever such a wrong choice has been made, and this poor fit will cause the unsuited base to be replaced by the fitting one.

This diagram shows the essential features of DNA replication, as envisaged by modern theory: each "daughter" chromosome is made up of one old and one new strand, and the specific base sequence in the two daughter molecules (II and III) is the same as in the original one.

Like every major discovery, the Watson Crick model has posed many more questions than it has answered, questions such as: Is the whole of the chromosomal DNA used for coding, or does a chromosome contain some "nonsense" sequences, which are used as full stops or commas or for some altogether different purpose?

Is the genetical information restricted to DNA, even in higher organisms in which the chromosomal DNA is always tightly bound up with protein and a certain amount of RNA? How is the code deciphered and translated into specific biochemical processes? RNA obviously plays a major role in this; but how exactly does it function?

THE MONOPOLY OF THE GENES IN INHERITANCE

The transmission of a character depends on genes, although its manifestation may be influenced by environment. Since the genes are in the nucleus, this type of inheritance is called "nuclear." The overwhelming majority of well studied cases of inheritance are nuclear, but this does not mean that "extranuclear" or "cytoplasmic" inheritance is impossible.

In fact, we have already encountered one case of extranuclear inheritance. Chlorophyll formation, as we saw, is under the control of a large number of genes; but these genes require the co-operation of the chloroplast. Changes may occur in a chloroplast that make it unable to respond to the controlling genes, and these changes may be transmitted from the first abnormal chloroplast to all its descendants.

Many cases of chloroplast inheritance are known, and since chloroplasts—like genes—reproduce by replication and division, this type of extranuclear inheritance is fairly well understood.

Other cases of cytoplasmic inheritance exist and can be detected by special methods. Two rules are used for their detection, one negative, one positive. The negative rule states that hereditary traits which are transmitted through the cytoplasm must not show Mendelian segregation in crosses. The positive rule is based on the fact that, at least in all higher organisms, the female gamete contains a large amount of cytoplasm, while the male gamete contains very little.

One would therefore expect that cytoplasmically inherited characters would follow the maternal line. This is, indeed, true for chloroplast inheritance.

In a cross between plants with normal and abnormal chloroplasts, *all* the progeny have normal chloroplasts if those of the mother plant were normal and abnormal ones if those of the mother plant were abnormal. Occasionally, some chloroplasts may be transmitted through the pollen, but this does not alter the essential difference between the progeny of the two "reciprocal" crosses.

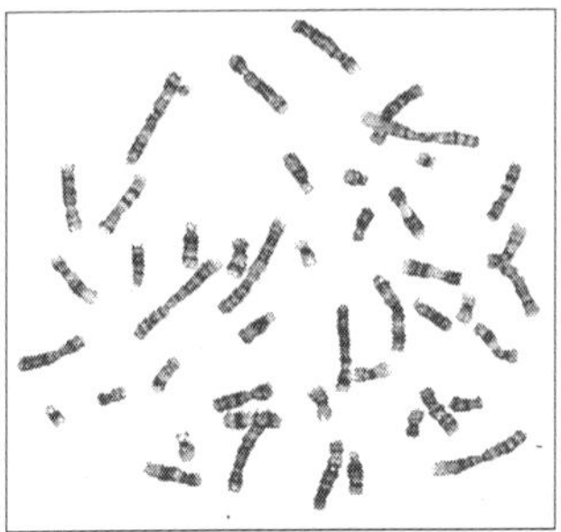

Fig. Genes of Inheritance

A comparison between the results of reciprocal crosses is the most useful tool in the search for cytoplasmic inheritance. But it is a tool that has to be used with caution, for differences between progenies of reciprocal matings may occur also when inheritance is purely nuclear.

The simplest example is sex-linked inheritance. A cross between a gold cock *ss* and a silver hen *S* produces silver cockerels *Ss* and gold pullets; the reciprocal cross between a silver cock *SS* and a gold hen *s* produces only silver offspring: the cockerels are *Ss*, the pullets are *S*. Such cases are easily distinguished from cytoplasmic inheritance.

Another cause for differences between reciprocal crosses is early gene action in the unfertilized ovum. In the flour moth Ephestia, the normal black eye colour of the caterpillars and moths develops through a series of biochemical steps, one of which is blocked when a gene *A* is replaced by its allelomorph *a*. *A* is dominant over *a*, and the eyes of *Aa* caterpillars and moths are black like those of homozygotes *AA*.

The eyes of *aa* animals are light because one of the necessary precursor substances is not produced in the absence of *A*. When a heterozygous black eyed male *Aa* is crossed with a light eyed female *aa*, the progeny segregates into 50 per cent *Aa* caterpillars with black eyes and 50 per cent *aa* caterpillars with light eyes, just as could have been predicted from Mendel's first law. The reciprocal cross gives a different result. Here all caterpillars have black eyes. That segregation into *Aa* and *aa* did, nevertheless, occur can be seen when the moths emerge from their cocoons; for half of them have black eyes and half have white eyes, exactly as in cross I.

Apparently the *A* gene in the mother produces sufficient precursor substance in the egg to last through the lifetime of the caterpillar; when the store is exhausted, no more can be formed in the absence of gene *A*, and no black eye pigment develops in the moth.

Again it is clear that we are dealing with nuclear and not with cytoplasmic inheritance. The distinction would have been less easy if the effect of the maternal gene had persisted throughout the life of the progeny. This, too, may happen and gives rise to cases that at first sight look like cytoplasmic inheritance.

Whether an individual will develop a left handed or a right handed coil depends on the arrangement of the first four cells in the young embryo, and this in turn depends on a pattern of asymmetry that is laid down in the unfertilized egg under the influence of the mother's genes.

One pair of genes controls this pattern, *D* the gene for right handed coiling being dominant over its recessive allele d for left handed coiling. A mother of genotype *DD* or *Dd* has only right handed offspring, independent of the genotype of the father.

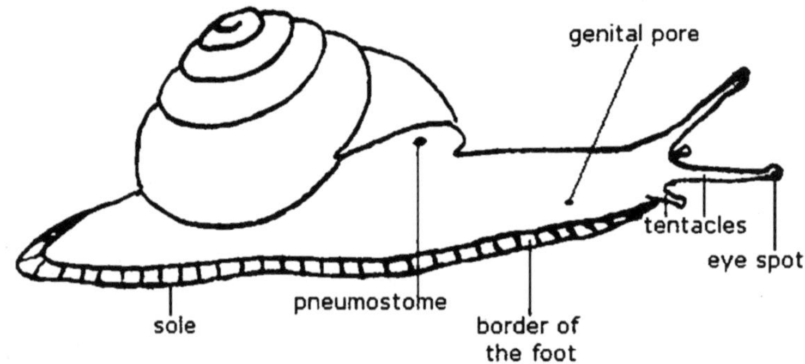

Fig. Pond Snail Limnaea

Similarly, a mother of genotype *dd* has only left handed offspring. In general, the direction of coiling in any individual is determined by the genotype of its mother, not by its own genotype. This can easily be followed in the breeding scheme, in which use has been made of the fact that snails are hermaphroditic, that is, they have both male and female sex organs and are capable of either self fertilization or cross fertilization.

If the breeding tests had not been continued beyond the first generation, this would have seemed a clear case of cytoplasmic inheritance according to our rules; for there was no segregation, and in reciprocal crosses all offspring had the maternal phenotype.

When all similar instances of "maternal inheritance" of nuclear effects are left aside, there remain a number of cases that must be attributed to cytoplasmic inheritance, or at least to inheritance by way of the cytoplasm. This is not a purely verbal distinction; it refers to the experience that several cases of what appeared to be cytoplasmic inheritance turned out to be transmission of a microorganism through the cytoplasm. Perhaps the most spectacular case is that of the "killers" and "mate killers" in the tiny fresh water animal Paramecium. This organism consists of one cell and produces clones by continued division into two; in addition, mating occurs and gives Mendelian segregation of genes. When different clones of this animal are kept together in a container, it may happen that all those of one clone are killed by a secretion from the other. For the killers themselves, this secretion is harm less. Different killer strains have different means of killing their victims. Most of them do not kill their mates; but there an some strains that, instead of killing from a distance by secretion, kill their mates through close contact. The killer character has a nuclear as well as a cytoplasmic basis. The situation is very similar for ordinary killers and mate killers; it will be described for the former. The nuclear basis is a pair of alleles *K* and *k*. Animals that are homozygous for *k* are sensitive to killing and cannot themselves become killers.

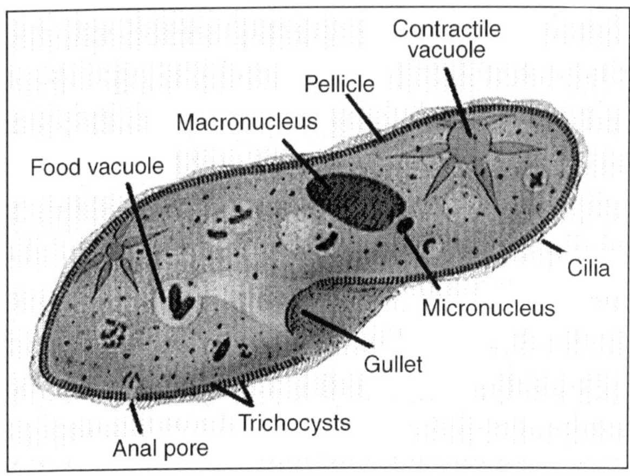

Fig. The Microscopic animal Paramecium.

Animals that carry the *K* homozygously or heterozygously in normal cytoplasm are potential killers; they are actual killers when their cytoplasm contains particles called "kappa," which in turn produce the lethal poison. In *KK* or *Kk* animals kappa particles are transmitted from cell to cell in the cytoplasm; once they have been lost from a cell they do not again develop by themselves. The large bean shaped body and the small round one adjoining it are nuclei. The star shaped structures are vacuoles which pump superfluous fluid out of the cell. The inheritance of kappa was at first considered a good example of cytoplasmic inheritance. Closer studies, however, showed that the kappa particles resemble bacteria, and their transmission from cell to cell in the cytoplasm is therefore more aptly compared with the transmission of a parasitic microorganism. The fact that kappa particles can be maintained only in animals with the gene K is no argument against this interpretation. It is a very common observation that resistance or sensitivity to a parasite may depend on the genotype of the infected host. In man there is evidence for genetically determined sensitivity to the polio virus, and bacterial strains may become resistant to a bacteriophage through mutation of one of their genes.

In Drosophila, certain animals are hypersensitive to the effects of carbon dioxide and are killed by doses that have only slight and transitory effects on normal flies. This property is inherited cytoplasmically; but, again, the responsible particles appears to be a virus rather than a normal component of cytoplasm. A decision between these alternatives is not easy, and geneticists are divided in their interpretation of this case and similar ones. But even if all of them are attributed to true cytoplasmic inheritance, their number among animals is exceedingly small and is not likely to be increased considerably by future discoveries. It is somewhat larger in plants, where reciprocal crosses often differ and the differences do not always regard the chloroplasts. One character that in many plants is inherited through the cytoplasm has a certain economic importance. This is male sterility, the inability to form functional pollen. Where, as in the breeding of hybrid corn, selffertilization has to be prevented, the introduction of the cytoplasmic factor for male sterility into the line used as female parent can save much work.

In summary, it seems safe to conclude that in the control of inheritance the monopoly of the genes, although not absolute, is very strong. Whether cytoplasmic inheritance plays a role in determining differences between widely differing organisms, belonging to different species or genera, is a much debated question. Unfortunately, it is not amenable to experimental solution, because crosses between species or genera are usually sterile or, if fertile, have sterile progeny (like the mule). We shall not consider cytoplasmic inheritance when, we shall throw a brief glance at the forces that have brought about evolution.

MUTATION

An observed mutation reflects a change in a gene. Presumably the mutated gene differs from the original one in chemical composition or structure. The new gene, like the old one, must be capable of accurate self replication, for a mutated gene replicates in the mutated form.

There is no recovery from mutation except in the very rare cases when a subsequent mutation restores the original gene. If our present view of the nature of the gene is correct, a mutation consists in a change in the number, types, or arrangement of nucleotides that "spell" a gene. On the Watson Crick model of DNA replication, such changes would be perpetuated. Mutation is not restricted to the germ cells; however, in sexually reproducing species only mutations that occur in germ cells are transmitted to the progeny.

In asexually propagated species, mutations that occur in ordinary body cells may become established as mutant strains: many varieties of fruit trees have originated as "bud sports." In a diploid species, recessive mutations can produce bud sports or mosaics only when the individual is already heterozygous for the gene in question. A snapdragon flower which is half purple and half lavender; it grew on a branch which, in addition to two or more such mosaic flowers, also bore several that were either wholly purple or wholly lavender. The plant itself was heterozygous for the recessive lavender gene, and in one of the cells from which the mosaic branch developed the normal allele had mutated to lavender. Dominant mutations may give rise to immediately observable bud sports or mosaics. Mutation takes place regularly in all species of organisms. Each individual gene mutates only very rarely, perhaps once in 100,000 or once in a million cells; but as the number of genes in most organisms is very high the over all mutation frequency per generation may be considerable. In man some mutations, for instance, that to hemophilia, occur at the relatively high frequency of about one in 50,000; this means that about one out of 50,000 gametes carries a newly arisen gene for hemophilia.

For various reasons, it is likely that the majority of human genes mutate less frequently, perhaps once in 100,000 gametes or even more rarely. If we make the conservative assumption that the 23 chromosomes in a human gamete between them carry 10,000 genes, then it would follow that one gamete out of ten carries a newly mutated gene. This is not more than a rough guess, but it is not likely to be too high. Only a small fraction of new mutations is immediately detectable as mutant offspring. A recessive mutation in the gamete of a diploid species will yield phenotypically normal offspring, unless the other gamete happens to carry the some recessive gene as mutation in the population. Apart from such cases, a newly arisen recessive gene may be carried in heterozygotes for many generations, until the mating of two such heterozygotes produces a homozygous offspring. An exception is formed by sexlinked recessive mutations; these will show up in the first XY individual with the mutated gene.

Even dominant or partially dominant mutations do not always give rise to detectably mutant progeny. Many of them produce effects that are too small to be observed or that might equally well be attributed to environmental influences. This applies particularly to mutations affecting quantitative characters, for instance, egg production in poultry, grain yield in corn, general vigour, or intelligence. Thus the realization that about one out of ten human gametes carries a new mutation does not mean that about one out of ten children will be in some way abnormal or unusual. It does, however, mean that in man, as in every other species, new mutations occur inexorably even without artificial causes such as exposure to X rays. These new mutations must not be thought of as new in the sense that they have not occurred before. A species has a limited number of genes, and each gene can mutate only to a limited number of alleles. Each mutation of a given gene to a given allele recurs at a certain low frequency, and in a species with a long evolutionary history every possible mutation is likely to have occurred repeatedly. In microorganisms, in which millions of cells can be screened for mutations, it is easy to see that the same mutations turn up again and again. This is an important fact. It explains why the majority of mutations are harmful. In a species that, through long periods of evolution, has become adapted to its mode of life all useful alleles that can arise by mutation will already have been incorporated into the genotype, so that all or most members of the species carry them in homozygous condition. Once this stage has been reached, only harmful or at least less valuable alleles can still occur as new mutations. Most desert animals have a yellow gray coat colour, which forms an effective camouflage. Evolution in the desert has favored animals with the right type of coat colour genes, and alleles arising from new mutations can only be to lighter or darker shades, away from the best possible one.

This is a simple example, but the same principle applies to more complicated cases in which many genes interact to produce the best adapted phenotype. Indeed, the more complex the genetic control of a character the more readily is it upset by mutation in one of its controlling genes. A primitive piece of machinery may continue to work when one of its pieces has been replaced by a slightly different one; a complicated apparatus will be wrecked by the slightest alteration in the dimensions or position of one of its cogs or levers. Yet even the most complicated man made machine is simple compared with the most primitive living organism.

The fact that the majority of mutations is harmful has recently developed into a matter of immediate and grave concern. As long as mutations occur only from natural sources, there is nothing we can do about controlling their frequency. Modern man, however, is making use of agents that artificially increase the frequency of mutations above its natural level.

Foremost among these agents are high energy radiations from X-ray machines, from fallout produced by nuclear tests, and from a number of other sources. The benefits to be derived from these devices have to be balanced against the harm we inflict on unborn generations by loading them with an increased burden of mutated genes. So much has been written about this urgent and difficult problem that it would be redundant to discuss it here.

Only one point may be mentioned. The question is often raised whether new mutations may not sometimes be beneficial. This is, indeed, true. Very occasionally beneficial mutations occur. Plant breeders have successfully used radiation or chemical treatment for obtaining useful mutations in cereals, ornamental plants, fruit trees, and other plants. Irradiation of molds like Penicillium has produced strains with increased yield of antibiotic.

But each such useful mutation has been obtained at the cost of about a thousand harmful or useless ones, which had to be discarded. Even in animal breeding such a procedure would be impossibly wasteful. For mankind any increase in mutation frequency remains highly undesirable in spite of the possibility that among a thousand harmful mutations there may be one that is beneficial. While the genetical dangers of radiation are among the most widely discussed topics of our time, much less attention is paid to the possibility that mutations may be produced by some of the many chemicals that civilized man uses in his food, his drugs, his cosmetics, his industrial processes. There is a good reason for the lack of scientific pronouncement on this question: our profound ignorance of the relevant facts. It is true that the past twenty years have seen the discovery of an ever increasing number of chemicals that produce mutations in Drosophila, in plants, and in microorganisms.

But while it may be confidently assumed that human genes, like those of every other tested species including mice, will mutate on exposure to ionizing radiation, the same cannot be taken for granted in regard to chemical "mutagens." X rays penetrate easily through living matter of every description. Chemicals may be retained by the outer layers of the body before they reach the germ cells; they may be excreted before they have had a chance to produce mutations; they may be trapped by cell constituents outside the nucleus; they may be metabolized into compounds with different chemical properties.

All these possibilities will differ between species, and a chemical that produces many mutations in one organism may be quite ineffective in another. Thus caffeine has produced mutations in certain bacteria and fungi. In Drosophila it has had a similar, but much weaker, effect. Its effect on the germ cells of mice, rats or guinea pigs remains to be tested by experiment. Whether it is a mutagen for people who imbibe it in coffee or tea cannot be decided experimentally. Data on mice and other laboratory mammals can provide no more than presumptive evidence, which different geneticists will interpret differently according to their scientific caution and their degree of addiction to coffee or tea. Yet, to obtain such data remains an urgent object for genetical research in the near future. Mutation research thus forms an important branch of applied genetics, both positively, as a tool for the production of improved strains of plants and microorganisms, and negatively, as a means for detecting possible harmful genetical effects of processes and chemicals used in human societies. For the research worker mutation work is one of the most valuable tools in the study of the genetic material. When in 1927, at the Fifth International Congress of Genetics in Berlin, H. J. Muller from the United States first reported the production of mutations in Drosophila by X rays he opened in the history of genetics. With the aid of X-ray machines, every laboratory worker could now produce unprecedented numbers of new mutations in whatever organism he was studying. These mutations, in turn, provided material for all types of genetical investigation, which without this might have had to wait for decades until the right kind of mutation turned up spontaneously. Nowadays the geneticist who wants some particular mutation for his experiments turns quite naturally to the X-ray machine to provide it. In research on microorganisms, ultraviolet radiation serves the same purpose more effectively. Most of the mutations for the study of gene action in microorganisms or of genetic recombination in bacteria were produced by exposure of cells to ultraviolet light. In higher organisms, too, ultraviolet light can produce mutations when it reaches the nucleus of the germ cells; but shielding through the outer layers of tissue is so great that special tricks have to be used to ensure penetration. Although excessive exposure of the human body to the ultraviolet light of the sun may cause nasty burns, we need not worry about its effect on the progeny.

The first chemicals with mutagenic action were discovered at the beginning of the Second World War. At present we know scores of such substances, belonging to very different chemical groups. Some, like mustard gas and certain related substances, have produced mutations in every kind of tested organism.

Others act only on certain types of organism or cell. Formaldehyde, for instance, produces mutations in Drosophila males but not in Drosophila females. In recent years much effort has been directed toward finding chemicals that are even more selective and act only on certain genes. Obviously the detection of such substances would be of tremendous importance for the theory and practice of genetics. For the research worker they would provide a tool for studying the chemical nature of individual genes; for the applied geneticist they would provide means for producing at will mutations of a desirable type. We are still far from having reached this goal, but there have been encouraging signs that it may not be unattainable.

It is already clear that chemicals may differ from radiation and from each other in the relative frequencies with which they produce different mutations. We still know little about the manner in which radiation or chemical mutagens produce their effects. X rays and similar types of radiation release large amounts of energy on their passage through living matter; when this happens in or near a chromosome it may result in chemical change of a gene. The much smaller amounts of energy released by ultraviolet radiation may act in a similar way; but possibly much of the mutagenic action of ultraviolet light is mediated by chemical effects of the radiation on the fluid surrounding the genes. To some extent this appears to be true even for high energy radiations like X rays. Mutagenic chemicals may react with the genes directly, or indirectly through reactions with the cytoplasm. It is also possible—and indeed may be a frequent cause of mutation—that a gene "makes a mistake" when it replicates. One can imagine many ways by which a mutagen may provoke such mistakes, Especially mutagens that are chemically related to the purines and pyrimidines in DNA are liable to act in this way.

Our ideas about what causes mutations to occur in nature are even more speculative. Certainly radiation from cosmic rays and from radioactive substances in the soil and the organisms themselves must contribute their share of mutations; but the amount of radiation from these sources is too low to account for more than a small part of all naturally occurring mutations. Mutagenic chemicals may be responsible for much or all of the remainder. Several known mutagens are natural products; quite recently, a very potent mutagen—heliotrin—has been found in the common ragwort. In bacteria, at least, a proportion of naturally occurring mutations seems to be due to mutagenic purines that are produced in the cells themselves. This can be concluded from experiments in which purines such as caffeine or related substances were used for producing mutations.

The mutagenic effect of the purines could be completely prevented when the bacteria were given certain other substances, which in some way acted as "antimutagens." When untreated bacteria were grown in solutions of these antimutagens they developed many fewer mutations than bacteria grown in ordinary nutrient solution.

As would be expected, the antimutagens did not interfere with the production of mutations by radiation. After these speculations we reach firm ground again when we consider a different kind of genetical change which occurs infrequently in nature and can be induced by radiation and many chemicals: chromosome breakage. In organisms with sufficiently large chromosomes it is easy to see breakage by X rays, mustard gas, or other mutagenic treatment; in other organisms breakage may be detected by genetical methods, that is, by examination of the progeny. The genetical consequences of chromosome breakage depend on the fate of the broken chromosomes. In some way which is not yet fully understood broken chromosome ends can stick together to form again permanent structures; but, as chromosomes in a living cell float freely in the nuclear sap, the new chromosomes are not always the same as the original ones.

Chromosomes may exchange pieces, fragments may be lost, and other kinds of "structural change" may take place. In the present context the most important change is a "deficiency," that is, loss of a fragment either from the end of a chromosome or, through rejoining between the wrong pieces, from one of its inner regions. A fragment that fails to rejoin usually is lost, and with it the genes carried on it.

In a haploid organism like Chlamydomonas or many strains of yeast and other fungi all but the smallest deficiencies result in death, the vast majority of genes are necessary for survival. Either deficiency kills homozygotes, that is, individuals lacking the same gene or genes in both partner chromosome; but, while the deficiency has no noticeable. Two breaks have occurred in the same chromosome. The broken ends have stuck together, the middle piece has been left out and is likely to be lost at one of the next cell divisions effect on the heterozygote, the deficiency makes even the heterozygote strikingly abnormal. Thus the first deficiency behaves like a recessive gene, the second like a dominant one.

Fig. Origin of a Chromosome Deficiency.

Cells with broken chromosomes, whether or not these rejoin into rearrangements, survive well as long as they do not divide, but for various reasons they usually die at the next division or soon after. This explains why dividing cells are much more sensitive to radiation than nondividing ones. The selective killing of dividing cells by X rays is utilized in the treatment of cancer. Tumors characteristically contain a much higher proportion of dividing cells than the surrounding normal tissue and are therefore selectively destroyed by radiation. Treatment with mutagenic chemicals is of more recent origin. These chemicals, like X rays, destroy tumor cells by chromosome breakage; they may give good results especially in cases where X rays cannot be used. The first chemical of this kind was nitrogen mustard.

Its value for the treatment of cancer was discovered in the United States at the beginning of the Second World War. Simultaneously and independently the first chemically induced mutations in Drosophila were found in Scotland in the progeny of flies that had been exposed to mustard gas or nitrogen mustard.

Also at the same time, and without knowledge of the American and British work, German scientists found that an entirely different substance—urethane—causes chromosome breakage and rearrangement in plants, and British clinicians tested urethane successfully on patients with leukemia. This story of apparent coincidences is an illustration of the truth, well known to students of the history of science, that certain times are "ripe" for the discovery of certain scientific facts. Mendel was ahead of his time, and his great discovery remained practically unknown until the time had become ripe for it. This kind of misfortune is a hazard that is peculiar to a genius. For the common run of scientists it is a humbling thought that in their absence their discoveries would probably have been made by someone else within the same decade. While the ability of certain mutagenic agents to destroy tumor cells may at least in part be understood as a consequence of their ability to break chromosomes, another connection between mutagenesis and cancer is less easily interpreted. It is well known that X rays not only can *cure* cancer but that continued exposure to low doses of X rays may *produce* cancer, as the early X-ray workers learned to their great detriment. The same parallelism has meanwhile been found for a number of chemical mutagens, including nitrogen mustard and urethane.

Bibliography

Ashok Kumar Sharma: *Biochemistry of Nucleic Acids*, Random Publication, Delhi, 2011.

B.B. Buchanan, Wilhelm Gruissem and Russell L. Jones: *Biochemistry and Molecular Biology of Plants*, I.K. International Publication, Delhi, 2007.

C S Patil and Prabhu M Biradar: *Molecular Biology*, APH Publication, Delhi, 2008.

Frank H Stephenson: *Calculations For Molecular Biology and Biotechnology*, Elsevier, 2004.

Franklin: *Biochemistry and Molecular Biology of Antimicrobial Drug Action, 6e*, Springer, 2001.

G.P. Garg: *Biochemistry of Nucleic Acids*, Discovery Publication, Delhi, 2010.

G.P. Jeyanthi: *Molecular Biology*, MJP Publication, Delhi, 2009.

H.P. Gajera, S.V. Patel and B.A. Golakiya: *Biochemistry, Molecular Biology and Biotechnology: Instant Notes*, New India Publishing Agency, Delhi, 2015.

Kevin Davies: *The Sequence : Inside the Race for the Human Genome*, Penguin Press, 2001.

Manish L Srivastava: *Molecular Biology*, Shree Publication, Delhi, 2008.

M'Charek: *Human Genome Diversity Project, The*, Cambridge University Press, New York, 2001.

Mousumi Debnath: *Cell and Molecular Biology*, Pointer Publication, Delhi, 2008.

N Vidyavathi and D M Chetan: *Molecular Biology*, I K International Publication, Delhi, 2007.

P Kumar: *Concepts in Molecular Biology*, Sarup and Sons Publication, Delhi, 2006.

Peter E Nielsen: *Peptide Nucleic Acids : Protocols and Applications*, Horizon Press/ Routledge, 2004.

Pragya Khanna: *Cell and Molecular Biology*, I K International Publication, Delhi, 2008.

Prakash S Lohar: *Cell and Molecular Biology* , MJP Publication, Delhi, 2007.

R Gupta: *Molecular Biology*, Pearl Books, Delhi, 2007.

Richard Durbin: *Biological Sequence Analysis: Probabilistic Models of Proteins And Nucleic Acids*, Cambridge University Press, New York, 2003.

S K Singh: *Cytology, Genetics and Molecular Biology*, Campus Books International, Delhi, 2008.

S Sundrara Rajan: *Cell and Molecular Biology*, Anmol Publication, Delhi, 2003.

S.P. Vyas and A. Mehta: *Cell and Molecular Biology*, CBS Publication, Delhi, 2011.

Santosh Nagar: *Cytogenetics and Molecular Biology*, Agrobios Publication, Delhi, 2013.

Shilpa Pandey: *Basics of Molecular Biology*, Anmol Publication, Delhi, 2008.

Surzycki: *Basic Techniques in Molecular Biology*, Springer, 2003.

Tao Jiang, Ying Xu and Q. Michael Zhang: *Current Topics in Computational Molecular Biology*, MIT Press, Delhi, 2004.

William H. Elliott and Daphne C. Elliott: *Biochemistry and Molecular Biology*, Oxford University Press, Delhi, 2005.

Index